MANAGING HUMAN RESOURCES
IN THE OIL & GAS INDUSTRY

STEVE WERNER • ANDREW INKPEN • MICHAEL H. MOFFETT

> **Disclaimer**
> The recommendations, advice, descriptions, and the methods in this book are presented solely for educational purposes. The author and publisher assume no liability whatsoever for any loss or damage that results from the use of any of the material in this book. Use of the material in this book is solely at the risk of the user.

Copyright© 2016 by
PennWell Corporation
1421 South Sheridan Road
Tulsa, Oklahoma 74112-6600 USA

800.752.9764
+1.918.831.9421
sales@pennwell.com
www.pennwellbooks.com
www.pennwell.com

Marketing Manager: Sarah De Vos
National Account Executive: Barbara McGee Coons

Director: Mary McGee
Managing Editor: Stephen Hill
Production Manager: Sheila Brock
Production Editor: Tony Quinn
Book Designer: Susan E. Ormston
Cover Designer: Charles Thomas

Library of Congress Cataloging-in-Publication Data

Names: Werner, Steve, 1962- author. | Inkpen, Andrew C., author. | Moffett, Michael H., author.
Title: Managing human resources in the oil and gas industry / Steve Werner, Andrew Inkpen, Michael H. Moffett.
Description: Tulsa, OK : PennWell, 2016. | Includes bibliographical references and index.
Identifiers: LCCN 2015032602 | ISBN 9781593703622
Subjects: LCSH: Personnel management. | Oil industries--Management. | Gas industry--Management.
Classification: LCC HF5549 .W4377 2016 | DDC 622/.3380683--dc23
LC record available at http://lccn.loc.gov/2015032602

All rights reserved. No part of this book may be reproduced, stored in a retrieval system, or transcribed in any form or by any means, electronic or mechanical, including photocopying and recording, without the prior written permission of the publisher.

Printed in the United States of America

1 2 3 4 5   20 19 18 17 16

Contents

Chapter 1. Overview of the Management of Human Resources

Strategic Implications of Human Resource Management 3
    Gaining and sustaining a competitive advantage 4
    Satisfying multiple stakeholders 6
Managing Human Resources 8
Staffing 9
    Key staffing decisions 11
    Maximizing the effectiveness of staffing 15
Training and Development 17
    Key training and development decisions 19
    Maximizing the effectiveness of training 20
Performance Management 22
    Key performance management decisions 23
    Maximizing the effectiveness of performance management 25
Compensation and Benefits 27
    Key compensation and benefits decisions 30
    Maximizing the effectiveness of compensation and benefits 33
Format of the Book 35
References 36

Chapter 2. The Global Oil and Gas Industry

Oil and Gas Industry Background 41
    Oil and gas reserves 43
Oil and Gas in the Global Economy 44
    Oil and gas supply 44
    Industry financial performance 46
    The role of OPEC 47
    The resource curse 49
Major Industry Players and Competitors 49
    Integrated oil companies 51
    National oil companies (NOCs) 53
    Independents 56
    Oil field service companies and other firms 57
The Oil and Gas Industry Value Chain 58
    Upstream: Exploration, development, and production 58
    Midstream: Trading and transportation 60

    Downstream: Oil refining and marketing 61
Beyond Oil 64
    Natural gas 64
    Petrochemicals 67
Evolution of the Industry 68
    Innovation and technology 68
    Mergers and acquisitions 68
    China and India 69
    Unconventional oil and gas 69
    Industry substitutes and alternative fuels 70
    What's next for the global oil industry? 70
Conclusion 70
References 71

Chapter 3. The Global Nature of the Oil and Gas Industry

The Global Industry 73
    Oil and gas resources 76
    Proliferation of multinational partnerships 78
**Managing Human Resources
in the Global Environment** 79
Staffing in This Global Environment 80
    Which labor market in the global context? 80
    Which recruiting method should be used in the global context? 84
    Which selection method should be used in the global context? 87
    Maximizing the effectiveness of staffing in the global context 90
Training in the Global Environment 91
    Who gets trained in the global context? 91
    What do we train in the global context? 92
    How do we train in the global context? 94
    Maximizing the effectiveness of training in the global context 97
Performance Management in the Global Environment 98
    How is performance measured in the global context? 99
    Who evaluates performance in the global context? 99
    What format should be used in the global context? 100
    Maximizing the effectiveness of performance management in the
        global context 101
Compensation in the Global Environment 102
    Determining pay level in the global context 103
    Determining pay mix in the global context 108

Determining the type of performance-based pay in a global
  context 109
Determining benefits in a global context 110
Maximizing the effectiveness of compensation in the global
  context 111
Conclusion. 113
References 113

CASE 1. Employment in an Upstream African Project: The Case of the Chad-Cameroon Petroleum Development Project

Reinvestment 121
Structure of Employment. 123
Conclusion. 126

Chapter 4. The Importance of Health and Safety in the Oil and Gas Industry

Introduction 127
**Occupational Safety and Health
 in the Oil and Gas Industry** 129
  Personal safety and process safety 132
  Security 136
  Occupational disease 139
Safety through Human Resource Management 140
Safety through Staffing 141
  Recruit applicants who tend to not take safety-related risks 143
  Select applicants who tend to not take safety-related risks 143
  Use a multiple-hurdles approach on safety-related selection tools .. 144
  Maximize the effectiveness of safety-related staffing 144
Safety through Training 145
  Who should receive safety training? 145
  The content of safety and health training 146
  Safety and health training methods 148
  Maximizing the effectiveness of safety and health training 149
Safety through Performance Management 150
  Measuring safety performance 150
  Who should be involved in measuring safety performance? 155
  The format for measuring safety performance 155
  Maximizing the effectiveness of safety performance management 156
Safety through Compensation 156
  Pay level and safety performance 156
  Pay mix and safety performance 157

    Types of performance-based pay and safety performance 157
    Benefits and safety performance 161
    Maximizing the effectiveness of compensation for safety
        performance 162
Conclusion. 162
References 162

Chapter 5. Projects in the Oil and Gas Industry

Introduction 167
Project Development in the Oil and Gas Industry. 170
    Phase 1: Project feasibility 172
    Phase 2: Project analysis 172
    Phase 3: Investment decision 173
    Phase 4: Project execution 173
Managing Human Resources in Project-Oriented Companies 174
Staffing Projects. 176
    Labor markets in project-oriented companies 176
    Recruiting for project-oriented companies 177
    Selecting in project-oriented companies 180
    Combining tools in project-oriented companies 182
    Maximizing the effectiveness of staffing in project-oriented
        companies 183
Training for Projects 184
    Who gets trained in project-oriented companies? 184
    What do we train for in project-oriented companies? 186
    How do we train in project-oriented companies? 189
    Maximizing the effectiveness of training in project-oriented
        companies 191
Performance Management in Projects 192
    How do we measure performance in project-oriented
        companies? 192
    Who is involved in performance management in project-oriented
        firms? 197
    What format should be used in project-oriented firms? 197
    Maximizing the effectiveness of project management in project-
        oriented firms 198
Compensation in Projects 199
    Pay level in project-oriented companies 199
    Pay mix in project-oriented companies 200
    Performance-based pay in project-oriented companies 200

Benefits in project-oriented companies..........................203
  Maximizing the effectiveness of compensation in project-oriented
     companies..204
Conclusion....204
References...205

Chapter 6. The Unconventional Workforce of the Oil and Gas Industry

The Workforce of the Oil and Gas Industry 211
The Prevalence of Contractors.......................................211
Construction Contracts for Oil and Gas Projects.....................214
  Contracts based on ownership 214
  Contracts based on management....................................215
Contractors and Safety 216
Rotators in the Oil and Gas Industry 217
  Rotators defined 217
  Current rotator employment challenges 219
Characteristics of the Current Workforce 221
  An aging workforce...221
  Male-dominated...222
  Difficult working conditions 227
  The skills gap 227
**Managing the Unconventional Workforce of the Oil and Gas
  Industry** 230
Staffing the Unconventional Workforce of the Oil and Gas Industry...230
  Labor markets 231
  Recruiting 234
  Selecting 236
  Combining tools 237
  Maximizing the effectiveness of staffing.........................237
**Training the Unconventional Workforce of the Oil and Gas
  Industry** 238
  Who gets trained?..238
  What gets trained? 239
  How do we train?...240
  Maximizing the effectiveness of training 241
**The Performance Management of the Unconventional Workforce
  of the Oil and Gas Industry** 241
  How do we measure performance?...................................242
  Who is involved and in what format? 242
  Maximizing the effectiveness of performance management 243

Compensating the Unconventional Workforce of the Oil and Gas Industry 243
    Pay level 243
    Pay mix 244
    Performance-based pay 245
    Benefits 245
    Maximizing the effectiveness of compensation 247
Conclusion 248
References 248

CASE 2. Skill Shortages in the Oil and Gas Industry: The Case of Australian LNG Development

The Labor Challenge 253
    Australia and project locations 254
    Staffing and skill requirements 257
    Gorgon's cost blowout 259
    Strategic promise and challenge 260
Employment Structures and Strategies 261
    Indigenous skills development 261
    Capex versus sustainable skills 262
    The FIFO workforce 262
    Project sponsor employment investments 263
    Nontraditional employment sources 263
    Other alternatives 263
Summary 265
References 265

Chapter 7. The Proactive Stakeholders of the Oil and Gas Industry

The Influence of Stakeholders in the Oil and Gas Industry 267
Stakeholders in the Oil and Gas Industry 268
    Communities and society 268
    Owners and investors 275
    Employees 276
    Customers 277
    NGOs 278
    Unions 279
Corporate Ethics and Accountability 285
Human Resource Management from a Stakeholder Perspective 287

Staffing from a Stakeholder Perspective 288
    Determining labor markets from a stakeholder perspective 288
    Recruitment from a stakeholder perspective 290
    Selection from a stakeholder perspective. 291
    Combining selection tools from a stakeholder perspective 293
    Maximizing the effectiveness of staffing from a stakeholder
    perspective 293

Training from a Stakeholder Perspective 293
    Who gets trained from a stakeholder perspective? 294
    What gets trained from a stakeholder perspective? 295
    How do we train from a stakeholder perspective? 296
    Maximizing the effectiveness of training from a stakeholder
    perspective 297

Performance Management from a Stakeholder Perspective 298
    How do we measure performance from a stakeholder
    perspective? 298
    Who is involved in performance management from a stakeholder
    perspective? 299
    What format should be used from a stakeholder perspective? 300

Compensation from a Stakeholder Perspective 302
    Pay level from a stakeholder perspective 302
    Pay mix from a stakeholder perspective 303
    Performance-based pay from a stakeholder perspective. 303
    Benefits from a stakeholder perspective. 303
    Maximizing the effectiveness of compensation from a stakeholder
    perspective 304

Conclusion. 305
References 305

Chapter 8. Government Involvement in the Oil and Gas Industry

The Influence of Governments in the Oil and Gas Industry 309
    The importance of oil and gas to governments. 309
    Government ownership of companies 311
    Government ownership of oil 319
    Regulations 322
    Government and private firm partnerships. 327

**Improving Government Relations through Human Resources
Management** 328

Improving Government Relations through Staffing 329
    Determining labor markets from a government relations
        perspective 332
    Recruiting for better government relations 333
    Selection for better government relations 334
    Combining selection tools for better government relations 334
    Maximizing the effectiveness of staffing for better
        government relations... 335
Training for Better Government Relations 335
    Who gets trained for better government relations?.................. 336
    What gets trained for better government relations? 336
    How to train for better government relations 337
    Maximizing the effectiveness of training from a government
        relations perspective 338
Performance Management for Better Government Relations 338
    How to measure performance for better government relations....... 339
    Who is involved in performance management for better
        government relations?.. 340
    What format should be used for better government relations? 341
    Maximizing the effectiveness of performance management for
        better government relations.................................... 342
Compensation for Better Government Relations 342
    Pay level for better government relations........................... 343
    Pay mix for better government relations............................ 343
    Performance-based pay for better government relations 343
    Benefits for better government relations 344
    Maximizing the effectiveness of compensation from a government
        relations perspective 345
Conclusion... 346
References.. 347

CASE 3. The Bakken Boom: Unconventional Oil and Employment

Unconventional Oil 351
    Unconventional oil in Bakken..................................... 352
    Unconventional oil and employment 353
Creating Boomtowns... 355
    Boomtown wages and prices...................................... 356
    Boomtowns and societal impact 358
Conclusion.. 360

Chapter 9. Final Thoughts on Managing Human Resources in the Oil and Gas Industry

Major Challenges Facing the Industry 361
    Continued crude oil and natural gas price volatility 361
    Resource access challenges for the majors
        and other exploration and production (E&P) firms 362
    Technological change 362
    Climate change action and legislation: Doing business in a greener
        world 363
    Continued cost pressure 363
    NOCs growing and evolving as global competitors 363
    Shrinking talent pools in the industrialized countries 364
    Risks and challenges enhanced by social media and information
        access 364
Skills Required to Deal with the Industry Challenges 364
Managing Human Resources in the Oil and Gas Industry 365
    Staffing in the oil and gas industry 365
    Training in the oil and gas industry 367
    Performance management in the oil and gas industry 368
    Compensation in the oil and gas industry 369
Conclusion 370
References 370

Index 371

About the Authors 391

1

Overview of the Management of Human Resources

The oil and gas industry is arguably the most important industry in the world. It provides energy for our vehicles, heat for our homes, and electricity to run our daily lives. It provides the raw material for countless products ranging from asphalt to zippers. A comprehensive list of all the products made from oil would include hundreds of products we use every day, including aspirin, balloons, clothes, dentures, eyeglasses, footballs, guitar strings, and hair spray.[1] The oil and gas industry also has important implications for national security, international relations, national and international politics, and the economic development of countries. It provides millions of jobs around the globe.

Unlike most industries, the oil and gas industry depends heavily on natural resources as well as capital resources. Like most industries, human resources are critical for the oil and gas industry to succeed. The people a company chooses as its employees and how they are managed will impact every aspect of the firm. The research on human resource management (HRM) is clear: How employees are selected, trained, compensated, and managed affects the bottom line of the firm as well as the productivity and well-being of the employees themselves.[2] Yet, because of the distinct differences of the oil and gas industry from other industries, the management of human resources in this industry also differs. This is where this book comes in. We look at the distinct differences between the oil and gas industry and other industries and show how these differences impact the management of human resources. These differences are outlined below:

1. **The global nature of oil and gas.** The oil and gas industry is one of the most global industries in the world. The products of the

industry are used in every country in the world, and most are producers. The key players in the industry operate in dozens of different countries and have employees of dozens of different nationalities. Company partnerships in oil and gas frequently span many different countries.

2. **The importance of health and safety.** There are many safety hazards in the oil and gas industry, from the danger of falling objects on a drilling rig to a potential explosion at a refinery to malaria threats in Sub-Saharan Africa. Many industry products are volatile, toxic, and environmentally sensitive. Work sites such as refineries and oil production platforms have many complex pieces of equipment and must be carefully managed by a highly skilled workforce. Heavy pieces of equipment, such as ships and gas turbines, are built in some of the world's largest and most complex fabrication yards that may employ tens of thousands of workers. Security hazards are great because of the many remote locations of employees.

3. **The importance of project management.** Projects are important part of all aspects of the industry. Projects in the oil and gas industry include the construction and development of facilities such as onshore oil and gas production facilities, offshore production facilities, onshore petroleum storage facilities, offshore storage facilities, oil tankers, liquefied natural gas (LNG) liquefaction facilities, pipelines, refineries, terminals, petrochemical plants, lubricant blending units, gas stations, and carbon capture and storage facilities.

4. **The unique workforce.** The workforce of the oil and gas industry has a number of characteristics that are unconventional as compared with most other industries. These characteristics include the prevalence of contractors, the use of *rotators* (rotational employees who commute to a work site on a fixed schedule, such as 28 days on and 28 days off), an aging workforce, a male-dominated workforce, difficult work conditions, and an impending skills gap.

5. **The proactiveness of stakeholders.** There are few industries that possess the complexity of stakeholder interests as that of the global oil and gas industry. The historically poor reputation of the industry has increased the visibility and scrutiny focused on firms operating in the oil and gas sector, evoking considerable criticism for any perceived negative outcomes. The broad reach of the industry means that it touches nearly every person on the planet as a customer or community member, and so everyone on earth is a stakeholder.

6. **The level of government involvement.** Government is involved in nearly every facet of the oil and gas industry. In most countries, the government owns the oil and gas resources, even those below privately held land. Governments own many of the largest oil and gas companies in the world. Governments regulate most aspects of the business including health, safety, and the environment; drilling; pricing; labor; and foreign practices. Governments are so involved because of the economic, social, and national security impact of oil and gas industry.

In this book, we will look closely at these differences and show how they affect the management of human resources. By understanding these differences and adapting to them, managers in the oil and gas industry can become much more effective.

Although exactly which functions fall within the field of the management of human resources is debatable, most scholars and human resources (HR) professionals agree that staffing, training, compensation, and performance management are the critical functions of HRM.[3]

Staffing is composed of all of the practices that determine a company's workforce, including recruiting, selection, and termination decisions. Because training includes both training and development, this function encompasses all of the practices that organizations use to improve employees' job performance and increase employee skills, talents, and competencies in the short and long term. Compensation comprises all of the practices that companies use to reward employees, including wages, performance-related pay, and benefits. Finally, performance management covers all of the practices that companies use to ensure that employees perform up to their potential, including performance evaluations, performance feedback, pay for performance, and setting performance goals. In this book, we look at the distinct differences between the oil and gas industry and other industries and show how these differences impact these four specific functional areas of HRM.

Strategic Implications of Human Resource Management

Human resource management is certainly important to the people in an organization. Clearly, how people are selected, trained, compensated, and managed is important to them, because it affects what they do, how they do it, and what they receive for getting it done.[4] But just as clearly, how people are managed is important for an organization. This is where the notion of strategic HRM comes in. Strategic HRM considers how managing people can help an organization achieve its goals. It also considers how it can help a company gain and sustain a competitive advantage. That is, how can a company, through its employees, do things that its competitors cannot do, now or in the future?

Gaining and sustaining a competitive advantage

So how does a firm gain and sustain an advantage over its competitors? Firms can attain a competitive advantage by having capabilities that their competitors do not have. To sustain this competitive advantage, these capabilities need to be valuable, reasonably rare, difficult to imitate, and difficult to replace with something else.[5] These capabilities can be related to any aspect of business: the firm's products, marketing, learning capabilities, hedging, HR, and so forth. For example, in the oil and gas industry, ExxonMobil is known for its process discipline and excellence, a corporate capability that has clearly helped the company achieve a competitive advantage. Statoil and Petrobras are known for their deep-water drilling expertise. Shell has been known for having a strong marketing capability and being willing to be the first with new technologies like the Prelude *floating LNG* facility.

Capabilities that create a sustainable, competitive advantage are an important part of a firm's specific corporate, business, and functional strategies. These strategies are a coordinated set of practices and actions tailored to take advantage of a firm's current capabilities and create new ones. Corporate strategies relate to actions and practices that cover the entire organization. In particular, the corporate strategy of a firm deals with the scope of businesses or industrial sectors in which a firm competes. For example, large *integrated oil companies* (IOCs), like ExxonMobil and Shell, compete across a wide range of industry sectors. The term *integrated* in IOC refers to a broad set of businesses from upstream (e.g., exploration) to downstream (marketing). In contrast, Apache is an independent oil and gas company that has for decades focused its activities on the smaller and lower-margin oil and gas properties that some of the majors are not interested in. Because Apache has a particularly lean operational structure, it is able to make money on properties that others could not.

Business strategies relate to the choices and actions of a specific business unit. For example, one of the historic core businesses of the IOCs is retail fuel sales through a combination of company-owned, leased, and franchised gas stations. Over the past few decades, supermarkets and nontraditional fuel retailers have entered the fuel retailing sector. As result of the new entrants, the IOCs have decided that in many markets they can no longer generate adequate returns through company-owned retailers. As a result, the IOCs have made a strategic choice to sell many of their company-owned and operated gas stations. This is a business strategy decision that deals with the competitive nature of the retail fuels industry. The IOCs may choose to remain in the sector as fuels suppliers and as licensors of their brands.

Functional strategies relate to the actions and practices of a specific function such as production, marketing, or HRM. For example, in the fuel retailing sector, companies must make decisions about how to market their products. Chevron has developed a marketing campaign around the proprietary fuel additive called Techron. The nature of the marketing campaign can be called an example of a functional strategy.

It is generally believed that to be effective, all these strategies need to fit well with the opportunities and threats of the firm's external business environment. That is, the strategies need to take into consideration any aspect of the firm's business environment that could impact the effectiveness of the strategy such as the legal environment, the political environment, the available technologies, and so forth. The effectiveness of the strategy also depends on its fit with the strengths and weaknesses of the firm. That is, the strategies also need to take into consideration aspects of the firm itself that could impact the effectiveness of the strategy such as its size, labor force, management, and the like. Finally, the corporate, business, and functional strategies should also fit with each other to be effective. This leads us to strategic HRM.

Human resource management strategies are the coordinated HR practices and actions that help the firm achieve competitive advantage. To be effective, HR strategies must fit with the company's external environment, the firm's strengths and weaknesses, and the other strategies of the firm.[6] Although there is no consensus, some common HR strategies that have been suggested include bundles of practices that foster (1) employee skills building, (2) employee motivation, (3) employee empowerment, (4) employee monitoring and control, and (5) cost savings.[7] Each of these strategies requires coordination among all HR practices to achieve an effective HR strategy. That is, the firm's staffing, compensation, training, and performance management practices must all be in sync with each other to achieve an overall HR strategy that supports the organization and business strategies. All these relationships are shown in figure 1–1.

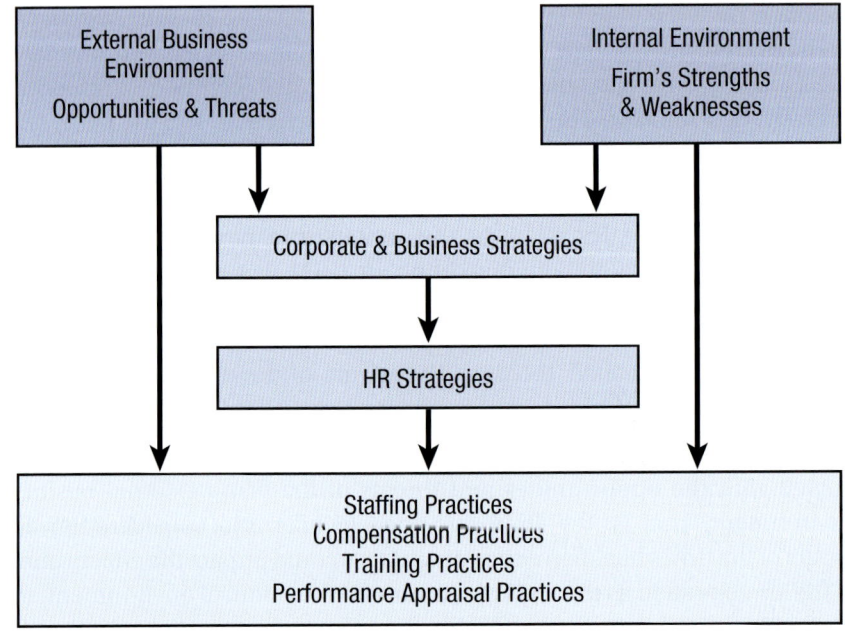

Fig. 1–1. Strategic HRM in context

The practices of these different areas of HR must also fit the firm's business environment. Because the business environment of the oil and gas industry is so unconventional, the strategic view dictates that firms in this environment need to go beyond conventional HR practices and tailor their HR practices to this specific unique business environment. In this book, we relate some of the major differences in the business environment of the oil and gas sector to each of the four major HR functions, consistent with this strategic view.

Satisfying multiple stakeholders

Earlier, we mentioned that being able to use HR strategically helps firms to achieve their goals, but what are those goals? Years ago, the prevailing view was that a firm's only goal was to return profits to the shareholders. Since then, this view has been debated, with a different view, one that believes that a firm has many important stakeholders, and that its success can be measured by how well it satisfies all of them.[8]

Nowadays many believe that a firm cannot be considered successful if there are stakeholders who are unsatisfied with what the firm is doing. Who are these multiple stakeholders? They are any organization or group of people that have a vested interest in the firm. They include employees (executives, managers, and nonmanagerial employees); customers; suppliers; joint venture partners; the community; governments; unions; consumer advocates; the media; and, of course, owners, shareholders, and suppliers of capital (banks). Each of these entities is affected by a company in different ways and thus wants different things from the organization. Owners want a good return on their investment. Customers want a good product at a good price. Employees want good working conditions and good compensation. Unions want a growing base of satisfied members. Governments want the firm to provide a growing tax base and jobs and abide by regulations. The community wants the firm to be supportive of the community and the environment. Because the goals of the stakeholders can work against each other (e.g., spending lots of money to be environmentally conscientious can hurt profits), trying to keep all these stakeholders satisfied can be a tricky endeavor.[9] Managing HR strategically can help achieve a balance that keeps all stakeholders at least relatively satisfied.

As with competitive advantage, the firm's HR practices within the various HR functions can independently, or in coordination, impact the firm's ability to satisfy its various stakeholders. The HR practices can impact stakeholders in many different ways.[10] First, HR- and labor-related costs can make up a large percentage of an organization's total costs. Because these costs have a direct effect on profits, the financial costs of labor and its management can have a significant impact on some stakeholders directly (e.g., owners and the profits they will receive) and indirectly (e.g., how many financial resources the firm will have to address societal concerns). Because payroll and benefits are such a large percentage of labor-related costs, compensation and benefits are particularly influential in this way.

Second, HR strategies and practices can enhance motivation and productivity. How people are treated and what is expected of them can greatly affect people's attitudes and motivation toward work. Performance management and compensation are the two HR functional areas that are most likely to affect the motivation of employees. All of the functional areas (staffing, training, performance management, and compensation and benefits) are likely to have an impact on productivity. Because employee motivation and productivity are strongly related to all organizational outcomes, this aspect can affect all stakeholders.

Third, HR strategies and practices can affect the organization's ability to attract and retain talent. Effective staffing practices and attractive compensation packages are likely to help attract and retain talented employees. Although all HR areas are likely to contribute to employees wanting to stay with a firm, compensation is probably the most important functional area with respect to retention. Because of the importance of talented employees to all organizational outcomes, attraction and retention can affect all stakeholders.

Fourth, HR strategies and policies can affect employee behaviors so that they support firm strategies. All HR functional areas can influence employees to work in ways that support the organization's strategy. For example, if the organization has a corporate strategy of high customer service, the staffing function can help hire people who intuitively treat customers well; the training function can educate employees on the fundamentals and importance of good customer service; the performance management function can monitor and provide goals and feedback with respect to the employees' level of customer service; and the compensation function can reward those employees who provide the expected level of customer service. Because the achievement of strategic goals is important to all stakeholders, shaping employee behaviors to support firm strategies can affect the satisfaction of all stakeholders.

We have shown that gaining and sustaining a competitive advantage and that satisfying stakeholders are critical for the success of any organization. Further, HR strategies and the functions of staffing, training, performance management, and compensation can directly and indirectly impact both competitive advantage and stakeholder satisfaction. Also, the unique aspects of the oil and gas industry (as compared with other industries) should be reflected in the HR strategies and functional practices to help achieve both. We now briefly discuss each of the functional areas, defining the area, discussing the key decisions in each area, and how to maximize the effectiveness of each.

Managing Human Resources

As previously mentioned, the functional areas of staffing, training, performance management, and compensation are seen as the most critical areas of HRM. They are critical in all industries, and the oil and gas industry is no exception. Although not as heavily studied as other industries, research clearly shows that these areas (and the management of HR in general) can have a big impact on the bottom line and various other outcomes of firms in the oil and gas industry. Research has shown that human resource strategies and practices have considerable influence on the implementation and success of the company strategy of oil firms.[11] The effectiveness of HRM practices have also been shown to be critical to the performance of firms in the oil and gas industry in many ways, including cutting down delays in implementing new projects, expanding the business in new areas, improving operational performance, and improving overall financial performance.[12] Additionally, research has shown that effective staffing, training, performance management, and compensation practices impact employees in the oil and gas industry in many positive ways, including their positive effects on employee commitment, satisfaction, and retention.[13] We now look at these four functional areas in detail.

Staffing

Staffing deals with all the decisions made that determine who will actually work for the company and what they will be doing. These decisions may be made by employees in the firm's HR department or by line managers. Clearly, this is a critical function because it will determine who the employees will be, which will affect every facet of the business. Think of how different a company would be if all of its employees were replaced by different people.

In the oil and gas industry, labor availability is seen as one of the most important business issues companies are currently faced with.[14] Staffing is currently a critical issue in the industry because of several factors.[15] First, as we will discuss in chapter 3, recruiting can be particularly difficult because jobs are located in locations around the world where working and living conditions are challenging. Second, the continued growth of the industry means that companies need more workers to keep up with increased demand. Figure 1–2 compares the job growth in several sectors of the US oil and gas industry in the last decade to all job growth in the United States.

Further, it has been estimated that the oil and gas industry will have created 1.4 million new jobs in the United States between 2012 and 2020.[17] Third, the aging workforce means that companies need to replace a large number of retiring workers. What is known as "the great crew change" is one of the most important human resource topics in the oil and gas industry today. This change will occur because the average age of workers in the energy industry is now over 50, and the US Department of Labor estimates that over half of the workforce will retire within the next 10 years.[18] We discuss the great crew change in detail in chapter 6.

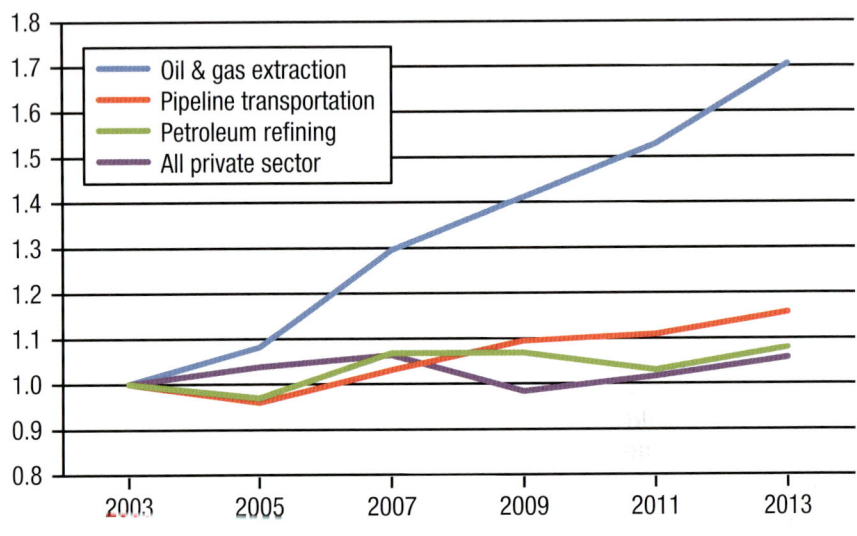

Fig. 1–2. Number of employees in US sector relative to 2003
Source: US Bureau of Labor Statistics, 2014.

Staffing broadly captures many decisions having to do with who to recruit, how to recruit them, and how to hire them:

- Will we hire them as employees or contractors?
- Will we hire them as full-time, part-time, or temporary employees?
- Where will we recruit potential applicants?
- How will we recruit potential applicants?
- Which selection tools will we use to determine the best applicants?
- How will scores on those tools be calculated?
- How will we determine when we should no longer retain an employee?

Of course, many of these decisions are quite complicated and entail a huge number of options. For those who would like to learn more about this, there are several good books about staffing in general, including books by Heneman, Judge, and Kammeyer-Mueller;[19] Gatewood, Feild, and Barrick;[20] and Phillips and Gully.[21] We focus on the key decisions in this area to provide a little more detail about what staffing entails.

Key staffing decisions

Staffing includes many important decisions that pertain to recruitment, selection, and continued employment. We narrow these down to four that we believe are particularly critical.

Which labor market? This question primarily affects recruiting and is particularly important in the oil and gas industry because of the industry's global nature. Where will we look to find capable and motivated employees? Do we try to attract potential applicants from an area close to the work site or from anywhere in the world? Historically, the labor market has been seen as the closest geographical area to the work site where you could find qualified employees. If we could find qualified employees in the neighborhood of the work site, that would be the labor market we focused on. If we had to expand our search for qualified employees across the entire state, that would be the labor market. As we'll show in chapters 3 and 8, this question is particularly complicated in the oil and gas sector because of its global nature and the strong involvement by foreign governments. Once companies decide where to focus their recruiting efforts, they must decide on what methods to use, which leads to the next key decision.

Which recruiting method? Many HR professionals in the oil and gas industry believe that strategic recruitment and attracting talent are the most important issues they are currently faced with.[22] Recruiting and selection are usually job-based. That is, companies try to attract and hire people with the skills needed to do a particular job. One of the differences of the oil and gas industry from other industries is that there are many jobs that are unique to this industry. Table 1–1 shows some of these jobs, some common tasks of the jobs, and the projected growth in the jobs from 2012 to 2022. These data come from O*net (www.onetonline.org), a website sponsored by the US Department of Labor that also provides information on the knowledge, skills, abilities, tools and technologies, and work activities needed for each job.

Table 1–1. Some common jobs in the oil and gas industry

Job	Major Tasks	Projected Growth (2012-2022)
Derrick operator	• Inspect derricks • Clean and oil derricks • Maintain drilling fluid • Repair pumps, mud tanks, and other equipment • Ensure proper operation of rig pumps and drilling mud systems	15%–21%
Gas plant operators	• Monitor transportation and storage of dangerous products to safety guidelines • Monitor, start, and shut down equipment • Control operation of compressors, scrubbers, evaporators, and refrigeration equipment • Record, review, and compile operations records, test results, and gauge readings. • Clean, maintain, and repair equipment	–3% or lower
Petroleum engineers	• Evaluate economic viability of potential drilling sites • Develop plans for oil and gas field drilling • Analyze data to rework processes and well modifications to enhance production • Interpret drilling and testing information • Specify and supervise well modification and stimulation programs	22% or higher
Petroleum pump system operators	• Monitor process indicators, instruments, and gauges • Regulate the flow of oil in pipelines and tanks • Control manifold and pumping systems • Operate control panels to regulate temperature, pressure, and flow rate • Plan movement of products	–3% or lower
Rotary oil and gas drill operators	• Train crews on safety and effectiveness • Control speed of rotary tables • Connect sections of drill pipe • Maintain and adjust machinery • Make drilling mud	15%–21%
Roustabout	• Clean up spilled oil • Unscrew or tighten pipes, casing, tubing, and rods • Locate leaks • Dismantle and repair oil field machinery • Move pipes	15%–21%
Wellhead pumper	• Monitor control panels during pumping operations • Operate engines and pumps • Repair gas and oil meters and gauges • Attach pumps and hoses to wellheads • Unload and assemble pipes and pumping equipment	15%–21%

Source: O*net (www.onetonline.org).

There are many different ways to attract applicants to your open jobs. Two basic ways are filling jobs with current employees (internal recruiting) or with new employees (external recruiting). There are a number of benefits to recruiting current employees rather than new employees, including better information about the applicant, employee familiarity with the organization, and generally lower selection and compensation costs. Also, employees prefer employers who promote career advancement from within the company, which enhances attracting the best talent and employee retention. However, internal recruiting has some disadvantages, including potential infighting and increased organizational politics by internal applicants for the same job, fewer new outside ideas, potentially not hiring the best person for the job, and many ripple effects by creating a new opening for the job that the internal hire left. Internal recruitment methods include informally "getting the word out" about an opening, posting openings on the firm's intranet, and emailing employees information about open jobs. Companies can also be more proactive and identify potential candidates by maintaining and using talent inventories (which is a database of current employees' skills and qualifications) and succession management plans.

Recruiting methods for external candidates include using employee referrals, the company website, advertising, job fairs, university recruiting, public and private employment agencies, search firms (headhunters), and Internet job boards. Indeed, Internet job boards make up one of the more common sources of new employees in the global oil and gas industry, as shown in figure 1-3.[23] Popular Internet job boards in the oil and gas industry include Pricelock.com, Oilonline.com, OilExec.com, Oilcareers.com, Rigzone.com, and drillers.com. Advertising on the company website, industry-specific advertising, and advertising to student career centers are also popular advertising methods used in the oil and gas industry.[24] One other method is to directly approach employees from competitors, known as poaching. Although this was historically taboo in the oil and gas industry, BP created the precedent in 2006 when it hired a refinery manager from Royal Dutch Shell to run its problematic plant in Texas City.[25] The effectiveness of these various methods depends on many factors, including the effectiveness of the employees doing the recruiting, the labor market, the company's characteristics, the nature of the job, and how well the method is implemented. Once organizations have a pool of viable potential employees, the next key question that needs to be answered is: How do we decide whom to hire?

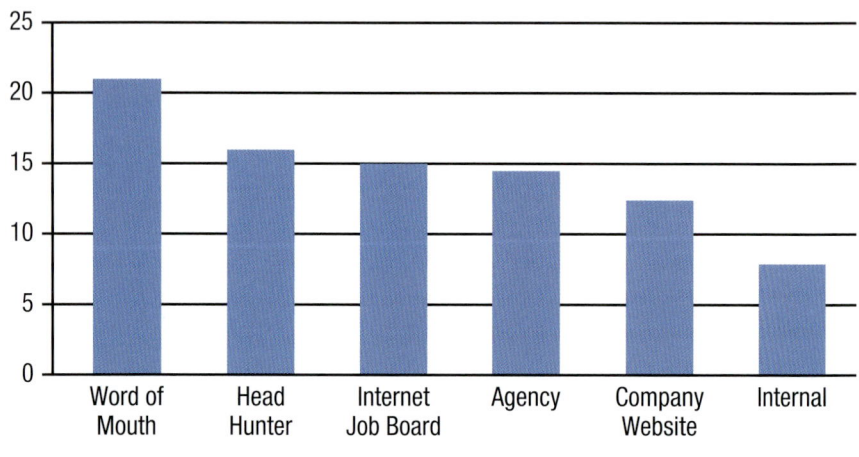

Fig. 1–3. Prevalent sources of recruits in the oil and gas industry
Source: *Hays Oil and Gas Global Salary Guide*, 2013.

Which selection tools? To help decide which applicant will end up being the best employee, organizations use a variety of tools that help predict future performance. These selection tools are generally tests that the applicant takes and can include ability tests, which measure the candidate's ability on some attribute that is needed on the job. Ability tests can be used to measure general mental ability, psychomotor ability, sensory ability, or physical ability. Other common tests include personality, integrity, job knowledge, situational judgment, and work sample tests. Although most selection tools are tests, the most common selection tool is the interview. Interviews can vary greatly across companies. Interviews can be very informal and unstructured or highly structured with the same questions being asked of all job candidates and the scoring clearly specified. Tests and interviews are commonly used to select employees, but other tools are used to screen applicants early in the process. Early screening methods include scrutinizing resumes and cover letters, job applications, and biographical data. Other screening methods that may be used later in the hiring process include medical and drug tests, background checks, and reference checks.[26]

How do we decide which selection tools to use? There are at least three clear criteria to consider: First, is the tool able to distinguish good future performers from poor future performers? This is also known as *validity*. If a test or interview is not valid, there is no point in using it. The validity of a test usually depends on the nature of the job. Physical ability is important for the job of roustabout, but not for a petroleum engineer. We find that structured interviews that are based on an accurate job description are always more valid than unstructured interviews. The second factor to consider is the cost of the test. Cost can include developmental costs as well as the costs of administering the test. Other things being equal, a cheaper test is always preferable to an expensive one. The third factor is adverse impact, which we will discuss shortly. In summary, the ideal selection tools to use are those that are valid, have lower costs, and create minimal *adverse impact*. Once we decide which tools to use, we must decide how to score candidates using those tools.

How should we combine our selection tools? Because we almost always use more than one selection method, we must decide how to combine the scores of different tools. There are two basic approaches. For the first, known as the *multiple hurdles approach*, candidates must score above some minimum threshold on every test (and interview) to be hirable. Thus, scoring below the cutoff on any one test eliminates an applicant's candidacy. For the second approach, known as the *compensatory approach*, all scores are averaged, so that scoring low on one test can be compensated for by scoring high on another. Companies can also use a combination of the two approaches. For example, for the job of wellhead pumper, you may have a cutoff for a mechanical knowledge test (a critical requirement for the job), and then use the compensatory approach for other tests on the candidates who made the first cut. Which approach to use should be based on a number of factors, including the importance of the skill being measured to do the job (more important skills should not be allowed to be compensated for), the ratio of applicants to openings (having a higher number of applicants allows organizations to be choosier using multiple hurdles), and the validity and costs of selection tools (expensive tools should be used after most applicants have been screened out by early hurdles).[27]

Maximizing the effectiveness of staffing

Determining the right labor market, using the best recruitment methods and selection tools, and combining the selection tools in the best way can go a long way toward attracting and hiring a high-quality workforce. However, there are some other aspects of the staffing process that managers need to be aware of to make sure that the staffing function is the best it can be: The organization's staffing decisions should be legally compliant, applicants should be treated fairly, and the process should be regularly evaluated and adjusted.[28]

Be legally compliant. There are many legal issues to be aware of during the recruitment and selection process. However, as mentioned earlier, one of key legal issues is the possibility of discriminatory effects caused by various selection tools. Discrimination may be intentional and blatant, known as *disparate treatment*. But it may also be unintentional and result from differences in scores on selection tests being related to some demographic characteristic, for example, women scoring lower than men on a physical ability test of lifting heavy pipes. This is known as *disparate impact* (or adverse impact). In the United States, federal antidiscrimination laws are meant to prevent selection from being influenced by gender, national origin, color, race, religion, disability, and age, so any test that is affected by these factors (e.g., a test in which women tend to score lower than men) raises issues of possible discrimination. Although adverse impact in tests is allowed if the test can be shown to be valid (e.g., the ability to lift heavy pipes is an essential requirement of a wellhead pumper), it is always better to use tests with less, rather than more, adverse impact, other things (particularly validity) being equal.

Legal compliance is important for many reasons. First, legal problems can be very expensive and time consuming. Legal expenses, fines, settlements, and judgments against a company can cost millions of dollars. For example, in 2011 Chevron became embroiled in a legal dispute with the Brazilian government over an offshore oil spill at the Frade field off Brazil's southeast coast. Brazilian prosecutors sought huge damages and filed criminal charges against Chevron. The criminal charges were eventually dropped. After almost two years of legal battles, Chevron agreed to pay about $42 million to settle the lawsuits. Second, being in the news frequently for legal violations or employee-driven lawsuits hurts a company's reputation. A poor company reputation can make future recruiting and selection much more difficult. For example, the decline in BP's reputation in the United States because of the Deepwater Horizon oil spill in 2010 in the Gulf of Mexico most certainly hurt its subsequent recruiting ability. Finally, frequent legal problems may indicate that the company does not generally treat its candidates and employees fairly, which leads us to the next topic.

Treat applicants fairly. There is a great deal of research that shows that perceptions of justice (employees believing that they were treated fairly) greatly influence the feelings, attitudes, and behaviors of applicants and employees. For employees, perceptions of fairness have been shown to be related to feelings of commitment toward the organization, job satisfaction, intentions to remain with the company, and performance.[29] The fair treatment of applicants (and subsequent employees) should occur in three ways. First, the outcomes should be fair. In the case of selection, that means that there are legitimate and reasonable justifications why one person is hired over another. Second, the process used to make the decisions should be fair. The process is much more likely to be fair if valid selection measures are used. Third, the personal interactions of the applicants with company employees should be fair. Applicants should be reasonably well informed of the process and should be treated respectfully throughout the process. All three aspects of fairness have been shown to be important in applicants' perceptions of firms.[30] Treating successful applicants fairly is particularly important because of the continued relationship the firm will have with them. In addition, treating rejected applicants fairly is important because they are likely to pass on their negative experiences to many of their friends and family, damaging the reputation of the firm.

Monitor, evaluate effectiveness, and adjust. When making so many decisions regarding who and how to recruit and select job candidates, it is unlikely that anyone would get all the decisions exactly right on the first try. Monitoring and evaluating the process and outcomes throughout will help determine if anything can be done better. Did we find that one recruitment method tended to supply us with all our successful applicants? Was there one particular selection test that seemed to clearly distinguish the good performers from the poor ones? Continually evaluating the process and the outcomes (the quality of the hires) will go a long way in improving the system so that it is as effective as it could possibly be, as long as reasonable adjustments are made based on those evaluations. In summary, to maximize the effectiveness of any recruiting and selection process, all those involved must strive to be in legal compliance; treat applicants fairly; and monitor, evaluate, and adjust the system continually.

Training and Development

Training and development are similar in that both involve activities provided to employees to improve their skills and competencies. Training tends to focus on improving performance in the short term, while development focuses on competencies that will affect longer term performance. Thus, training tends to focus on improving employees' performance in their current jobs, while development prepares them for the future.[31] It is important to note that training and staffing are interdependent. The more money invested in staffing, where the organization spends considerable resources to make sure it hires people with all the skills they need to do the job, the less training is needed. Conversely, if little is invested in staffing, and the new hires have few of the skills the company needs, then considerable investment in training is needed. Nevertheless, even when companies do a thorough job in their staffing process and hire well-qualified employees,

employees will still need some training in specific areas of their current jobs and development to prepare them for future jobs.

As with all functional areas of HRM, training and development can have strategic implications for an organization, given that they can influence customer service, help overcome staffing difficulties, improve efficiency and productivity, improve an organization's competitiveness, ease the implementation of new technology, and improve how knowledge is captured and shared throughout an organization.[32] Training and development can also have strategic implications for an organization through their effects on employee retention, motivation, and attitudes. Employees who are given training and development opportunities in the oil and gas industry are more committed to the organization, more optimistic, and more satisfied with their jobs.[33] In the oil and gas sector, some common types of training are on-the-job, technical, safety, interpersonal skills, business skills, country culture, and language training.[34] Table 1-2 shows the types of training and length of training for the industry-specific oil and gas jobs shown earlier.[35]

Table 1–2. Training requirements for oil and gas industry jobs

Job	Type of Training	Length of Training
Derrick operator	Informal training with experienced workers On the job	A few days to a few months
Gas plant operators	Informal training with experienced workers On the job Apprenticeships	One to two years
Petroleum engineers	Work-related experience On the job Vocational training Bachelor's degree or higher	Several years
Petroleum pump system operators	Working with experienced workers On the job Apprenticeships	A few months to one year
Rotary oil and gas drill operators	Working with experienced workers On the job Apprenticeships	A few months to one year
Roustabout	Informal training with experienced workers On the job	A few days to a few months
Wellhead pumper	Working with experienced workers On the job Apprenticeships	A few months to one year

Source: O*net (www.onetonline.org).

Key training and development decisions

Companies and managers must make many important decisions that pertain to the training and development of employees. We narrow these down to three that we believe are particularly critical.

Who gets trained? Every employee in a company is likely to benefit from some type of training (or development). However, who gets trained usually comes down to these business decisions: How much will it cost, and how much will the company benefit? How much the company will benefit can be partially determined by the need for training. The need for training can be established at a number of different levels. It can be established at the organization level, by evaluating the need for training given the company's strategy, culture, and current business environment. For example, because of the current shortage of fully qualified workers in the oil and gas sector, companies have found they must have programs in place that assess the deficiencies of new hires and address those deficiencies quickly through standardized, smaller, and more focused training units.[36] The need for training can also be determined at the job level, by looking at the specific skills, knowledge, abilities, and behaviors needed to perform a job at the highest level. For example, an examination of the job of hydrologist may reveal that an important part of the job is water analysis. It may then be determined that all hydrologists should be trained on the use of a new water-quality-measurement instrument. Last, the need for training can also be determined at the person level. Person-level needs analysis looks at the gaps between the knowledge, skills, and abilities (KSAs) of individual employees and those KSAs needed to perform the job at a high level. This can be part of the performance management process, which we will discuss later in this chapter. Thus, given these different needs analyses, companies can determine that training should be given to the entire workforce or any number of employees, including just one employee.

What do we train them to do? The various needs analyses can reveal many different areas where training is necessary and warranted. Thus, it follows that there are a large number of possible content areas for training and a wide variety of different training goals for these content areas. The objectives can include improving general or specific knowledge, improving or developing general or specific skills, changing employee attitudes, changing behaviors, and promoting ethical behavior. Improving or developing skills can include technical, managerial, interpersonal, and language skills. As mentioned previously, some common training content in the oil and gas industry encompasses on-the-job, technical, safety, interpersonal skills, business skills, country culture, and language training.[37] Oil and gas companies anticipate skills shortages in the near future for technical, management, marketing, and leadership skills.[38] For the long term, and related to the anticipated shortages of qualified workers, oil and gas companies are increasingly partnering with universities, governments, and nonprofit associations to promote awareness and provide oil-and gas-specific knowledge for the future labor market.[39] Oil and gas companies are also focusing on leadership development because they anticipate a shortage of qualified leaders.[40] Given that we have now determined who gets training and the content of that training, the only key question remaining is how.

How do we train them? There are many different ways to train and develop employees. These include on-the-job training, on-site but off-the-job training, and off-site training methods. Many different training methods are used in the oil and gas sector including on-the-job training, university courses, workshops, conferences, videos, mentoring, teleconferencing, simulations, and interactive multimedia (web-based learning).[41] Each of these many methods has strengths and weaknesses. Some are more effective than others, some are less expensive than others, and some are more likely to actually transfer to the job than others. Cost and effectiveness are important, but if the training does not transfer to the job, then the company will not receive any benefits from it. Therefore, which one to use depends on its effectiveness, cost, and transferability, but also on who is receiving the training and the content.[42]

Maximizing the effectiveness of training

Choosing the right audience, the right content, and the best method for training can go a long way to achieving a highly capable workforce. However, there are some other aspects of the training process that managers need to be aware of to make sure that the training function is operating at the highest level. For example, the company's training should maximize learning, be transferable to the job, and be regularly monitored, evaluated, and adjusted.

Maximize learning. We have briefly discussed the who, what, and how of training, but the conditions in which learning occurs can have a big impact on how much trainees actually learn. These conditions include factors related to the organization, the training location, the trainer, and the training session. Some of the organizational factors that improve the effectiveness of learning include a culture that accepts and rewards training, supervisors who reward rather than punish employees who undertake training, adequate technological support, and opportunities for the trainees to use what they have learned. Regarding the training location, factors that improve the effectiveness of learning include a location that is convenient, which encourages attendance; a setting that is comfortable, quiet, and private; a workspace that facilitates learning; and the availability of resources that provide trainees with everything they need to actively participate. Some of the trainer and training session factors include clear communication; culturally relevant material; trainers who make the material meaningful; and sessions that allow for practice, provide feedback, and stimulate a feeling of accomplishment by trainees.[43] Table 1-3 summarizes these factors. The next aspect that helps your company's training be successful is transferability to the job.

Table 1–3. Factors that maximize learning in training

Organizational Factors	Location Factors	Trainer Factors	Training Session Factors
A culture that accepts and rewards training	Convenience to encourage attendance	Clear communicator	Culturally relevant material
Supervisors that reward rather than punish employees who seek training	Comfortable, quiet, and private setting	Trainer makes material meaningful	Sessions allow practice
Adequate technological support	A workspace that facilitates learning	Trainer provides constructive feedback	Sessions that provide feeling of accomplishment
Giving trainees the opportunity to use what they've learned	Resources to encourage active participation	Respected and knowledgeable trainer	Sessions that motivate trainee to learn

Facilitate training transfer and performance maintenance. Even if trainees thoroughly learn the training session content, it will not benefit the company unless the learning is transferred to behaviors on the job. Therefore from the company's view, creating an organizational and job environment that helps trainees use the learned material on the job and maintain high performance levels using the learned material is critical for training to be effective. Getting trainees to initially and regularly apply the learned content to their jobs can be accomplished in a number of ways, including setting up specific goals that require the use of the new skills, abilities, or knowledge; reinforcing the use of the new content through positive feedback; measuring performance; and providing regular feedback. These can be accomplished through a good performance management system, which we will discuss later in this chapter.

Monitor, evaluate effectiveness, and adjust. When making so many decisions regarding who, what, and how to train, it is unlikely that anyone would get all the decisions exactly right on the first try. Monitoring and evaluating the process and outcomes throughout will help determine if anything can be improved. Questions that should be asked about the training system include the following:

- Did we find that one type of training method tended to work particularly well for those in one particular job, but not for others?
- Was there one particular method that clearly led to better job performance for those that took the training?
- Were the trainees satisfied with the training?
- Did they believe that they learned something?
- Did trainees actually learn skills?
- Did they use them on the job?
- Has productivity improved?

Continually monitoring training will go a long way toward improving the system so that it is as effective as it could possibly be, as long as reasonable adjustments are made based on those evaluations. In summary, to maximize the effectiveness of any training process, learning should be maximized; the transfer of training and the maintenance of performance should be facilitated; and the system should be monitored, evaluated, and adjusted regularly.

Performance Management

Performance management is the process used to measure, evaluate, and influence employee performance. Years ago we thought of this function as performance appraisal, and it was just the measurement and feedback of employees' performance.[44] Performance management is now thought of as broader, encompassing not only measurement and feedback but also setting goals for employees, tying rewards to those goals, and long-term performance aspects such as career development. Performance management has strategic implications for companies because it can enhance motivation and productivity, support strategic objectives, and help detect performance problems.[45] Performance management can improve employee performance in two specific ways. First, it can help motivate employees by clarifying performance expectations, specifying outcomes related to those expectations, and providing feedback of performance. Second, it can reveal performance problems of employees (possibly through no fault of their own) and allow the organization the opportunity to fix the problems (through training, job restructuring, etc.).

Performance management is related to the other HR functions, such as staffing, by defining what high performance on the job means. Because companies tend to hire based on the job, and they are trying to predict future job performance, they need to know how to define good job performance—this is generally taken from the job description. Performance management is usually involved in promotion or termination decisions as well.

Performance management is related to training in several ways. First, performance management can reveal performance problems that can be fixed through training. Second, the long-term career-planning aspect of performance management should be tied to a long-term development plan. Third, the performance management process is complicated and subject to biases and error. Training those doing performance evaluations tends to improve the process.

Key performance management decisions

Companies and managers must make many important decisions that pertain to the performance management of employees. We narrow these down to the three that we believe are particularly critical.

How do we measure performance? This question seems trivial and obvious, but it is critical and complicated. To evaluate the quality of someone's performance, we must determine what performance is. Although performance on a task is generally specified by the job description, sometimes job descriptions are out of date or too general. Task performance is only one part of a more holistic view of performance in general. Employee behaviors that help the organization but are not necessarily task-related can also be considered part of performance. This is known as *contextual performance* and can include such facets as having a good attitude, helping fellow workers, volunteering for difficult assignments, following company rules, and being a positive representative of the company. Finally, another aspect of performance that is not necessarily task related consists of counterproductive work behaviors that negatively affect the organization, such as theft, sabotage, and the like.

The defining of performance leads to a decision of how to measure it. Performance can generally be measured by evaluating an employee's traits, behaviors, or outcomes. Examples of traits include dependability, loyalty, and being team player. Traits are the least effective way of measuring performance, since trait descriptions tend to be too general and more difficult to relate to the job. Behaviors are more specific and easier to relate to the job, and thus tend to be a good measure of performance, although there is some subjectivity in their use. Finally, outcomes can be good measures because they can be directly related to the job, can be specific, and are objective, but for some jobs the outcomes can differ greatly through no fault of the employee.[46] You can see that how we measure performance includes determining not only which of these types to use but also how many and the specific measures for each (e.g., which outcomes?). This is likely to depend on many factors, such as the company's strategy. One survey of oil company managers identified 21 different indicators of managerial performance including measures related to efficiencies of internal processes, meeting stakeholder (customers, employees, government, and society) needs, the effective management of employees, and the quality of services.[47]

Who should be involved? Historically, an employee's supervisor was the only person involved in the performance management process (which was usually just a performance appraisal). However, research has found that performance management works better when the supervisor is an administrator of the process who puts together information from many different sources.[48] Therefore, useful information regarding the performance of an employee can come from many different sources including the supervisor, subordinates, the employee, co-workers, customers, suppliers, and outcome measures. More sources are usually better than fewer sources, for the following reasons. First, different sources see different aspects of behavior, so a more holistic picture of performance can be painted with many sources. Second, when measuring anything, more data points are better than fewer, because biases or errors in any one data point will be counted less. Finally, biases or errors will tend to cancel each other out when there are more sources, resulting in a more reliable measure. Anonymity tends to get more honest evaluations when using sources other than the supervisor. The process of evaluating employees with many different sources of information is known as *360 degree feedback*. In 360 degree feedback, how the sources are combined can vary widely, but no one source should have an overpowering influence on the overall rating.

What format should be used? The formats of the actual performance appraisal can vary widely. Rating scales are common for rating traits or behaviors. However, ratings scales themselves can vary widely. The number of items, the number of scale points, the type of anchors and scale points (numbers or words), and the specificity of the anchors and scale points can all vary. There are two specific types of rating scales for behaviors: (1) the *behaviorally anchored rating scale* (BARS) and the *behavioral observation scale* (BOS). In BARS the scale points are specific examples of behaviors, while in BOS the scale points are frequencies of specific behaviors.

One critical issue regarding the format of evaluations is: Are employees rated on an absolute level or relative to each other (e.g., ranking or forced distribution)? Forced distribution is sometimes used to overcome leniency or central tendency errors by managers, who usually otherwise do not give employees low ratings. Forced distributions also tend to make decisions about employees easier and may help motivate employees. However, forced distributions have several drawbacks, such as that managers and employees tend to not like them, they can undermine teamwork, and they can create errors if the performance of the employees does not happen to follow a normal distribution.[49]

Maximizing the effectiveness of performance management

Choosing an accurate measure of performance, a broad group of knowledgeable sources, and the right format can go a long way toward attaining the benefits of a good performance management system. However, managers need to be mindful of the following to ensure that the performance management function is working at the highest level: provide appropriate feedback; reduce rating errors to improve rater accuracy; set goals and give rewards; and regularly monitor, evaluate, and adjust the process.

Provide appropriate feedback. An important aspect of performance management is providing employees with appropriate feedback about their performance. Just handing them a piece of paper with their performance evaluation is not enough. What makes feedback appropriate? First, it should be well-timed, usually meaning it should be immediate. Second, supervisors should be prepared for the feedback session, with ample evidence to support the evaluation. Third, supervisors should focus the session on identifying the factors that affected the employee's performance. Supervisors should understand that the employees may have different explanations for the causes of their performance, and removing roadblocks to good performance should be a priority. Finally, feedback should be a continuous process, with supervisors frequently following up on agreements made during the feedback sessions. With good feedback and follow up, the performance evaluation can help identify and recognize good performance and correct poor performance, maximizing its effectiveness. For example, a study in Nigeria's oil industry found that performance deficiencies were the most common reason for additional employee training, and that employee performance significantly improved after training.[50]

Improve evaluation accuracy and reduce rating errors. An important goal of every performance evaluation is accuracy. The performance evaluation conclusions should accurately reflect actual performance. Research has found a number of factors that help improve evaluation accuracy. First, precise rating scale formats tend to be more accurate than general ones. More precise rating scales assess single job activities rather than groups of them, separately rate different performance dimensions, and avoid the use of ambiguous terms like "motivated." Second, memory aids such as behavioral diaries, critical incident files, or performance management software help to ensure that all relevant information is used in the evaluation, increasing its accuracy. Third, supervisors should be rewarded for timely and accurate performance appraisals. This can be done by making timely and accurate performance appraisals part of supervisors' performance. Finally, rater training (e.g., frame of reference training) can help improve evaluation accuracy by teaching raters how to improve the observation skills of relevant performance and ways to reduce rating errors (such as being too lenient or strict).

Set goals and give rewards. Performance management is much more effective when it includes goal setting and rewards tied to those goals. Both have been shown to be highly effective in increasing motivation to perform.[51] To maximize the effectiveness of goal setting, goals should be challenging and specific, and employees should accept them as reasonable and perceive them as attainable. Employees are more likely to find goals reasonable and perceive them as attainable if the employees participate in the goal-setting process. Tying rewards to goal attainment usually has very positive effects on motivation, which we will discuss below in the compensation and benefits section.

Monitor, evaluate effectiveness, and adjust. When making so many decisions regarding performance management, it is unlikely that anyone would get all the decisions exactly right on the first try. Monitoring and evaluating the process and outcomes throughout will help determine if anything can be improved. Continually evaluating the process will go a long way toward improving the system so that it is as effective as it could possibly be, as long as reasonable adjustments are made based on those evaluations. Questions that should be asked about the system include the following:

- Do employee surveys reveal that employees think the process is unfair?
- Do some managers rate all of their employees at the highest level?
- Do the evaluations not distinguish between employees who clearly differ in performance?
- Are some employees not evaluated?

In summary, to maximize the effectiveness of any performance management system, appropriate feedback should be provided to employees; evaluations should be accurate and timely; rewards should be tied to specific, challenging, but accepted goals; and the system should be monitored, evaluated, and adjusted regularly.

Compensation and Benefits

Compensation and benefits are all the pay and benefits employees receive. Lately, compensation professionals have expanded this concept into "total rewards" and "the employee value proposition," which broaden the notion to include anything perceived as beneficial by the employee. This includes nonmonetary things such as the 12-hour shift, which is popular among employees in the petroleum industry because it provides large blocks of time off and reduces commuting time and costs.[52] Compensation is composed of the monetary pay employees receive, including base salary and any additional performance-based pay such as individual-based bonuses, recognition awards, group based incentives, and commissions. Benefits are noncash payments employees receive, such as retirement savings plans, health care benefits and services, paid leave (vacation days, holidays, sick leave, etc.), work-life benefits (such as flexible work arrangements, child care services, etc.), and others (personal services, business travel assistance, etc.).[53]

Compensation and benefits are closely tied to the other functions of HR.[54] Compensation and benefits are related to staffing in several ways. First, compensation and benefits appear to be an important factor in the recruitment of potential applicants. The greater the compensation and the more benefits a company offers, the larger the applicant pool. Second, compensation can be directly linked to recruitment through sign-on bonuses and referral bonuses and to retention through retention bonuses. Third, compensation, benefit level, and certain benefits can affect employee retention (e.g., number of vacation days related to employee seniority or retirement plans that do not carry over to other employers).

Compensation and benefits are also related to training in several ways. First, most employees see training and development opportunities as a benefit. Second, employers are likely to pay new hires less if they need a lot of training compared to those who already have all the skills needed for the job. Finally, although pay in most companies is based on the job, in some companies it is based on the skills the employees have. This is known as *skill-based pay*. When companies offer skill-based pay, training is closely integrated with employee pay. The more training they receive, the more skills they acquire, and the more pay they get. Pay could also be tied to *competencies* (which are a broader set of skills) in a similar fashion. Finally, pay is closely tied to performance management through any pay-for-performance aspects of the company's pay package.

Compensation and benefits have many strategic implications. First, both cost firms a great deal of money, and so they can have a major impact on the bottom line. For some organizations, labor costs can be more than 50% of total costs (although this tends to be less for oil and gas companies). Benefits are about 30% of labor costs. Second, as mentioned previously, compensation and benefits can affect employee attraction and retention. Third, compensation can affect motivation; high pay levels and pay for performance can both have strong motivating influences and thus improve productivity. Finally, compensation (specifically, pay for performance), together with a good performance management program, can direct employee behaviors toward strategic goals.[55]

Historically, the oil and gas industry has been a high-paying industry. Figure 1–4 shows the average hourly wage in a number of sectors of the oil and gas industry.[56] Note that they are all much higher than the average US hourly wage. The oil and gas industry pays well for several reasons. First, the industry has large capital expenses—the costs of land leases, machinery, transportation vehicles, and so on, are substantial. So labor costs are a relatively low percentage of total costs. This allows firms to pay more for labor without raising total costs much. Second, the industry has many high-knowledge jobs that require skills that must be compensated for. For example, many of the discipline areas in the oil and gas industry, such as geosciences, quality control, structural engineering, reservoir engineering, operations management, and mechanical engineering, require undergraduate or graduate degrees and subsequently pay much more than industries with fewer educational requirements. Figure 1–5 shows the annual wage for some typical jobs in the oil and gas sector.[57] Third, the working conditions are frequently difficult and may be in locations that are remote, so travel premiums and hardship premiums are common. Fourth, unlike some other industries (e.g., the airline industry), the oil and gas industry has been relatively profitable (particularly when looking at the absolute level of profits rather than profits vs. sales), so companies have had the ability to pay relatively high wages.

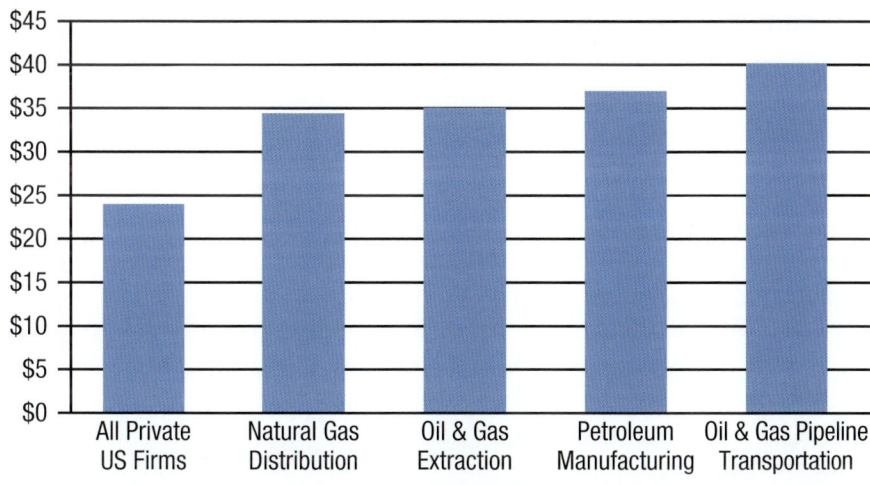

Fig. 1–4. Hourly wage of select sectors in United States
Source: US Bureau of Labor Statistics.

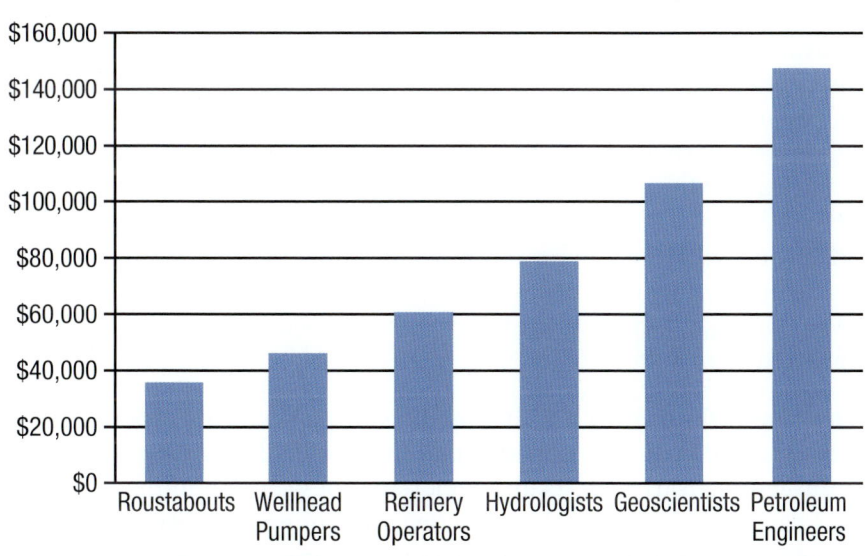

Fig. 1–5. Annual wage of select occupations in the United States (May 2012, private firm average annual wages = $49,200)
Source: US Bureau of Labor Statistics.

Key compensation and benefits decisions

Companies and managers must make many important decisions that pertain to compensation and benefits. We narrow these down to four that we believe are particularly critical.

What should the pay level be? This is a critical question that is the core of any compensation strategy. Companies can choose a lag, match, or lead strategy relative to the pay of their competitors.[58] A lag strategy means that you will, on average, pay less than other organizations are paying for the same job. Of course, how much less would have to be determined. A lag strategy lowers your labor costs relative to the other companies that are paying more. However, a lag strategy comes with some disadvantages. Paying less than your competitors will make it harder to attract and retain workers. It will also negatively affect motivation, because many will see it as unfair (in fact, they are likely to reduce their effort with the justification that their pay is reduced). The opposite is true for a lead strategy. Firms with a lead strategy will have higher labor costs than their competitors, but they will attract many more workers, and will more easily retain them. Further motivation is enhanced, because workers feel fairly treated. Coupled with a good performance management system, workers are motivated to perform because losing the job would likely mean having to take another one that pays less somewhere else. Firms with a lead strategy usually implement their strategy as paying a certain percentage above the industry average (5% or 10% are common standards) or paying at a certain percentile of firms (10th or 25th percentile is common, meaning that only 10% or 25% of the firms are paying more). A match strategy is the most common, publicly referred to as "competitive pay," with average labor costs and average attraction, retention, and motivational effects.

What should the pay mix be? The pay mix is the percentage of various aspects of compensation and benefits that make up the total rewards package. Pay mix usually considers the relative importance of salary, performance-based pay (also known as incentives), and benefits. For example, salary could be 70% of total rewards, benefits 30%, and no performance-based pay. Or salary could be 50%, incentive pay 30%, and benefits 20%. Salary and benefits are guaranteed, while performance-based pay is not, since it depends on the level of performance. Performance-based pay is also known as *variable pay*. The use of variable pay has been increasing worldwide the last few decades. Variable pay when tied to individual performance is highly motivational, but requires a well-designed incentive system that includes good performance measures, a reasonable link between pay and performance, and meaningful rewards. Variable pay when tied to group or organization performance loses some of its motivational effects, but is an effective way to lower labor costs during difficult economic times. In the United States, benefits tend to be about 30% of the pay mix, but could be as low as 8% or as high as 50%. Because employees receive benefits regardless of their level of performance, they are not seen as particularly motivational. However, benefits can shape employee behaviors (e.g., offering the benefit of reimbursing educational expenses will lead to more employees taking classes), can signal what the company values, and can support a company's culture.

The pay mix should be determined with the company and HR strategy in mind. HR strategies that are more paternalistic (and employee centered) would have a high mix of benefits and salary. HR strategies that are empowering and performance centered would have a high mix of incentives. HR strategies that focus on cost containment would have a low mix of benefits. Pay mix is somewhat tied to pay-level strategy in that firms tend to focus on salary when determining the pay-level strategy. However, the pay-level strategy should consider the pay mix and the level of total compensation when communicating about compensation to employees.

Which type of performance-based pay should be used? All firms use some type of performance-based pay for at least some of their employees. Today, performance-based pay is likely to be a component of the pay mix of most employees, with its importance increasing at higher levels of the organization.[59] Incentives are believed to be more effective at controlling costs, aligning employee behaviors with the firm culture and strategies, and improving financial performance than base salary. However, the different types of performance-based pay vary greatly. The most common types are merit pay, bonuses, recognition pay, and commission.[60] Table 1–4 depicts the differences in these types of plans.

Table 1–4. Four types of performance-based plans

Type of Plan	Performance Measure	Reward	Time Frame	Applicable Employees
Merit plan	Performance appraisal	Raise to salary	Annual	Non-managers, all
Bonus plan	Objective outcomes	Monetary lump sum	Short-term to long-term	All, higher level
Recognition plan	Exemplary performance	Small cash or non-cash	Short-term	All
Commission plan	Sales	Monetary lump sum	Short-term to long-term	Salesforce

Merit pay is an increase in salary based on performance appraisals. Average performers get an average raise, while top performers get bigger raises. Poor performers get no raises. Sometimes, to save the expense of compounding raises, firms use lump-sum merit, where the reward is the amount of the annual raise, but is given all at once and then not added to the base salary. Lump-sum merit is similar to a bonus, except lump-sum merit is based on a performance appraisal, while bonuses are based on some objective, predetermined performance measure. Some common performance measures for bonuses are sales numbers, quality indicators, production levels, and the like. Bonuses can be long term or short term depending on how far out the timeline for the performance measure is. In the oil and gas industry, about 43% of the employees receive a bonus, which equals 14% of their total compensation package on average.[61]

Recognition awards tend to be low-value monetary awards with the purpose of recognizing good performance. Common recognition awards include spot awards (small awards given to employees on the spot when outstanding performance is noted by their supervisor), *peer-to-peer awards* (small awards given to employees based on comments by their co-workers), and *suggestion awards* (small to large awards given when employee suggestions are successfully used by the company). Recognition awards are relatively inexpensive for the company to use, yet seem to have good motivational effects, which explains their growing popularity. Finally, commissions are generally applied to salespeople, where they retain some portion of their sales. Commissions can represent a small or large percentage of the total pay mix and tend to be based on total sales, sales milestones, new revenue, or gross profits. In the oil and gas industry, about 7.5% of employees receive a commission, which equals 10% of their total compensation package on average.[62]

Which benefits should be used? Some benefits are legally required, while many are voluntarily provided by employers. In the United States, legally required benefits make up about 8% of total labor costs,[63] although this number is expected to rise in the next few years with the implementation of mandated health care benefits. Currently, legally required benefits in the United States are FICA (Social Security and Medicare taxes), unemployment insurance, worker's compensation insurance (for work-related injuries or illnesses), disability insurance, and family medical leave (unpaid leaves of absence for workers with family-related needs). Health insurance (as specified by the Patient Protection and Affordable Care Act) could also be considered a mandated benefit, because not providing it leads to large fines. Nevertheless, a large number of other benefits can be offered by employers. The Society for Human Resource Management issues an annual benefit report that lists more than 200 different benefits provided by organizations.[64] These benefits can be classified as benefits related to retirement savings plans, health care, paid leave, and work-life services. Figure 1–6 shows the most prevalent benefits in the global oil and gas sector.[65]

With so many possible benefits to offer, how do companies decide which to provide? Many factors go into this decision, including employee needs, HR and company strategy, industry factors, company ability to pay, the offerings of competitors, and perceived value of the benefit.[66] Some companies try to better meet the needs of their employees by offering *flexible benefit plans*. Flexible benefit plans allow employees to choose which benefits they want to receive from a menu of possibilities. There may be a core group of benefits that all employees receive. Some of the benefits of flexible benefits plans are that they allow employees to receive the benefits they want most, they encourage employees to gain familiarity with their benefits, and they can save the company money since it does not have to offer all benefits to all employees. However, flexible benefit plans have their disadvantages too: Their administration is more complex, employees may make unwise choices, and the tax implications can be substantial.

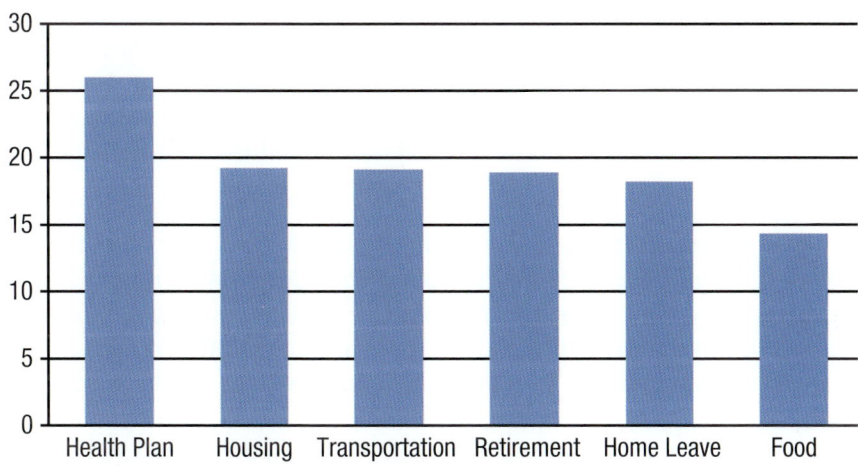

Fig. 1–6. Most prevalent benefits in global oil and gas
Source: *Hays Oil and Gas Global Salary Guide*, 2013.

Maximizing the effectiveness of compensation and benefits

Choosing an appropriate pay level, pay mix, type of pay for performance, and mix of benefits can help to get the most value from a company's labor expenses. However, there are some other aspects of the compensation and benefits system that managers need to be aware of to make sure that it is highly effective: Gain employee acceptance; be legally compliant; and regularly monitor, evaluate, and adjust the system.

Gain employee acceptance. Companies tend to put a great deal of effort into making sure their pay systems are fair. For example, a survey of human resource managers in the Canadian petroleum industry found that the two main concerns influencing their wage-setting practices were the fairness and equity of the pay related to other companies and the fairness and equity of the pay related to other plants within in their company. Internal and external equity were more important than all other considerations including their company's profitability, recruitment considerations, turnover considerations, and productivity effects.[67]

Although most companies try to make their pay system fair, employees' feelings of fairness regarding their compensation and benefits are largely driven by their perceptions. Employees are more likely to believe that the pay system is fair if they or someone representing them has participated in its design. Thus, taking employee input into consideration either directly or through advisory committees can help gain acceptance. Employee input during the design and implementation period of pay systems can also help catch any potential problems early. Acceptance can also be gained through communication and education. When employees' understand how their pay level and their bonuses are determined, they are more likely to perceive them as fair. Benefits can now be easily explained through personalized benefits statements.

Be legally compliant. Legal compliance is important for a compensation system to be effective. A company cannot consider its system successful if it causes many fines, lawsuits, or expensive settlements. Legal difficulties can also affect the firm's reputation. There are many different legal issues related to compensation, but the most common include discrimination issues, overtime pay, minimum wage, living wage laws, and benefit eligibility.[68] Of course, these laws vary by country, and in the United States, they even vary by state and municipality. Training on these various laws can keep managers from making legally questionable compensation decisions. Our earlier discussion of adverse impact of selection decisions also applies to pay decisions. Thus, pay systems should be regularly checked to see if there are any significant differences in pay based on various demographic factors.

Monitor, evaluate effectiveness, and adjust. There is probably no HR function where monitoring, evaluating, and adjusting are more important than for compensation. Monitoring and evaluating the system and related outcomes (employee turnover, absenteeism, increased grievances, etc.) regularly will help determine if anything can be improved. Continually evaluating the pay system may reveal much needed adjustments—Do employee surveys reveal that employees think the pay system is unfair? Are top performers leaving? However, regardless of the outcomes of system evaluations, pay systems regularly need adjusting. Changes in market pay levels, performance expectations, and cost of living can all greatly effect pay expectations and so adjustments are usually needed regularly (at least annually). In summary, to maximize the effectiveness of any compensation system, the system should be accepted by employees; should be legally compliant; and should be monitored, evaluated, and adjusted regularly.

Format of the Book

This chapter has described the strategic role of HRM. It has also described the four main areas of managing HR, the key decisions managers must make within each, and some issues that must be kept in mind to maximize the effectiveness of the function. In the next chapter we provide a brief overview of the oil and gas sector, focusing on the major components of the industry and specifying how the industry differs from most others. These differences are the global nature of oil and gas, the importance of health and safety, the importance of project management, the unique workforce, the proactiveness of stakeholders, and the level of government involvement. Each following chapter then describes each difference in detail and systematically shows how these differences affect each of the four main areas of HRM. Our last chapter also discusses how the four areas of HRM can help prepare organizations for the future of the industry. Each chapter uses many examples from the oil and gas sector and includes at least two "Current Challenges" that describe a current HR challenge in the oil and gas industry that should be on the minds of managers.

References

1. *Things Made From Oil That We Use Daily*. PBS. Retrieved from www.pbs.org/independentlens/classroom/wwo/petroleum.pdf.
2. See, for example, the following sources:
   Huselid, M. A. (1995). The impact of human resource management practices on turnover, productivity, and corporate financial performance. *Academy of management journal, 38*(3), 635-672.
   Huselid, M. A., Jackson, S. E., & Schuler, R. S. (1997). Technical and strategic human resources management effectiveness as determinants of firm performance. *Academy of Management journal, 40*(1), 171-188.
   Bowen, D. E., & Ostroff, C. (2004). Understanding HRM–firm performance linkages: The role of the "strength" of the HRM system. *Academy of management review, 29*(2), 203-221.
   Purcell, J., & Hutchinson, S. (2007). Front-line managers as agents in the HRM-performance causal chain: Theory, analysis and evidence. *Human Resource Management Journal, 17*(1), 3-20.
3. Wright, P. M., & Boswell, W. R. (2002). Desegregating HRM: A review and synthesis of micro and macro human resource management research. *Journal of Management, 28*(3), 247-276.
   Jackson, S. E., Schuler, R. S., & Werner, S. (2011). *Managing human resources*. CengageBrain.com.
4. Guest, D. E. (1999). Human resource management—the workers' verdict. *Human Resource Management Journal, 9*(3), 5-25.
   Guest, D. (2002). Human resource management, corporate performance and employee wellbeing: Building the worker into HRM. *Journal of Industrial Relations, 44*(3), 335-358.
   Allen, D. G., Shore, L. M., & Griffeth, R. W. (2003). The role of perceived organizational support and supportive human resource practices in the turnover process. Journal of management, 29(1), 99-118.
5. Barney, J. B. (1997). *Gaining and sustaining competitive advantage*. Reading, MA: Addison-Wesley.
   Hitt, M. A., Ireland, R. D., & Hoskisson, R. E. (2012). *Strategic management cases: Competitiveness and globalization*. CengageBrain.com.
   Hitt, M. A., Keats, B. W., & DeMarie, S. M. (1998). Navigating in the new competitive landscape: Building strategic flexibility and competitive advantage in the 21st century. *The Academy of Management Executive, 12*(4), 22-42.
6. Schuler, R. S., & Jackson, S. E. (1987). Linking competitive strategies with human resource management practices. *The Academy of Management Executive, 1*(3), 207-219.
   Wright, P. M., & McMahan, G. C. (1992). Theoretical perspectives for strategic human resource management. *Journal of management, 18*(2), 295-320.
7. Wright, P. M., & Boswell, W. R. (2002). Desegregating HRM: A review and synthesis of micro and macro human resource management research. *Journal of management, 28*(3), 247-276.
8. Donaldson, T., & Preston, L. E. (1995). The stakeholder theory of the corporation: Concepts, evidence, and implications. *Academy of management Review, 20*(1), 65-91.
   Sundaram, A. K., & Inkpen, A. C. (2004). The corporate objective revisited. *Organization science, 15*(3), 350-363.
   Freeman, R. E., Wicks, A. C., & Parmar, B. (2004). Stakeholder theory and "the

corporate objective revisited." *Organization Science, 15*(3), 364–369.
Laplume, A. O., Sonpar, K., & Litz, R. A. (2008). Stakeholder theory: Reviewing a theory that moves us. *Journal of management, 34*(6), 1152–1189.

9. Freeman, R. E. (2010). *Strategic management: A stakeholder approach.* Cambridge, United Kingdom: Cambridge University Press.
10. Jackson, S. E., Schuler, R. S., & Werner, S. (2011). *Managing human resources.* CengageBrain.com.
11. Ritson, N. (1999). Corporate strategy and the role of HRM: Critical cases in oil and chemicals. *Employee Relations, 21*(2), 159–176.
12. Parast, M. M., Adams, S. G., & Jones, E. C. (2011). Improving operational and business performance in the petroleum industry through quality management. *International Journal of Quality & Reliability Management, 28*(4), 426–450.
    Pieterse, H., & Rothmann, S. (2009). Perceptions of the role and contribution of human resource practitioners in a global petrochemical company. *South African Journal of Economic and Management Science, 12*(3), 370–384.
    Mohanty, R. P., & Deshmukh, S. G. (1997). Evolution of a decision support system for human resource planning in a petroleum company. *International journal of production economics, 51*(3), 251–261.
13. Dhiman, G. R., & Mohanty, R. P. (2010). HRM practices, attitudinal outcomes and turnover intent: An empirical study in Indian oil and gas exploration and production sector. *South Asian Journal of Management, 74*(4), 74–104.
14. Ernst & Young, (2011). *Human resources in Canada's oil and gas sector: A snapshot of challenges and directions.* Retrieved from http://www.ey.com/Publication/vwLUAssets/Human-resources-in-Canada-oil-gas-sector/$FILE/Human-resources-in-Canada-oil-gas-sector.pdf.
    Hays Oil and Gas. (2013). *The oil and gas global salary guide, 2013.* Retrieved from http://hays.clikpages.co.uk/Oil_and_Gas_Salary_Guide_2013/
15. US Department of Labor. (2013). *High growth industry profile—Energy.* Retrieved from http://www.doleta.gov/brg/indprof/energy_profile.cfm
    Meinert, D. (2013). Hiring frenzy: The North Dakota oil boom has fueled uncontrolled growth and HR headaches. *HR Magazine,* June, 31–35.
16. US Bureau of Labor Statistics. Retrieved from www.bls.gov.
17. International Labor Office. (2012). *Current and future skills, human resources development and safety training for contractors in the oil and gas industry* (Sectoral Activities Dept. Issues Paper). Geneva, Switzerland: International Labor Organization.
18. US Department of Labor. (2013). *High growth industry profile—Energy.* Retrieved from http://www.doleta.gov/brg/indprof/energy_profile.cfm
19. Heneman, H. G., Judge, T. A., & Kammeyer-Mueller, J. D. (2012). *Staffing organizations* (7th ed.). New York, NY: McGraw-Hill/Irwin.
20. Gatewood, R. D., Feild, H. S., & Barrick, M. R. (2010). *Human resource selection.* Mason, OH: Cengage Learning.
21. Phillips, J. M., & Gully, S. M. (2012). *Strategic staffing* (2nd ed.). Upper Saddle River, NJ: Prentice Hall.
22. Ernst & Young. (2011). *Human resources in Canada's oil and gas sector: A snapshot of challenges and directions.* Retrieved from http://www.ey.com/Publication/vwLUAssets/Human-resources-in-Canada-oil-gas-sector/$FILE/Human-resources-in-Canada-oil-gas-sector.pdf.
23. Hays Oil and Gas. (2013). *The oil and gas global salary guide, 2013.* Retrieved from http://hays.clikpages.co.uk/Oil_and_Gas_Salary_Guide_2013/.

24. Ernst & Young, (2011). Human resources in *Canada's oil and gas sector: A snapshot of challenges and directions*. Retrieved from http://www.ey.com/Publication/vwLUAssets/Human-resources-in-Canada-oil-gas-sector/$FILE/Human-resources-in-Canada-oil-gas-sector.pdf.
25. Phillips, J. M., & Gully, S. M. (2012). *Strategic staffing* (2nd ed.). Upper Saddle River, NJ: Prentice Hall.
26. Gatewood, R. D., Feild, H. S., & Barrick, M. R. (2010). *Human resource selection*. Mason, OH: Cengage Learning.
27. Heneman, H. G., Judge, T. A., & Kammeyer-Mueller, J. D. (2012). *Staffing organizations* (7th ed.). New York, NY: McGraw-Hill/Irwin.
28. Jackson, S. E., Schuler, R. S., & Werner, S. (2011). *Managing human resources*. CengageBrain.com.
29. Cohen-Charash, Y., & Spector, P. E. (2001). The role of justice in organizations: A meta-analysis. *Organizational behavior and human decision processes, 86*(2), 278–321.
30. Ployhart, R. E., & Ryan, A. M. (1997). Toward an explanation of applicant reactions: An examination of organizational justice and attribution frameworks. *Organizational Behavior and Human Decision Processes, 72*(3), 308–335.
31. Noe, R. A. (2013). *Employee training and development*. Boston, MA: McGraw-Hill/Irwin.
32. Noe, R. A. (2013). *Employee training and development*. Boston, MA: McGraw-Hill/Irwin.
    Jackson, S. E., Schuler, R. S., & Werner, S. (2011). *Managing human resources*. CengageBrain.com.
    Ozigbo, N. C. (2012). The implications of human resource management and organizational culture adoption on knowledge management practices in Nigerian oil and gas industry. *Communications of the IIMA, 12*(1): 91–104.
    Ahmed, H. (2007). Improved operations through manpower management in the oil sector. *Journal of Petroleum Science and Engineering, 55*, 187–199.
33. Al-Emadi, M. A. S., & Marquardt, M. J. (2007). Relationship between employees' beliefs regarding training benefits and employees' organizational commitment in a petroleum company in the State of Qatar. *International Journal of Training and Development, 11*(1): 49–70.
    Darkwah, A. K. (2013). Keeping hope alive: An analysis of training opportunities for Ghanaian youth in the emerging oil and gas industry. *International Development Planning Review, 35*(2): 119–134.
34. Pyron, D. (2008). Solutions to the recruitment and retention challenge. *Talent & Technology, 2*(2): 4–6.
    Williams, B. (2007). HR Council tackling Canadian petroleum staffing challenge. *Offshore, 2*, 8–10.
35. O*net. Retrieved from www.onetonline.org.
36. Ryder, J.A. (2007). Complex human resource challenges: Call for new approaches. *Talent & Technology, 1*, 14–16.
37. Pyron, D. (2008). Solutions to the recruitment and retention challenge. *Talent & Technology, 2*(2): 4–6.
    Williams, B. (2007). HR Council tackling Canadian petroleum staffing challenge. *Offshore, 2*, 8–10.
38. Energy Institute, Deloitte, and Norman Broadbent. (2008). *Skills needed in the energy industry: A report on the initial findings of three surveys*. Retrieved from http://www.energyinst.org/documents/5.

39. Williams, B. (2007). Oil & gas UK accelerating staffing initiatives for UK offshore oil and gas industry. *Offshore, 3*, 4–9.
    Anderson, A. (1984). Training for the offshore petroleum industry in Atlantic Canada. *The Journal of Canadian Petroleum*, July–August, 70–72.
40. Pyron, D. (2008). Solutions to the recruitment and retention challenge. *Talent & Technology, 2*(2), 4–6.
41. Aibieyi, S. (2012). The impact of post-training on job performance in Nigeria's oil industry. *Educational Research Quarterly, 35*(3), 3–32.
    Cullen, E.T. (2011). Effective training: A case study from the oil & gas industry. Professional Safety, March, 40–47.
    Greaves, W., & Heideman, K. (2000). Interactive multimedia and internet training for technical professionals in the oil and gas industry. *Computers and Geosciences, 26*, 713–717.
    Koottungal, L. (2011). Training gets real. *Oil & Gas Journal, 109*(5), 12.
42. Noe, R. A. (2013). *Employee training and development*. Boston, MA: McGraw-Hill/Irwin.
43. Noe, R. A. (2013). *Employee training and development*. Boston, MA: McGraw-Hill/Irwin.
    Jackson, S. E., Schuler, R. S., & Werner, S. (2011). *Managing human resources*. CengageBrain.com.
    Cullen, E. T. (2011). Effective training: A case study from the oil & gas industry. *Professional Safety*, March, 40–47.
44. Aguinis, H. (2013). *Performance management* (3rd ed.). Upper Saddle River, NJ: Prentice Hall.
45. Jackson, S. E., Schuler, R. S., & Werner, S. (2011). *Managing human resources*. CengageBrain.com.
46. Aguinis, H. (2013). *Performance management* (3rd ed.). Upper Saddle River, NJ: Prentice Hall.
47. Faghihi, A., Afsharnezhad, A., & Kheirandish, M. (2012). An empirical study on performance management: A case study of national Iranian oil production distribution company. *Management Science Letters, 2*, 2435–2440.
48. Jackson, S. E., Schuler, R. S., & Werner, S. (2011). *Managing human resources*. CengageBrain.com.
49. Aguinis, H. (2013). *Performance management* (3rd ed.). Upper Saddle River, NJ: Prentice Hall.
    Jackson, S. E., Schuler, R. S., & Werner, S. (2011). *Managing human resources*. CengageBrain.com.
    Berger, J., Harbring, C., & Sliwka, D. (2013). Performance appraisals and the impact of forced distribution—An experimental investigation. *Management Science, 59*(1), 54–68.
50. Aibieyi, S. (2012). The impact of post-training on job performance in Nigeria's oil industry. *Educational Research Quarterly, 35*(3), 3–32.
51. Mento, A. J., Steel, R. P., & Karren, R. J. (1987). A meta-analytic study of the effects of goal setting on task performance: 1966–1984. *Organizational Behavior and Human Decision Processes, 39*(1), 52–83.
52. Northrup, H. R. (1989). The twelve-hour shift in the petroleum and chemical industries revisited: An assessment by human resource management executives. *Industrial and Labor Relations Review, 42*, 640–648.
53. Milkovich, G. T., Newman, J. M., & Gerhart, B. (2014). *Compensation* (11th ed.). Burr Ridge, IL: Irwin/McGraw-Hill.

54. Jackson, S. E., Schuler, R. S., & Werner, S. (2011). *Managing human resources.* CengageBrain.com.
55. Martocchio, J. J. (2015). *Strategic compensation: A human resource management approach* (8th ed.). Upper Saddle River, New Jersey: Pearson Education.
56. Bureau of Labor Statistics. (n.d.). *Overview of BLS wage data by area and occupation.* Retrieved from http://www.bls.gov/bls/blswage.htm.
57. Bureau of Labor Statistics. (n.d.). *Overview of BLS wage data by area and occupation.* Retrieved from http://www.bls.gov/bls/blswage.htm.
58. Martocchio, J. J. (2015). *Strategic compensation: A human resource management approach* (8th ed.). Upper Saddle River, New Jersey: Pearson Education.
59. Milkovich, G. T., Newman, J. M., & Gerhart, B. (2014). *Compensation* (11th ed.). Burr Ridge, IL: Irwin/McGraw-Hill.
60. Jackson, S. E., Schuler, R. S., & Werner, S. (2011). Managing human resources. CengageBrain.com.
61. Hays Oil and Gas. (2013). *The oil and gas global salary guide, 2013.* Retrieved from http://hays.clikpages.co.uk/Oil_and_Gas_Salary_Guide_2013/.
62. Hays Oil and Gas. (2013). *The oil and gas global salary guide, 2013.* Retrieved from http://hays.clikpages.co.uk/Oil_and_Gas_Salary_Guide_2013/.
63. Bureau of Labor Statistics. (n.d.). *National compensation survey.* Retrieved from http://www.bls.gov/ncs/.
64. Society for Human Resource Management. (2013). *2013 Employee benefits.* Retrieved from www.shrmstore.shrm.org.
65. Hays Oil and Gas. (2013). *The oil and gas global salary guide, 2013.* Retrieved from http://hays.clikpages.co.uk/Oil_and_Gas_Salary_Guide_2013/
66. Martocchio, J. J. 2014. Employee Benefits (5th ed.). New York, NY: McGraw-Hill/Irwin.
67. Taras, D. G. (1997). Managerial intentions and wage determination in the Canadian petroleum industry. *Industrial Relations: A Journal of Economy and Society, 36*(2), 178-205.
68. Milkovich, G. T., Newman, J. M., & Gerhart, B. (2014). Compensation (11th ed.). Burr Ridge, IL: Irwin/McGraw-Hill.
    Jackson, S. E., Schuler, R. S., & Werner, S. (2011). *Managing human resources.* CengageBrain.com.

2

The Global Oil and Gas Industry

As discussed in chapter 1, the oil and gas industry is one of the largest, most complex, and most important global industries. The industry touches everyone's lives with products such as fuel for transportation, heating and electricity fuels, asphalt, lubricants, propane, and thousands of petrochemical products from carpets to eyeglasses to clothing. The industry impacts national security, elections, geopolitics, and international conflicts. The price of crude oil is the most closely watched commodity price in the global economy. In recent years, the industry has seen many tumultuous events, including the continuing efforts from oil-producing countries like Kazakhstan, Russia, and Nigeria to exert greater control over their resources; major technological advances in deepwater drilling and shale oil and gas; Chinese oil and gas firms and international acquisitions; ongoing strife in Sudan, Nigeria, Iraq, and other oil-exporting nations; continued heated discussion about global warming and nonhydrocarbon sources of energy; and major movements up and down in crude prices. All of this comes amid predictions that the global demand for energy will increase by 30%–40% by 2040.

Oil and Gas Industry Background

When Colonel Edwin Drake struck oil in northwestern Pennsylvania in 1859, the first phase of the oil industry began. John D. Rockefeller emerged in those early days as a pioneer in industrial organization. When Rockefeller combined Standard Oil and 39 affiliated companies to create Standard Oil Trust in 1882, his goal was not to form a monopoly, because these companies already controlled 90% of the kerosene market. His real goal was to achieve economies of scale by combining all the refining operations under a single management structure. In doing so, Rockefeller set the stage for what historian Alfred Chandler called the "dynamic logic of growth and competition that drives modern capitalism."[1]

With the discovery of oil in the Spindletop salt dome in East Texas in 1901, a new phase of the industry began. Before Spindletop, oil was used mainly for lamps and lubrication. After Spindletop, petroleum would be used as a major fuel for new inventions, such as the airplane and automobile. Ships and trains that had previously run on coal began to switch to oil. For the next century, oil, and then natural gas, would be the world's most important sources of energy.

Since the beginning of the oil industry, there have been fears from petroleum producers and consumers that eventually oil reserves would be depleted. In 1950, the United States Geological Survey estimated that the world's conventional recoverable resource base was about one trillion barrels. Fifty years later, that estimate had tripled to three trillion barrels. The peak oil theory is based on the fact that the amount of oil is finite. In recent years, the concept of peak oil has been much debated.

Regardless of whether the peak has or has not been reached, oil and natural gas are an indispensable source of the world's energy and petrochemical feedstocks, and will be for many years to come. The difficulty in determining oil and gas reserves is that "true reserves" are a complex combination of technology, price, and politics. While technical change continues to reveal new sources of oil and gas, prices have demonstrated more volatility than ever, and governments have sought more control than ever over resource information and access. As prices rise, reserves once considered not economic to develop may become feasible.

As illustrated by figure 2–1, crude oil prices never rose above $15 a barrel over the first century of the industry's existence. Then a series of shocks resulted in sudden and dramatic price changes in global prices. First the Arab oil embargo of 1974, then events in Iran and Iraq, and the subsequent OPEC (Organization of the Petroleum Exporting Countries) price shock of 1979 drove prices and potential industry profits upwards. Shifts in global supply and demand for oil during the 1980s and 1990s kept further price shocks at bay until the global run-up in the post-2001 era. First at $80 per barrel, and then hovering above $100 per barrel, the industry seemed to be reaching new ever-higher plateaus that drew more and more people and more and more organizations into the global industry. The rapid drop of prices to below $50 per barrel in early 2015 illustrated that price volatility continues to be a reality of the industry that all oil companies must deal with.

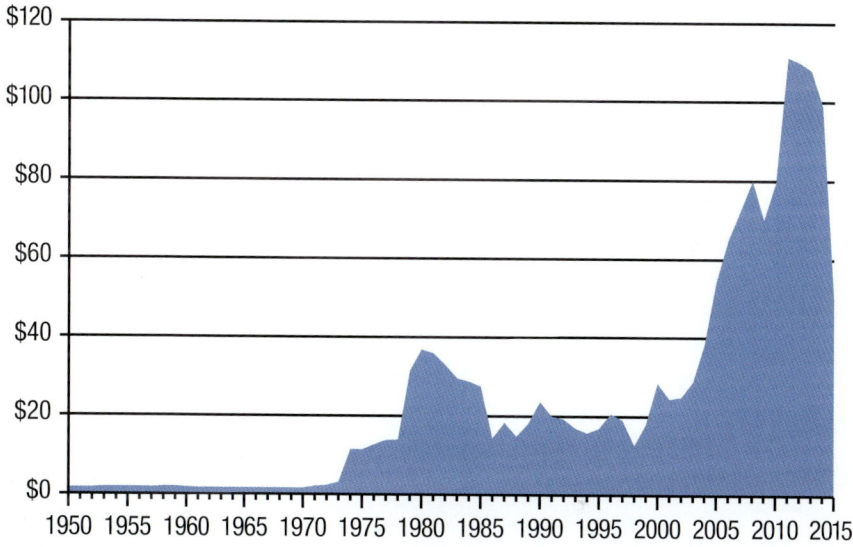

Fig. 2–1. The price of oil, 1950–2015 (US dollars per barrel)
Source: Based on *BP Statistical Review of World Energy, 2013*. 1984–2014 is Brent crude. 2015 is price through Feb 5, 2015.

Oil and gas reserves

Discovering new oil and gas reserves is the lifeblood of the industry. Without new reserves to replace oil and gas production, the industry would die. However, measuring and valuing reserves is a scientific and business challenge because reserves can only be measured if they have value in the marketplace. The oil sands of Alberta, Canada, are a good illustration of how difficult it is to accurately measure oil and gas reserves. Oil sands are deposits of bitumen, a molasses-like viscous oil that will not flow unless heated or diluted with lighter hydrocarbons. Although the oil sands in Alberta are now considered second only to the Saudi Arabia reserves in the potential amount of recoverable oil, for many years these were not viewed as real reserves because they were not economical to develop. In 2014 the main town in the oil sands region, Fort McMurray, was in the midst of a boom not unlike the gold rush booms of the 1800s. Housing and labor were scarce and the infrastructure was struggling to keep pace with the influx of people, companies, and capital. The development of the oil sands occurred because of a combination of rising oil prices and technological innovation. Oil sands production was predicted to rise to 3 million barrels per day (b/d) by 2020 and 4.5 million b/d by 2040. Williston, North Dakota, was also seeing a boomtown mentality because of the increasing production from shale oil.

From 2005 to 2014 oil production in North Dakota increased from 105,000 b/d to about 1 million b/d.

Oil and Gas in the Global Economy

Oil and gas play a vital role in the global economy. The International Energy Agency predicts that energy demand will rise significantly over the next three decades, with most of the increase coming from developing countries. Total energy demand in 2040 will be about 40% higher than in 2010. Most of the world's growing energy needs through 2040 will continue to be met by oil, gas, and coal. With increased energy efficiency, energy as a percentage of total gross domestic product (GDP) has fallen and is expected to continue to fall.

Oil and gas supply

Critical to understanding industry supply and demand is the fact that all countries are consumers of products derived from the oil and gas industry, but only a small set of nations are major producers of oil and gas. Over the past decades, the large, developed economies of the world have become net importers of oil and gas, giving rise to challenging geopolitical issues involving a diverse set of oil consumers and producers. Table 2–1 shows the major oil- and gas-producing nations and their change in output over a decade. Angola, Brazil, and Kazakhstan have made their way into the top tier of oil producers. In natural gas, almost all of the major producers have seen large increases in output as gas increasingly replaces fuels such as coal and nuclear. The United States has risen to the top of all gas-producing nations in recent years as a result of the exploitation of shale gas resources. This gas production boom has been a direct result of advancements in technology—specifically horizontal drilling and hydrofracturing (fracking)—and continuous improvement in drilling and completion processes that have lowered costs per well.

Table 2–1. Major oil- and gas-producing nations

Country	Oil Producing Nations		
	Oil Production 2012 (million bbls/d)	Percent of World Production, 2012	Output Change Since 2000
Saudi Arabia	11.530	13.3%	21.8%
Russia	10.643	12.8%	61.7%
United States	8.905	9.6%	15.1%
China	4.155	5.0%	27.6%
Canada	3.741	4.4%	38.4%
Iran	3.680	4.2%	−4.5%
United Arab Emirates	3.380	3.7%	27.1%
Kuwait	3.127	3.7%	39.4%
Iraq	3.115	3.7%	19.2%
Mexico	2.911	3.5%	−15.8%
Venezuela	2.725	3.4%	−12.0%
Nigeria	2.417	2.8%	11.9%
Brazil	2.149	2.7%	69.0%
Qatar	1.966	2.0%	130.4%
Norway	1.916	2.1%	−42.7%
Angola	1.784	2.1%	139.2%
Kazakhstan	1.728	2.0%	133.5%
Algeria	1.667	1.8%	7.6%
Libya	1.509	1.7%	2.3%
United Kingdom	0.967	1.1%	−64.1%
Subtotal	74.015	85.6%	
Total World	86.152	100%	14.9%
OPEC	37.405	43.4%	20.2%
Non-OPEC	48.747	56.6%	−2.0%

Gas Producing Nations			
Country	Gas Production 2012 (million bbls/d)	Percent of World Production, 2012	Output Change Since 2000
United States	681.4	20.4%	25.4%
Russia	592.3	17.6%	12.1%
Iran	160.5	4.8%	166.4%
Qatar	157.0	4.7%	562.7%
Canada	156.5	4.6%	−14.1%
Norway	114.9	3.4%	131.0%
China	107.2	3.2%	294.2%
Saudi Arabia	102.8	3.0%	106.4%
Algeria	81.5	2.4%	−3.5%
Indonesia	71.1	2.1%	9.0%
Malaysia	65.2	1.9%	44.1%
Turkmenistan	64.4	1.9%	51.3%
Netherlands	63.9	1.9%	9.9%
Egypt	60.9	1.8%	189.9%
Mexico	58.5	1.7%	52.3%
Uzbekistan	56.9	1.7%	11.5%
United Arab Emirates	51.7	1.5%	34.6%
Australia	49.0	1.5%	57.4%
Nigeria	43.2	1.3%	266.4%
Trinidad & Tobago	42.2	1.3%	171.9%
Total	2,781.1	82.7%	
Total World	3,363.9	100.0%	

Source: BP Statistical Review of World Energy 2013.

Industry financial performance

The oil and gas industry is highly cyclical, and the cycles can last many years.[2] In the 1990s, crude oil prices fell steadily, and in the new millennium, the first few years saw steadily rising prices. The Great Recession put a damper on some experts' prediction of $200 per barrel prices. Although the oil industry is highly profitable in some years, its long-term profitability is not much higher than average profitability across many industries. In the United States, the oil and gas industry has earned return on sales (net income divided by revenue) of about 8%, compared to an average of about 6% for all US manufacturing, mining, and wholesale trade corporations.

As evidence of the cyclical nature of the industry, some years ago *Fortune* reported that the oil industry ranked 30th out of 36 industries in return to investors over the 1985–95 period, 34th out of 36 US industries in return on equity in 1995, and 32nd in return on sales.[3]

The role of OPEC

The oil and gas industry has seen a remarkable bevy of government regulations and interventions over the past century, from heavy taxation of petrol in Europe to US price controls on domestic production in the 1970s. The creation of the OPEC represents government intervention on a global scale. OPEC, founded in Bagdad in 1960, has worked continuously to shift power, primarily bargaining power over prices, away from the large oil companies to the producing countries. Outside of the founding five members (Iran, Iraq, Kuwait, Saudi Arabia, and Venezuela), OPEC's membership has varied over time. Membership currently stands at 12, with Angola's entry in 2006 the first new entrant (a number have exited) since the early 1970s.

OPEC's official mission is "to coordinate and unify the petroleum policies of Member Countries and ensure the stabilization of oil prices in order to secure an efficient, economic, and regular supply of petroleum to consumers, a steady income to producers, and a fair return on capital to those investing in the petroleum industry."[4] Despite being a cartel, OPEC's ability to control prices is questionable. Surging oil prices in the 1980s resulted in energy conservation and increased exploration outside OPEC. Maintaining discipline among OPEC members has been a major problem (as is typical in all cartels). Massive cheating was blamed for the oil price crash of 1986, and in the 1990s, Venezuela was considered one of the bigger OPEC cheats in regularly producing more than its quota. Figure 2–2 shows the global oil production by non-OPEC and OPEC members since the 1960s.

OPEC's influence extends beyond present production into the future, as OPEC members continue to hold the dominant proportion of proven oil reserves, as illustrated in figure 2–3. Although there has been a continuous and committed effort to explore for decades by thousands of market participants, that which "has been found" continues to fall largely within the control of the OPEC nations.

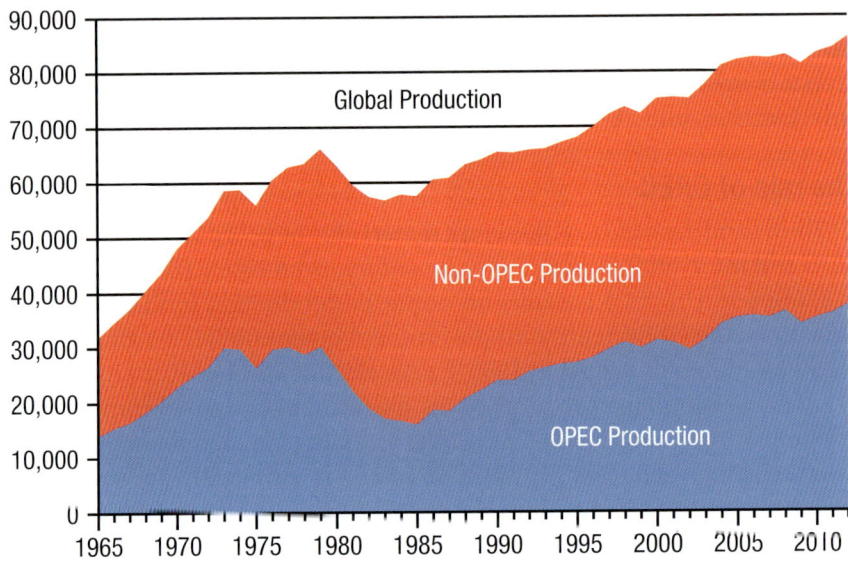

Fig. 2–2. Global oil production (thousands of barrels per day)
Source: Constructed by authors based on data drawn from *BP Statistical Review of World Energy, 2013.*

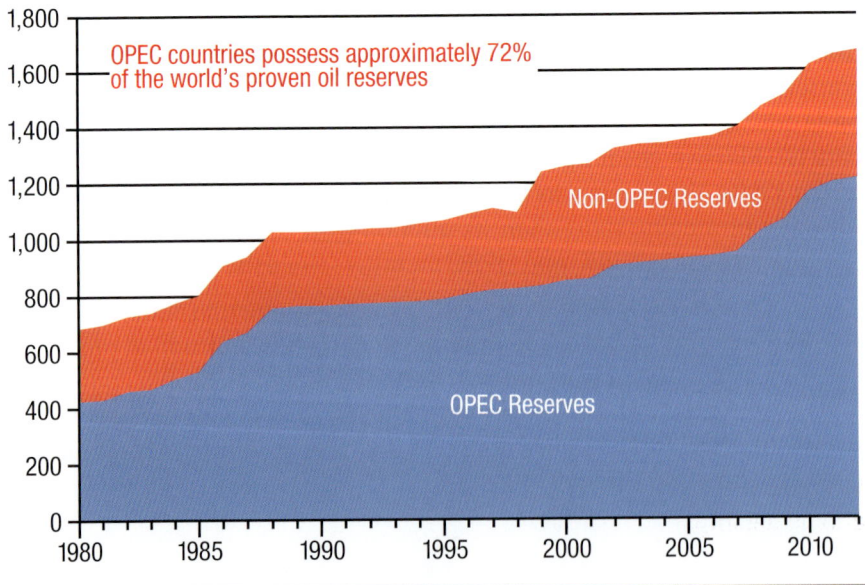

Fig. 2–3. Global oil reserves (billion barrels)
Source: Constructed by authors based on data drawn from *BP Statistical Review of World Energy, 2013.*

The resource curse

The *resource curse* is a paradox of the oil and gas industry. Despite high resource prices, the living standards in many oil-producing countries are low. This condition has led to the inability of countries rich in natural resources to use that wealth to strengthen their economies and, counterintuitively, to have lower economic growth than countries without an abundance of natural resources.[5] When times are good and oil prices are high, oil-rich countries may prosper. When oil prices fall, as they inevitably do, an overreliance on the oil sector can leave a country in a perilous situation. Moreover, the oil industries of the petroleum-nationalistic countries often suffer from a lack of investment and from heavily subsidized domestic petroleum products. For example, although Iran has huge oil reserves, the country's oil industry is in a shambles. Iran's oil production, although up from a decade ago, was less than the level reached under the government of the former Shah of Iran in 1979. Iran imported about 40% of its gasoline and is unable to produce sufficient crude to meet its OPEC quota.

Mexico also has declining production and significant imports of refined products. Until recently, the Mexican constitution did not allow foreign direct investment in the oil and gas industry. After many years of underinvestment and of Mexican governments using the oil industry as their primary source of revenue, the industry is in dire straits. Without major investment and new technology, Mexico's oil production is poised to fall. For example, production at the Cantarell oil field, one of the largest fields in the world, fell from more than 2 million b/d in 2004 to less than 400,000 b/d in 2014. To stem the decline in production, the Mexican government passed legislation in 2013 to change the constitution and allow foreign participation in the oil industry.

Major Industry Players and Competitors

The organizations that dominate the global oil and gas industry have changed dramatically over time—who they are, what they do, and, of critical significance for the future of the industry, how they compete. Table 2–2 lists what are termed the *global supermajors*, a diverse cross-section of countries and ownership.

Table 2–2. Top 10 global super majors

Company	Home Country	2012 Revenue (billion US$)	2012 Production (million bbls/day)	Reserves (billion bbls)	2012 Market Value (billion US$)	Reserve Life (years)	2012 Employees
Saudi Aramco	Saudi Arabia	$311	12.7	307	not available	66.2	54,000
Gazprom	Russia	$153	8.4	112	$92	36.5	393,000
NIOC	Iran	$110	6.1	311	not available	139.7	41,000
ExxonMobil	United States	$482	4.1	25	$417	16.7	75,000
PetroChina	China	$365	3.6	23	$239	17.5	553,000
Kuwait Petroleum	Kuwait	$114	3.3	112	not available	93.0	18,000
Royal Dutch Shell	UK/Netherlands	$467	3.3	8	$218	6.6	85,000
Pétroleos Mexicanos	Mexico	$129	3.2	11	not available	9.4	138,000
BP	United Kingdom	$376	3.0	7	$130	6.4	86,000
Chevron	United States	$231	2.9	9	$244	8.5	58,000

Integrated oil companies

The term *integrated oil companies* (IOCs) refers to companies that operate in many industry segments from exploration to refining, marketing, and retail. In the early days of the industry, there was true vertical integration in which producers refined most of their production and then marketed refined products through their company-owned retail outlets. In the modern industry, the IOCs operate in many segments, but also buy and sell oil and gas to and from other firms. (Somewhat confusingly, the term IOC can also mean international oil company.)

For many years, the largest IOCs (also known as oil majors) were the *Seven Sisters*, and included:

1. Standard Oil of New Jersey (Esso), which later became Exxon and then merged with Mobil to create ExxonMobil.
2. Royal Dutch Shell.
3. Anglo-Persian Oil Company, which became British Petroleum, then BP Amoco following a merger with Amoco (which was formerly Standard Oil of Indiana). The company is now known as BP.
4. Standard Oil of New York (Socony) became Mobil, which merged with Exxon.
5. Standard Oil of California (Socal) became Chevron.
6. Gulf Oil, most of which became part of Chevron.
7. Texaco, which merged with Chevron in 2001.

Figure 2–4 lists the largest oil and gas companies by stock market capitalization today. It is evident that the industry is dominated by a mix of global IOCs and *national oil companies* (NOCs). Based on market capitalization, the largest publicly traded (and in some cases, government-controlled) companies include a diverse and global set of firms such as PetroChina (China), Gazprom (Russia), Sinopec (China), Petrobras (Brazil), Total (France), and Eni (Italy).

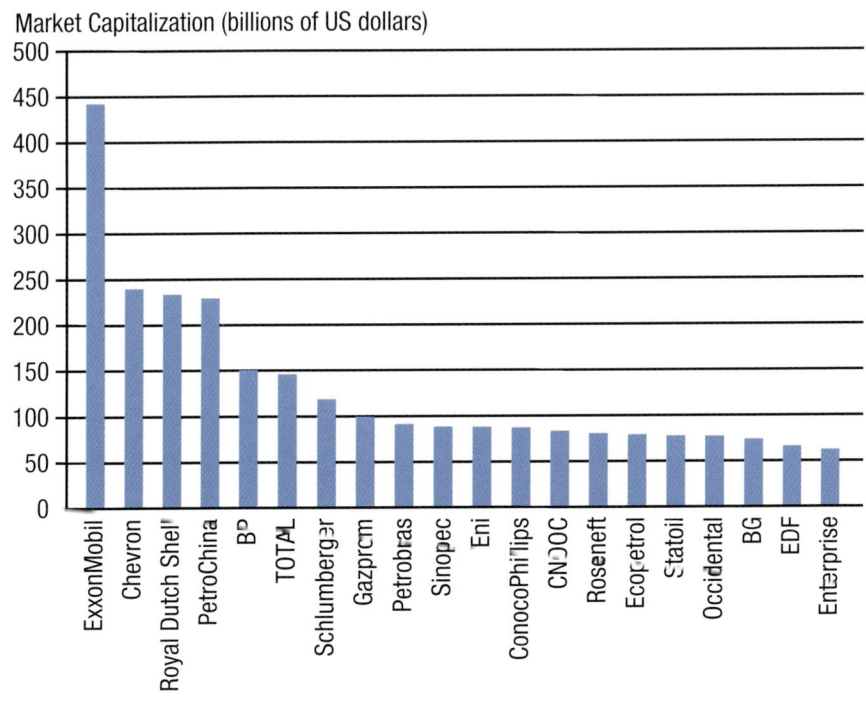

Fig. 2–4. Top 20 publicly traded oil and gas companies
Source: IHS. Market capitalization = Share price × Shares outstanding.

Given the long product life cycles and the huge capital investment required in the oil industry, the large IOCs were often described as stodgy and conservative. Before bankruptcy, Enron executives regularly derided the oil majors as dinosaurs that were too slow moving and that would eventually become extinct. The reality, of course, is very different. Oil majors like BP, ExxonMobil, and Shell (and their predecessor companies) have been around for more than a century. Through experience that is occasionally painful, the IOCs have learned how to deal with the enormous financial and political risks of the oil and gas industry. The IOCs take a long-term view and recognize that cycles and uncertainty are an inherent part of the industry.

On the surface, the IOCs look similar in terms of the activities they perform. All appear to be vertically integrated from exploration to retail distribution. However, there are fundamental cultural, organizational, and financial differences among the firms. The IOCs use various organizational designs to deal with vertical integration. The IOCs have different portfolios of projects around the world, and over the years, they have developed different relationships with various governments and NOCs.

National oil companies (NOCs)

One of the most important trends of past few decades has been the growing importance of NOCs. Although BP, ExxonMobil, and Shell are among the largest publicly traded companies in the world, they do not rank in the top 10 of the world's largest oil and gas firms measured by reserves. The largest oil and gas firms based on reserves are, by a large margin, NOCs partially or wholly state-owned. This is readily apparent from table 2–3. The NOCs control about 90% of the world's oil and gas, and most new oil is expected to be found in their territories.

Table 2–3. The 50 largest publicly traded oil and gas companies in the world

2013 Rank	2012 Rank	Company	Market Capitalization (billion US$)	Primary Business	HQ Country
1	1	ExxonMobil	442.1	Integrated IOC	United States
2	4	Chevron	240.2	Integrated IOC	United States
3	3	Royal Dutch Shell	233.8	Integrated IOC	Netherlands
4	2	PetroChina	229.4	Integrated NOC	China
5	5	BP	150.7	Integrated IOC	United Kingdom
6	8	TOTAL	145.9	Integrated IOC	France
7	13	Schlumberger	118.7	Oilfield & drilling services	United States
8	9	Gazprom	99.2	Integrated NOC	Russia
9	7	Petrobras	91.0	Integrated NOC	Brazil
10	11	Sinopec	88.2	Integrated NOC	China
11	14	Eni	87.6	Integrated IOC	Italy
12	17	ConocoPhillips	86.6	E&P	United States
13	10	CNOOC	83.0	Integrated NOC	China
14	12	Rosneft	80.2	Integrated NOC	Russia
15	6	Ecopetrol	78.8	Integrated NOC	Colombia
16	15	Statoil	77.2	Integrated NOC	Norway
17	18	Occidental	76.7	E&P	United States
18	20	BG	73.3	Integrated IOC	United Kingdom
19	36	EDF	65.5	Electricity	France
20	25	Enterprise	61.9	Midstream/infrastructure	United States
21	23	GDF Suez	56.9	Gas/utilities	France
22	16	China Shenhua	53.8	Coal	Malaysia
23	19	LUKOIL	52.6	Integrated IOC	Russia
24	21	Suncor	52.2	Integrated IOC	Canada

2013 Rank	2012 Rank	Company	Market Capitalization (billion US$)	Primary Business	HQ Country
25	26	Duke Energy	48.7	Electricity	United States
26	22	Reliance	46.8	R&M	India
27	38	Phillips 66	46.2	R&M	United States
28	39	EOG Resources	45.8	E&P	United States
29	40	Halliburton	43.0	Drilling & oilfield services	United States
30	33	ENEL	41.2	Electricity	Italy
31	27	ONGC	40.0	Integrated NOC	India
32	31	Anadarko	39.9	E&P	United States
33	44	Iberdrola	39.9	Low carbon power	Spain
34	45	Dominion	37.5	Gas/utilities	United States
35	32	Imperial Oil	37.5	Integrated IOC	Canada
36	28	Kinder Morgan	37.3	Midstream/infrastructure	United States
37	30	E.ON	37.0	Gas/utilities	Germany
38	35	NOVATEK	36.8	E&P	Russia
39	42	Canadian Natural	36.7	E&P	Canada
40	29	Southern	36.2	Electricity	United States
41	34	Enbridge	36.2	Midstream/infrastructure	Canada
42	43	Apache	34.3	E&P	United States
43	46	National Oilwell Varco	34.0	Equipment & EPC	United States
44	60	Endesa	34.0	Low carbon power	Spain
45	24	BHP Billiton*	33.1	Other	Australia
46	54	Repsol	32.9	Integrated IOC	Spain
47	37	TransCanada	32.3	Midstream/infrastructure	Canada
48	52	Sasol	31.8	Integrated IOC	South Africa
49	47	Husky	31.2	Integrated IOC	Canada
50	41	Surgutneftnegaz	30.7	Integrated IOC	Russia

Source: IHS. IOC is integrated oil company; NOC is national oil company. Note: BHP Billiton is ranked based on a value of 19% of the company's total market capitalization, representing the contribution of its petroleum segment to total EBIT in the 12 months ended 6/30/2013.

Viewed from an ownership and corporate objective perspective, NOCs lead a complicated existence. As controlled by the state, they must serve a variety of different commercial and political objectives. (It is important to note that many NOCs are also publicly traded, but the governing government share—the *golden share*—remains with the government.) NOCs are often required to provide a secure supply of oil and gas for the country; provide jobs for citizens; generate cash flows in the form of taxes, duties, and other payments to the state (which in turn is often very dependent on those same revenues for its government's funding); as well as "profit" and "returns" if publicly traded. In short, a nearly impossible list of objectives.

Viewed from a business perspective, NOCs have a mixed reputation in terms of discipline and efficiency. For example, Venezuela nationalized its oil industry in the 1970s and created Petróleos de Venezuela S.A. (PDVSA). PDVSA developed a reputation for professionalism and competence and was relatively free from the corruption and cronyism that pervaded, and continues to pervade, so many of the NOCs.[6] By 1998, 36 foreign oil firms were operating in Venezuela, and PDVSA had ambitious expansion plans. In 1999, Hugo Chávez became president and almost immediately began to question the management and autonomy of PDVSA. After a bitter strike in 2002, PDVSA lost about two-thirds of its managerial and technical staff, in many cases to competitors outside Venezuela. From a peak of 3 million b/d in 1998, exports have declined by half, and the company imported a significant amount of gasoline. As a company, PDVSA is indistinguishable from the government. The company is required to spend a 10th of its investment budget on social programs, which included sending low-cost heating oil to poor South Americans. Company hiring is based on social and political goals; for example, candidates from larger families are given priority. In 2006, the Venezuelan Congress approved new guidelines to turn 32 privately run oil fields over to state-controlled joint ventures.

According to many analysts, nationalization has failed to live up to expectations almost everywhere. NOCs often suffer from excessive and misguided government intervention. Many NOCs operate as the de facto treasury for the country. In Nigeria, for example, oil revenues represented more than 90% of hard currency earnings and about 60% of GDP. Nigeria's economic and financial crimes commission estimated that more than $380 billion of government revenues had been stolen or misused since 1960.[7] Some of the Middle Eastern NOCs are required to hire large numbers of locals, leaving them heavily overstaffed. Others, for example, in India and Russia, must sell their products at subsidized prices. Underinvestment in the downstream is a chronic problem for many NOCs, resulting in countries like Mexico and Iran with huge reserves having to import petroleum products. Monopoly positions held by many NOCs contribute to underinvestment. In Russia, Gazprom controls the pipeline network, making it difficult for other Russian gas producers to expand their production. Russia has also used its NOCs as agents of foreign policy, going so far as to occasionally threaten to shut off gas supplies to recalcitrant neighbors. These actions have prompted discussions in Western Europe about how to diversify gas supply away from Russian gas.

Some NOCs are well-run and profitable enterprises. Statoil of Norway is considered to be among the best of the NOCs. The NOCs of Brazil, Malaysia, and Qatar are also viewed as reasonably well-run companies. Petrobras (Brazil) has developed leading technology in deepwater drilling, Petronas (Malaysia) has expanded into multiple countries, and Qatar has built a world's leading position in liquefied natural gas (LNG).

The role that NOCs will play in the future is not clear. Some analysts see NOCs as inefficient and corrupt arms of government that will never compete in a true economic sense. Other analysts raise different issues, suggesting that the NOCs are in a period of transition and will become competitive forces to be reckoned with. Regardless of what happens, the NOCs and their sovereign owners control the vast majority of the world's oil and gas reserves.

Independents

Independents are the non-government-owned companies that focus on either the upstream or the downstream. Figure 2–5 lists the top independent refining and marketing firms. Many of these firms are sizable players and rank in the top 50 of all non-government-owned oil and gas companies. In the downstream refining and marketing area, the largest independents are scattered about the world's largest energy-consuming countries. The downstream independents tend to have lower market capitalizations than the upstream independents.

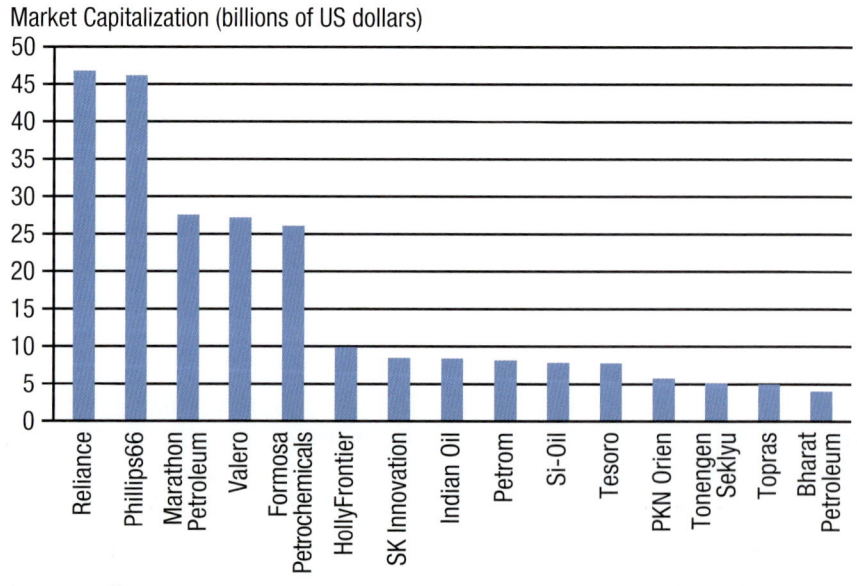

Fig. 2–5. Top independent refining and marketing firms
Source: "HIS Energy 50: The Definitive Annual Ranking of the World's Largest Listed Energy Firms," HIS, January 2014.

Oil field service companies and other firms

In addition to the IOCs, NOCs, and independents, the oil and gas industry includes a huge number of other firms that perform important functions. Oil field service firms play a critical role throughout the exploration, development, and production phases. The three largest of these are Schlumberger (123,000 employees), Halliburton (77,000 employees), and Baker Hughes (60,000 employees), as shown in table 2-4. These firms provide both products and services that, according to Baker Hughes website, help oil and gas producers "find, develop, produce, and manage oil and gas reservoirs." Because the oil field service firms do not seek ownership rights to oil and gas reserves, many analysts predict that their role will become increasingly important in the future as partners of the NOCs.

Table 2–4. Top 15 oilfield and drilling services firms

2013 Rank	2012 Rank	Company	Market Capitalization (billion US$)	Average Price/Earnings	Debt/Capital Ratio	HQ Country
1	1	Schlumberger	118.7	20	25%	United States
2	2	Halliburton	43.0	18	38%	United States
3	3	Baker Hughes	24.5	22	20%	United States
4	4	Seadrill	19.1	8	64%	Norway
5	5	Transocean	17.8	11	40%	Switzerland
6	7	COSL	15.7	19	49%	China
7	6	Ensco	13.4	10	28%	United Kingdom
8	10	Weatherford	11.9	29	52%	Switzerland
9	9	Noble	9.5	14	37%	United Kingdom
10	11	H&P	9.0	15	4%	United States
11	13	Core Laboratories	8.7	38	58%	Netherlands
12	8	Diamond Offshore	7.9	12	24%	United States
13	—	Oil States	5.6	16	27%	United States
14	14	Nabors	5.0	19	41%	Bermuda
15	12	CGG	3.1	33	36%	France

Source: "HIS Energy 50: The Definitive Annaual Ranking of the World's Largest Listed Energy Firms," IHS, January 2014.

Thousands of other firms provide a vast array of services and products for the industry. For example, gas utilities such as Gaz de France and Tokyo Gas are major customers for gas producers. Pipeline companies distribute gas, crude oil, and petroleum products. The firms involved in drilling and seismic services provide drilling rigs and expertise for onshore and offshore wells.

The Oil and Gas Industry Value Chain

In every industry, there are various activities that transform inputs of raw materials, knowledge, labor, and capital into end products purchased by customers. A *value chain* helps to identify the independent, economically viable segments of an industry.[8] The oil and gas industry value chain is shown in figure 2–6.

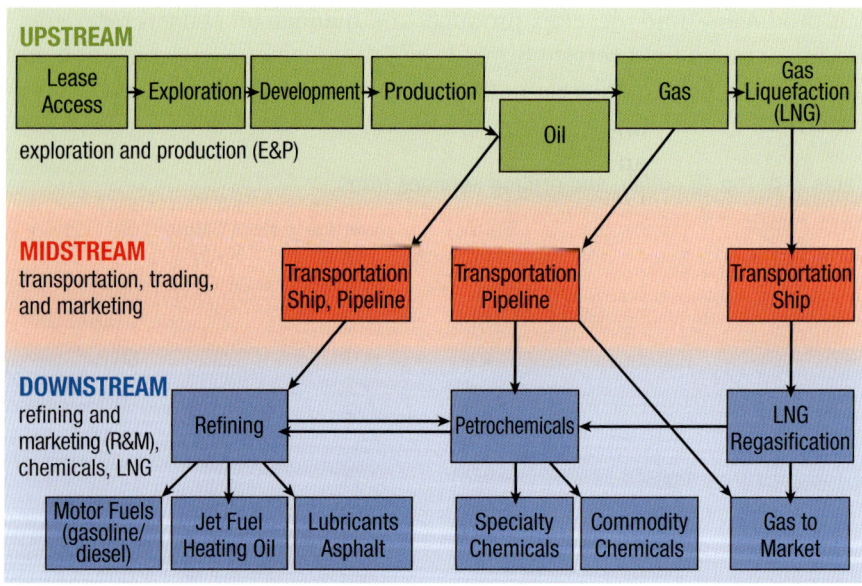

Fig. 2–6. The oil and gas value chain

Value refers to what customers are willing to pay for, so the value chain helps to identify specific activities that create value throughout the chain. Companies can use value chains to determine where they are strong and where they have limited competitive strength. All industries have *upstream* (close to raw materials and basic inputs) and *downstream* (close to the customer) segments. In the oil and gas industry, the terms *upstream*, *downstream*, and *midstream* are important descriptors of the industry activities.

Upstream: Exploration, development, and production

Upstream activities include *exploration, development,* and *production.* In simple terms, after a lease is obtained, oil and gas are discovered during exploration; the discovery requires development; and production is the long-term process of drilling and extracting oil and gas. Since exploration and development must take place where resources are located and most oil ownership regimes are based on state sovereignty, companies have to deal with complex government policies and regulations. Most countries grant oil and gas development rights to private companies through a process of either negotiation or bidding. The main aim of the private company is to maximize profit, whereas the host country government is interested in maximizing revenue. Not surprisingly, these two aims often conflict. The agreements between oil companies and governments are called *fiscal regimes*. The most common fiscal regime is a production-sharing agreement.

The method used to bid for, grant, and then renew or extend oil and gas rights varies from country to country. Once the rights to explore are acquired, a well is drilled. A financial analysis is a determining factor in the classification of a well as an oil well, natural gas well, or dry hole. If the well can produce enough oil or gas to cover the cost of completion and production, it will be put into production. Otherwise, it is classified as a dry hole, even if oil or gas is found. The percentage of wells completed is used as a measure of success. Immediately after World War II, 65% of the wells drilled were completed as oil or gas wells. This percentage declined to about 57% by the end of the 1960s. It then rose steadily during the 1970s to reach 70% at the end of that decade, primarily because of the rise in oil prices. This was followed by a plateau or modest decline through most of the 1980s. Beginning in 1990, completion rates increased dramatically to 77%. The increases of the 1990s had more to do with new technology than higher prices.[9]

Most upstream projects are done in some type of partnership structure. For example, a production-sharing agreement for the Azeri, Chirag, and Gunashli development in Azerbaijan was signed in September 1994. BP is the operator, with a 34.1% stake; the partners were Chevron, with 10.3%; Socar, 10%; INPEX, 10%; StatoilHydro, 8.56%; ExxonMobil, 8%; TPAO, 6.8%; Devon, 5.6%; Itochu, 3.9%; and Hess, 2.7%.

Reservoir management. For companies involved in the upstream, reservoir management is an essential skill. Reservoir management involves ensuring that reserves are replaced and that existing oil and gas fields are efficiently managed. Asset acquisition, divestiture, and partnering are key aspects of reservoir management. Upstream companies try to replace more than 100% of the oil and gas produced. Determining the level of *proved reserves* (the amount of oil and gas the firm is reasonably certain to recover under existing economic and operating conditions) is a complex process. Matthew Simmons, founder of the energy-focused investment bank Simmons and Company, commented that "95% of world 'proven reserves' are in-house guesses," "most reserve appreciation is exaggerated," and "95% of the world's 'proven reserves' are unaudited."[10] The pressure to replace reserves has on occasion resulted in some unintended behaviors. In 2004, Shell's CEO left earlier than anticipated after revelations that the company had overstated its reserves by nearly 25%.

Upstream profitability. Profitability is a function of costs and commodity prices. One way to consider profitability is with the break-even price for crude oil, which is the price that equals the cost of producing the oil. In Saudi Arabia and Kuwait, the break-even price was between $10 and $20 per barrel. New oil sands projects in Alberta had a break-even price of about $80 per barrel. For US shale oil projects, estimates of break-even prices ranged from $50 to 80 per barrel. In the Gulf of Mexico the break-even price was about $55–$70.

The term *break-even price* is also used to understand the relationship between oil-exporting nations and their fiscal management. In other words, the oil price that a country requires in order to match oil revenues to planned government expenditures is its fiscal break-even price. Many oil-exporting nations have seen their fiscal break-even prices rise significantly over the past few years. For example, it was estimated that Russia's fiscal break-even price in 2014 was getting close to crude market prices, which created a vulnerability to the steep decline in oil prices that occurred a few months later.

Midstream: Trading and transportation

The midstream in the value chain comprises the activities of storing, trading, and transporting crude oil and natural gas. As shown in figure 2–6, once oil and gas are in production, there is a divergence in the value chain. Crude oil that is produced must be sold and transported from the wellhead to a refinery. Natural gas must be moved to markets via pipeline or ship; we provide an overview of the gas business in a later section.

Crude oil has little or no value until it is refined into products such as gasoline and diesel. Thus, producers of crude oil must sell and transport their product to refineries. The market for crude oil involves many players, including refiners, speculators, commodities exchanges, shipping companies, IOCs, NOCs, independents, and OPEC. Market-making activities in the oil business have become front-page news, and the daily price of crude oil is as frequently reported in the news just as the weather is.

The ease by which liquids can be transported is a key reason why crude oil has become such an important source of energy. Although pipelines, ships, and barges are the most common transportation platforms for crude oil, railroads and tank trucks are also used in some parts of the world. In recent years railroads have made a resurgence in the United States and Canada because of the rapid growth in production of oil in North Dakota and Alberta and a shortage of pipeline capacity. The shipping industry is very fragmented and, because oil tankers travel for the most part in international waters, largely unregulated. New technologies in ship building in recent decades have allowed ships to become larger and safer.

Pipelines in Alaska, Chad, Russia, and other countries have enabled the transport of oil from very remote locations to markets. The construction and management of pipelines is fraught with geopolitical challenges, which means that the pipeline development process takes many years or even decades. Pipelines that cross national borders are enormously complex to negotiate and build. Countries with pipelines that cross their land have been known to use them as bargaining chips. Terrorists often sabotage pipelines, and in some countries, such as Nigeria and Iraq, oil theft from pipelines and the associated environmental and safety issues are daily occurrences.

Downstream: Oil refining and marketing

The refining of crude oil produces a variety of products, including gasoline, diesel fuel, jet fuel, home heating oil, and chemical feedstock. In the United States, about 60% of refinery product volume is gasoline. Derivative products are sold directly to end users through retail locations, directly to large users, such as utilities and commercial customers, and through wholesale networks.

The financial performance of the refining sector has always been volatile. The primary measure of industry profitability is the refining margin, which is the difference between the price of crude oil and that of the refined products. Crude prices can fluctuate for many reasons. Weather in the Gulf Coast states, political instability in oil-producing countries, and OPEC actions can all impact the price of crude oil. These fluctuations are not always accompanied by matching changes in the price of finished products, leading to large expansions or contractions in the refining margin. Refiners also get squeezed between the commodity markets for crude oil (crude is the largest cost to a refiner) and commodity markets for refined products like gasoline.

To put the downstream business in perspective, Lee Raymond, former ExxonMobil CEO, said in 1997, "I've been pessimistic on refining for 30 years, and I've run the damn places."[11] Shell's head of downstream operations described the business as "grubbing [i.e., begging] for pennies in a street. . . . If this industry, and especially the downstream, were to let its cost base slip, then we're going to have difficulty getting through those down-low cycles."[12] Refining margins for the US Gulf Coast, Northwestern Europe, and the Singapore markets are shown in figure 2–7.

The profitability of refining is driven primarily by the following factors
1. The costs of crude oil (by far the largest cost)
2. The cost of energy to run the refinery
3. The supply and demand for refinery products (i.e., if refining capacity is tight, refining margins usually rise)
4. Refinery product prices, which are set by a combination of the supply and demand of refinery products and crude oil prices
5. Refinery location and operational skills

After a so-called golden age of refining from 2002 to about 2007 (as seen by the large and positive spikes in refining margins in figure 2–7), refining entered a new era of change and consolidation. By 2014 many US and European refineries were either shut down or on the verge of closure. A report by A.T. Kearney concluded that by 2021, every refinery in Western Europe and North America would have to restructure, strategically reposition their assets, or leave the market.[13] Interestingly, despite the closure of various refineries in North America, total refining capacity continued to rise through de-bottlenecking and expansions to existing sites.

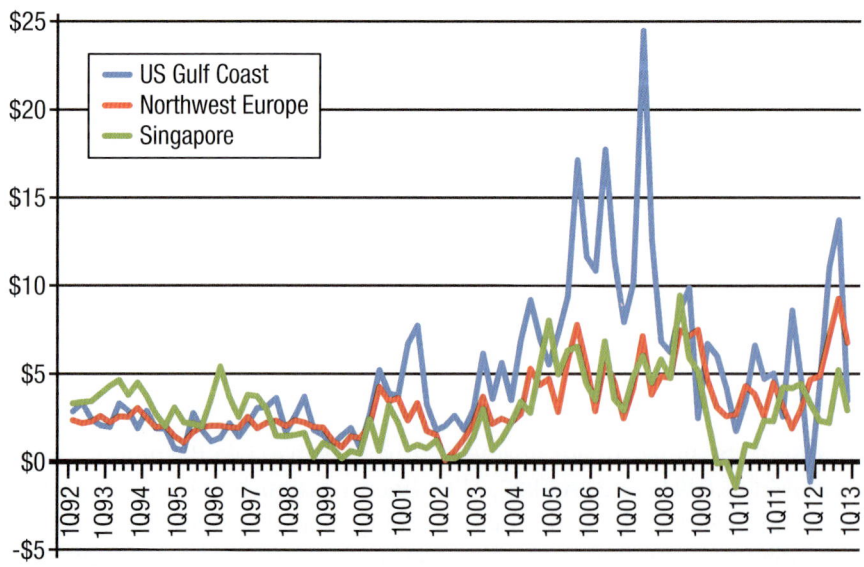

Fig. 2–7. Refining margins (US dollars per barrel)
Source: Constructed by authors based on data drawn from *BP Statistical Review of World Energy, 2013.*

Although no new greenfield refineries had been built in the United States for many decades, in Asia, Eastern Europe, and the Middle East, the industry has recently engaged in aggressive retooling, reinvestment, and expansion of refining capacity. In 2009, Reliance Industries completed the world's largest refinery complex at Jamnagar in India. The Jamnagar complex has a capacity of 1.24 million b/d. In the near term, Jamnagar is expected to focus on export markets. The largest market for Jamnagar is in the Middle East, followed by Africa, Europe, and the United States. Shipping costs are only pennies per gallon for finished products, even from India to the United States. A new state-of-the-art refinery and chemical plant complex is under construction in Malaysia, planned for completion in 2017.

Various questions arise in thinking about the future of refining:

- In 2011 the United States became a net exporter of refined products for the first time since 1949. Will the trend continue?
- Will the demand for electric and hybrid cars have a major impact on refined product demand?
- Where is global biofuel (mainly ethanol) demand going?
- How much refining capacity will close in Europe, Japan, and the United States?
- Will natural gas gain traction as a transportation fuel?
- Will there be more integration between refining and petrochemicals assets for the large NOCs?

Finally, it was clear that there was no clear "best" competitive model in refining. In North America and Western Europe, the oil majors were divesting or closing refineries. ConocoPhillips and Marathon were splitting into upstream and downstream companies. In contrast, Petrobras, one of the largest NOCs, was expanding refining capacity. Middle East NOCs were also increasing refining capacity. Russian upstream firms were looking to acquire downstream assets.

Transportation fuels retailing. In the transportation fuels retail sector, competition is intense, and margins have eroded over the past 15–20 years. For the oil majors, returns on capital employed are generally lower in retail than in other business areas. The entry of hypermarkets or supermarkets into retail gasoline and diesel sales in Western Europe and other markets had displaced small dealer networks, and national players found they could make good money from convenience store sales.

In the United States, supermarket and "petropreneur" entry into gasoline sales was also occurring, although not with the same speed as in Europe.

In most countries, transportation fuel was seen as a commodity product, which meant spending money on brand development had questionable results. The weakness of brands favored the entry of supermarkets because they compete on price and proximity and can sell fuel as a loss leader. With traditional retail barriers to competition gone, the majors were selling most of their company-owned stores in countries around the world. The buyers are a mix of convenience store specialists, such as Couche-Tard of Canada (more than 4,000 fuel stores in Canada and the United States); franchisers; distributors; and independent dealers.

Beyond Oil

The industry goes beyond crude oil and gasoline. An important component of the oil and gas industry is made up of natural gas and petrochemicals. We provide a brief overview of each below.

Natural gas

In recent years natural gas has played a much more important role in the global energy mix. Although the industry has always been referred to as "oil and gas," the focus on real development has always been oil. But several major technical innovations in recent years have revolutionized the natural gas sector, and today it represents the focus of much economic and employment development globally.

Liquefied natural gas (LNG). The first technical advancement is the continued growth of the LNG supply. For many years, natural gas was a niche product because, unlike crude oil, natural gas is not easily transported. Without a pipeline infrastructure, natural gas in its gas form cannot be transported far from its source. In some parts of the world, such Western and Central Europe and North America, a network of pipelines allows gas to be produced and distributed efficiently. In the United States gas pipeline companies operate more than 285,000 miles of pipe. In other parts of the world, such as offshore Africa or Qatar, pipelines to customers are not feasible. To transport "stranded" gas, it must be converted to liquid—LNG. To liquefy natural gas, impurities such as water, carbon dioxide, sulfur, and some of the heavier hydrocarbons must be removed. The gas is then cooled to about $-259°F$ ($-162°C$) at atmospheric pressure to condense the gas to liquid form. LNG is then transported by specially designed cryogenic sea vessels and road tankers to destination ports and facilities where it is "re-gased" to its natural gaseous form for use.

Historically, the costs of LNG treatment and transportation were so huge that development of gas reserves was slow. In recent years, LNG has moved from being a niche product to a vital part of the global energy business. As more players take part in investment, both in the upstream and downstream, and as new technologies are adopted, the prices for construction of LNG plants, receiving terminals, and ships have fallen, making LNG a more competitive energy source. LNG ships are also getting much larger and more efficient. In addition, natural gas to liquid (GTL) technology provides an alternative to LNG and converts gas to liquid products, such as fuels and lubricants, which can be easily transported. Questions remained about the economic viability of GTL technology, and only a few major projects had been completed, including the largest one, Shell's approximately $20 billion Pearl project in Qatar.

Major technological and structural changes continue to occur in the LNG business. The floating liquefied natural gas (FLNG) vessel is a technology that allows producers to commercialize offshore gas deposits without pipelines and onshore infrastructure. FLNGs create opportunities to commercialize gas fields that would otherwise be untouched. Another innovation is the floating natural gas liquefaction, regasification, and storage unit (FLRSU) vessel, which moves the various industrial processes offshore and makes the equipment available for redeployment at the end of the resource life.

Changes in the LNG market and in LNG shipping have increased flexibility for producers and consumers, and shorter contracts have been negotiated. The agreement to develop the huge Qatargas 2 project, jointly owned by ExxonMobil and Qatar Petroleum, was finalized without contracts for gas sales in place. An LNG ship can deliver its gas anywhere there is an LNG terminal, making LNG almost as flexible in delivery as crude oil. There is also speculation that the rapid growth in Middle East LNG supply could lead to a global convergence in gas pricing and markets, with LNG becoming a traded commodity. As well, buyers and sellers have been taking on new roles. Buyers have been investing in the upstream, including liquefaction plants. Producers, such as BP and Shell, have leased capacity at terminals and are extending their role into trading. New buyers have been emerging, including independent power producers.

Shale gas. The second factor that helps explain the increased importance of gas is *shale gas*. The impact of shale gas on US and global gas markets has resulted in what has been referred to as a game changer for US energy supply. As recently as 2003 the consensus was that the United States would have to import large quantities of LNG to satisfy gas demand. Less than a decade later, US production can easily meet domestic gas demand and several dormant LNG import terminals are being considered for conversion to LNG exports facilities. As demonstrated by figure 2–8, the United States is today the world's largest gas-producing nation.

Although the rest of the world has lagged behind the US shale gas experience, many other countries have the potential to develop shale gas resources. According to the US Energy Information Administration, several countries, such as France, Poland, Turkey, Ukraine, South Africa, Morocco, and Chile, could significantly reduce gas imports if they develop their shale gas resources.

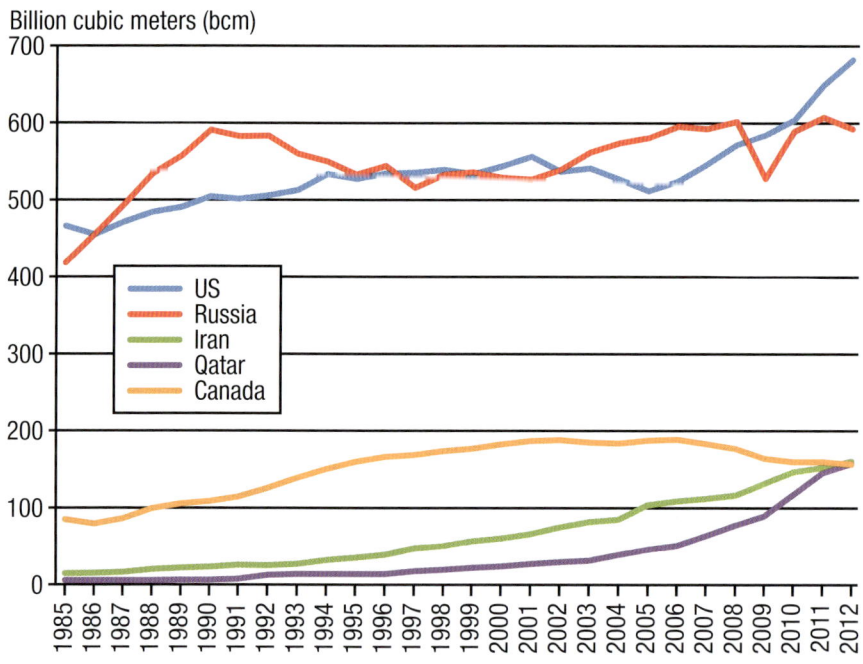

Fig. 2–8. Major gas-producing nations
Source: Constructed by authors based on data drawn from *BP Statistical Review of World Energy, 2013*.

Gas turbine technology. The third influential technical change driving the growth in the use of natural gas has been the development of the extremely efficient *combined cycle gas turbine* (CCGT). Unlike petroleum, which is used primarily for transportation fuels, natural gas has often been used for generating electricity and in commercial and residential heating applications. Technical advances in the 1990s resulted in CCGT technology, which allowed many utility and industrial power producers to use gas turbines for electrical power generation. With low capital and operating costs, gas turbines have become a focal point for a variety of electrical applications, resulting in a rapid growth in the demand for natural gas.

Natural gas use in electrical power has been particularly pronounced in the Asian marketplace. Economies like that of Japan, Taiwan, and Korea rely on the importation of natural gas for their power production. Although constituting only a portion of their energy mix, natural gas constitutes a reliable and, in many cases, a safer fuel (particularly since the Fukushima nuclear plant accident) choice for electricity. Although natural gas is a fossil fuel, it is considered the cleanest of the fossil fuels, with lower levels of carbon dioxide emissions than oil or coal.

Petrochemicals

Producing petrochemicals is the furthest downstream activity in the value chain. Although all of the major IOCs were involved in chemicals to some degree, they had different strategic approaches. ExxonMobil Chemical, one of the world's largest chemical businesses, produced both cyclical-commodity-type products, such as olefins and polyethylene, as well as a range of less cyclical specialty businesses. Many of ExxonMobil's refineries and chemical plants were co-located, providing opportunities for shared knowledge and support services and the creation of product-based synergies. In the past, BP and Shell had chemical businesses that were among the largest in the world. In 2005, BP decided that its chemical business was noncore and divested the majority of the business. BP's remaining chemicals businesses became part of the refining and marketing division and were no longer considered a separate corporate division. Shell also downsized its chemicals business. The rising players in chemicals are in the Middle East and Asia, and include NOCs such as Sabic (Saudi Arabia) and Sinopec (China), as well as non-state-owned companies, such as Reliance (India).

The commodity side of the petrochemical sector is capital-intensive and deeply cyclical. Margins and profitability for commodity chemicals depend on scale, capacity utilization, operating cost discipline, and access to low-cost feedstocks. The specialty market is at the other end of the spectrum. Specialty chemicals are sold on the basis of their performance in customer applications, not chemical composition. Patented products or technologies can enhance the value of specialty chemicals. Product differentiation may be the result of proprietary technologies such as unique catalysts or chemical processes, and it can also be the result of branding, marketing, customer service, and delivery.

Evolution of the Industry

A number of seismic forces have been at work driving change in the global oil and gas industry.

In this section we discuss how the industry has evolved into its present condition. Important changes in the industry have occurred because of innovation and technology, mergers and acquisitions, the growing importance of China and India for the industry, the growth of unconventional oil and gas, and the emergence of industry substitutes and alternative fuels. We conclude this section with a brief glimpse of the future of the industry.

Innovation and technology

Innovation plays a key role across the oil and gas value chain. Innovations in areas such as deepwater drilling and LNG shipping were discussed earlier. In the upstream, many important technologies have been developed in the past few decades, including increased use of 3-D seismic data to reduce drilling risk, and directional and horizontal drilling to improve production in reservoirs.[14] Innovations in financial instruments were used to limit exposure to resource price movements. In oil field management, wireless technologies allowed for faster and cheaper communication than the traditional wired underground infrastructure. In refining, nanotechnology has enabled refiners to tailor refining catalysts to accelerate reactions, increase product volumes, and remove impurities, which has led to increased refining capacity. In retailing, innovations such as unstaffed stations have reduced retail costs.

Mergers and acquisitions

Mergers and acquisitions have been an important element in the oil and gas industry since its inception. Although the mega-mergers, such as BP and Amoco, Total and PetroFina, Chevron and Texaco, and Exxon and Mobil, receive much of the press, there are also many smaller deals. In recent years NOCs have been making many acquisitions to gain access to resources and new technology. Private equity–backed acquisitions have also become more prevalent.

In looking at the mega-mergers done over the past few decades, one might conclude that eventually there will only be a handful of oil companies in the world. The reality is different. Research shows that the oil industry is much less concentrated today than it was 50 years ago.[15] There are opportunities for new entrants despite the huge size of the largest IOCs and NOCs. In the downstream, new entrants have had a significant impact on industry structure. In chemicals, Ineos, the privately held British company, grew through a series of related acquisitions to become one of the world's largest chemical companies. In the upstream, the huge financial scale of projects such as Gorgon, Kashagan, Sakhalin I and II or Qatargas 2 make it unlikely that a new entrant could challenge the majors in the largest and most technological projects. However, if NOCs in China and India continue to acquire and grow, they may develop the technological and financial skills to compete for large complex upstream projects.

China and India

In 1998, China became a net importer of oil for the first time. In 2006, China overtook Japan to become the world's second largest importer. By 2030, China will likely be importing about 80% of its oil. Clearly, China and Chinese companies are going to be major players in the oil and gas industry. Thousands of gas stations are being built, and Chinese companies are aggressively investing in upstream projects around the world. Unlike the United States and Europe, China has no qualms about allowing its oil companies to invest in countries like Sudan and Iran. Chinese companies have also been actively buying assets outside China, including the $15 billion purchase of the Canadian company Nexen.

India is also a force to be reckoned with in the global oil and gas industry. India, the fifth largest oil consumer, needs energy to feed its rapidly growing and industrializing economy. Companies such as Reliance are moving aggressively into the upstream, and stodgy state-owned companies such as ONGC, Oil India Limited, and Gas Authority of India are slowly becoming more productive. Like China, India is far from self-sufficient in energy and must find new energy sources.

Unconventional oil and gas

Growth in unconventional oil and gas production will have a profound impact on the world's energy supply over the next few decades. In fact, what we currently call *unconventional* will likely lose that label in the coming years. Unconventional gas, mainly *shale gas*, *coal bed methane*, and *tight gas* (gas locked in impermeable hard rock) will constitute the majority of the growth in natural gas production over the next few decades.

The production of *unconventional oil*, which refers primarily to the crude produced from deepwater, oil sands, and shale, will also grow substantially. The exploitation of unconventional resources is the result of technologies such as hydraulic fracturing and horizontal drilling as well as the entrepreneurial initiatives of industry participants doing what has been done for more than a hundred years—searching for innovative ways to economically create value from scarce resources.

Industry substitutes and alternative fuels

Various factors have contributed to a large investment flow into alternative fuel projects, including the rapid rise in oil and gas prices in recent years, concerns about global climate change, perceived competitive opportunities by energy companies (new entrants and entrenched players), and government subsidies. Despite these investments and the often strong public support for them, hydrocarbons will continue to be the world's primary energy source for years to come. By 2040 wind-powered energy is predicted to grow by seven times but will account for only about 7% of global electricity supply. Solar power is also expected to increase significantly but will probably only account for about 2% of global electricity supply in 2040.[16]

What's next for the global oil industry?

A few predictions seem fairly safe: The global demand for oil and gas will continue to rise over the next few decades; NOCs will continue to expand beyond their home markets; finding new conventional sources of oil and gas will get harder and require innovative new technologies; unconventional oil and gas production will grow substantially; investment in nonhydrocarbon energy sources will continue; the oil and gas industry will remain one of the most vital for the global economy; and despite the high prices of recent years, the industry will continue to go through up-and-down cycles. Finally, oil and gas firms, and especially the majors, will continue to do what they have done for more than a century: take a long-term view, invest for the future, push the boundaries of technology, and seek new resources and markets in every corner of the world.

Conclusion

In this chapter we provided a brief description and background of the oil and gas industry and demonstrated the importance of oil and gas in the global economy. We described the major players in the industry, including IOCs, NOCs, independents, and oil field service companies. We explained the oil and gas industry value chain, showing how upstream, midstream, and downstream all contribute value. We have also shown how the industry has gone beyond oil, particularly in natural gas and petrochemicals. We illustrated a number of factors that differentiate the oil and gas industry from others; for example, compared to others, it is extremely global, must frequently deal with governments, and is often project-oriented. This chapter concluded with the recent evolution of the industry. Next, we will look at the major differences between the oil and gas industry and other industries, and show the implications of those differences for the aspects of human resources discussed in chapter 1. We will begin with the global nature of the industry.

References

1. Chandler, A. D. (1990). The enduring logic of industrial success. *Harvard Business Review, 68*(2), 130–140.
2. *Fox News*. (2005, October 17). ExxonMobil's Lee Raymond [transcript]. Retrieved from http://www.foxnews.com/story/2005/10/17/transcript-exxon-mobil-lee-raymond.html.
3. The Fortune 500 medians. (1996, April 29). *Fortune*, 23–25.
4. http://www.opec.org/opec_web/en/about_us/23.htm.
5. Auty, R. (1993). *Sustaining development in mineral economies: The resource curse thesis*. London, UK: Routledge.
6. Special Report, National Oil Companies. (2006, August 12). *The Economist*, 55–57.
7. Mahtani, D. (2007, January 11). Nigeria struggles to eliminate corruption from its oil industry. *Financial Times*, p. 8.
8. The value chain concept was developed by Harvard Professor Michael Porter and is the main theme of his book *Competitive Advantage: Creating and Sustaining Superior Performance* (New York: Free Press, 1985). The concept was used by Porter to explain how firms created competitive advantage. Porter's generic value chain included primary and support activities. Primary activities included inbound logistics, operations (production), outbound logistics, marketing and sales (demand), and services (maintenance). Support activities included administrative infrastructure management, human resource management, technology (R&D), and procurement. The extension of the firm value chain to the industry is logically consistent, especially in the oil and gas industry where the IOCs compete across most of the major industry segments.
9. *Oil price history and analysis*. (n.d.). Retrieved from WTRG economics website: http://www.wtrg.com/prices.htm.
10. http://www.simmonsco-intl.com/About-Us/Our-Founder/Legacy-Presentations/.
11. Teitelbaum, R. (1997, April 28). Exxon: Pumping up profits. *Fortune*, 22–23.
12. Crooks, E. (2006, October 20). Interview: Rob Routs: You have to keep changing. *Financial Times, Special Report: Energy*, p. 10.
13. *Refining 2021: Who will be in the game?* (2012) AT Kearney, Retrieved from https://www.atkearney.com/paper/-/asset_publisher/dVxv4Hz2h8bS/content/refining-2021-who-will-be-in-the-game-/10192.
14. Oil history and analysis. Retrieved from WTRG Economics website: http://www.wtrg.com/prices.htm.

15. Ghemawat, P., & Ghadar, F. (2000). The dubious logic of global megamergers. *Harvard Business Review, 78*(4), 65–72.
16. ExxonMobil. (2013). *The outlook for energy: A view to 2040.* Retrieved from: http://cdn.exxonmobil.com/~/media/global/Reports/Outlook%20For%20Energy/2015/2015-Outlook-for-Energy_print-resolution.

3

The Global Nature of the Oil and Gas Industry

The Global Industry

As we discussed in the previous chapter, the oil and gas industry is one of the most global industries. The industry's products are used in every country in the world. Although oil and gas are not produced in every country, many countries that have not produced oil and gas in the past now have the potential to become producers, because of the development of new technologies for the extraction of shale oil and shale gas. In addition, after a decade of fairly steady oil prices (with the exception of a few big swings), exploration efforts have increased around the world. For example, in Africa, virtually every country either is an oil and gas producer or is engaged in exploration activities. Various countries, such as Kenya, Uganda, Ghana, Tanzania, and Mozambique, have recently become new oil producers, joining major producing countries such as Algeria, Angola, Equatorial Guinea, Libya, and Nigeria.

The extremely global nature of the oil and gas industry is clearly illustrated by looking at the key players in the industry. The largest oil and gas companies are true global players. ExxonMobil has about 75,000 employees and assets in about 100 countries. ExxonMobil produces oil and gas in many countries, with major production output from Abu Dhabi, Angola, Australia, Canada, Equatorial Guinea, Malaysia, Nigeria, Norway, the UK, and the United States. Figure 3–1 shows the countries where ExxonMobil operates.[1] About two-thirds of ExxonMobil's oil and gas production comes from outside the United States, and more than two-thirds of new management and professional employees are hired outside the United States.

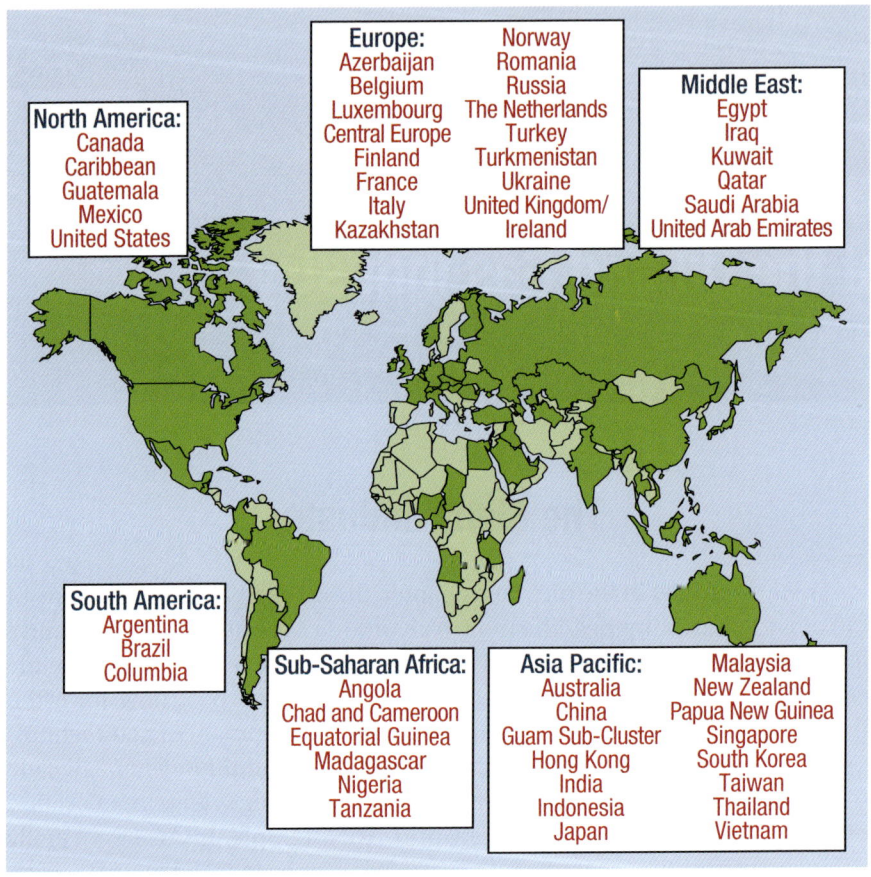

Fig. 3–1. ExxonMobil's global operations
Source: ExxonMobil.com.

BP has about 90,000 employees operating in 80 countries. Although nearly 40% of its employees are in Europe, the rest are spread throughout the world, including:

- North America (24%)
- Asia (19%)
- South and Central America (8%)
- Middle East and North Africa (6.8%)
- Sub-Saharan Africa (3%)
- Russia (1%)

The French company Total has nearly 100,000 employees and operates in more than 130 countries worldwide. Unique among the integrated oil companies (IOCs), Total has a huge presence in Africa in the downstream, with more than 3,600 owned or franchised gas stations. Royal Dutch Shell operates in more than 70 countries with 92,000 employees. In recent years, more than 30% of Shell's new hires who were university graduates came from universities outside Europe and the Americas. Figure 3–2 shows the geographical breadth of Shell employees.

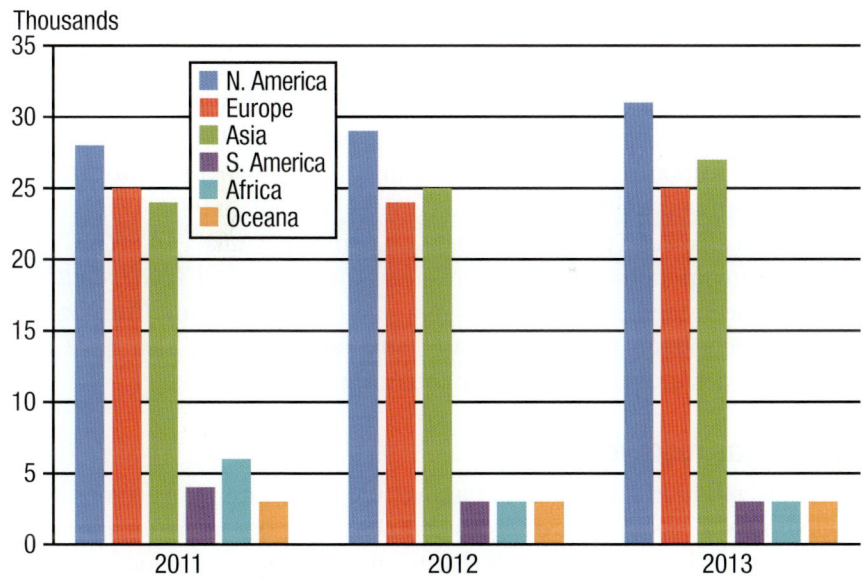

Fig. 3–2. Royal Dutch Shell employees by country
Source: Royal Dutch Shell.

Many of the national oil companies (NOCs) are also global companies. Petronas (Malaysia) has more than 30,000 employees operating in 32 countries. Statoil is headquartered in Stavanger, Norway, and has about 23,000 employees worldwide working in 33 countries. In recent years, the two largest Chinese NOCs, PetroChina and Sinopec, have embarked on aggressive global expansions involving acquisitions and partnerships. These two Chinese firms are much bigger than their largest IOC counterparts—PetroChina has more than 500,000 employees, and Sinopec has about 370,000 employees. Saudi Aramco (Saudi Arabia), the world's largest oil producer, produces roughly one in every eight barrels of crude oil the world consumes on any given day. Saudi Aramco has a network of affiliates in Asia, Europe, and North America to support its oil-exporting operations. The company also has interests in refining and marketing companies in the United States, South Korea, Japan, and China. Given that Saudi Aramco operates in far fewer countries than the largest IOCs, the company has one of the most diverse workforces in the industry—54,000 employees from 77 countries, including the United States, the Philippines, India, Great Britain, Canada, Jordan, Pakistan, South Africa, France, Trinidad and Tobago, Russia, Brazil Croatia, Singapore, Sweden, Kazakhstan, Macedonia, Oman, Sierra Leone, and Zimbabwe. Another very diverse company based in Saudi Arabia is Saudi Basic Industries Corporation (SABIC), the world's second largest diversified petrochemical company. SABIC, 70% owned by the Saudi Arabia government, has 40,000 employees operating in more than 40 countries.

Although many NOCs operate globally, quite a few are almost purely domestic companies with a largely national workforce. For example, Pemex (Mexico) has about 150,000 employees, and almost all of them work only in Mexico. Petróleos de Venezuela S.A. (PDVSA) has about 94,000 employees, and almost all oil and gas production is done in Venezuela. PDVSA has some downstream operations outside Venezuela, including the CITGO retail fuels chain in the United States and 50% of a refinery in Louisiana. The Nigerian National Petroleum Corporation and Sonangol (Angola) also have large domestic workforces and almost no international operations.

Oil and gas resources

As discussed in chapter 2 and above, the largest non-government-controlled oil and gas firms are highly global in their operating locations and workforces. A key reason why the firms are so global is the location of oil and gas resources. Oil and gas resources are not distributed equally across all nations—all countries are consumers of oil and gas, but not all countries are producers. For example, Japan and France are two of the largest consumers of hydrocarbon products, but they have almost no domestic resources. To supply their countries' energy needs, they must import large quantities of product. French and Japanese oil and gas companies involved in the upstream sector have no choice but to compete globally.

Figure 3–3 shows the leading oil-producing nations. Six countries produce 50% of the world's oil. Two of the leading producers, China and the United States, are also the world's largest importers because domestic production is insufficient to support the demand for oil. In 2013 China surpassed the United States to become the world's largest importer of oil, which means that Chinese oil and gas companies must search the world for new resources. Some analysts have predicted that with the growth in US oil production, the United States could be become energy self-sufficient by 2020.

Figure 3–4 shows the seven largest oil producers in the Gulf region (either the Persian or the Arabian Gulf, depending on one's perspective). All of the countries are major exporters, although Oman is now faced with declining reserves. Three of the countries, Kuwait, the United Arab Emirates (UAE), and Qatar, are among the world's richest nations per capita as a result of their large oil and gas exports and relatively small populations. Figure 3–4 shows Qatar with production of oil and other liquid hydrocarbons of almost 2 million barrels per day. Qatar is also the world's largest exporter of liquefied natural gas (LNG), so combined with liquids production is probably the richest country in the world per capita. By virtue of their large populations, the other large Gulf producers (Iran, Iraq, and Saudi Arabia) are much smaller in per capita income.

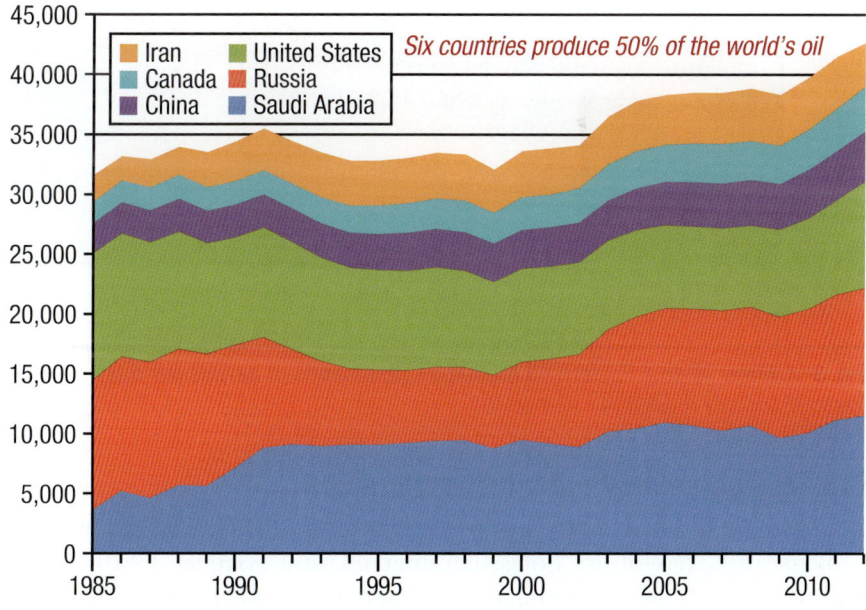

Fig. 3–3. Leading oil-producing nations (thousands of barrels per day)
Source: Constructed by authors based on data drawn from *BP Statistical Review of World Energy, 2013.*

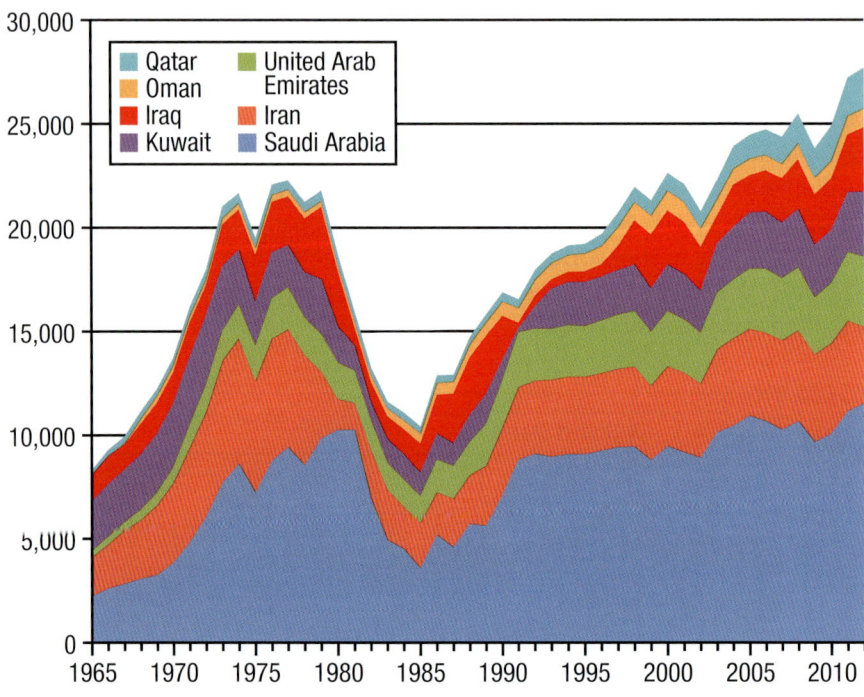

Fig. 3–4. Major Middle Eastern oil producers (thousands of barrels per day)
Source: *BP Statistical Review of World Energy, 2013.*

Proliferation of multinational partnerships

The oil and gas industry involves various risks: safety, environmental, geologic (e.g., risk of dry holes) commercial, and political. Partnering to spread risk and costs has been a key element of the oil and gas industry for more than a century. The creation of partnerships can occur at any stage in the value chain. Although more common in the highly capital-intensive upstream sector, downstream partnerships are also formed. For example, ExxonMobil and PDVSA have a 50-50 owned refinery in Chalmette, Louisiana.

In the upstream, large projects executed by a single company are a thing of the past. As projects get larger and more complex, no single company wants to have full risk exposure, and the resource owner or NOC partner often does not have the technology to access the oil and gas in its country. Most upstream projects are done in some type of partnership structure that often involves partners from different countries. For example, a production sharing agreement (PSA) for the Azeri, Chirag, and Gunashli development in Azerbaijan was signed in September 1994. BP is the operator with a 34.1%

stake, and the partners were Chevron with 10.3%; Socar (*State Oil Company of Azerbaijan*), 10%; INPEX (Japan), 10%; Statoil (Norway), 8.56%; ExxonMobil, 8%; TPAO (Turkey), 6.8%; Devon (US), 5.6%; Itochu (Japan), 3.9%; and Hess (Japan), 2.7%.

Another large multinational partnership is the Frade Project in the Northern Campos Basin, approximately 370 kilometers offshore Rio de Janeiro, Brazil, in 1,100 meters of water depth. The field was discovered in 1986 by Petrobras. Texaco negotiated a partnership with Petrobras and received Brazilian government oil and gas regulatory assignment in 2000. Conceptual engineering studies and the acquisition of 3-D seismic data were completed in 2000 by Texaco. Chevron inherited the project after its merger with Texaco in 2001, and Chevron is the project operator with a 51.74% interest. Petrobras has a 30% interest and Frade Japao Petroleo Limitada (FJPL), a Japanese consortium, has 18.26%. FJPL is made up of Japanese companies INPEX with 37.5%, government-owned Japan Oil, Gas & Metals National Corp. with 50%, and Sojitz with 12.5%.

There are various factors that help make complex multinational partnerships like Frade and Azeri, Chirag, and Gunashli work successfully. One of these factors is the quality of people involved in managing the partnership. Partnerships with many partners from different countries require managers with a global mind-set and the ability to deal with ambiguity, especially with respect to the evolving relationships between partners.

Managing Human Resources in the Global Environment

In chapter 2 we introduced the concept of the value chain. Global issues and events impact all of the activities in the value chain. In the earlier discussion in this chapter we emphasized the global nature of the key players in the industry. So as would be expected, the extremely global nature of the industry affects human resources (HR) strategy and all of the HR functions. It does so in two distinct ways.[2] First, being in a global industry means that firms have to conduct business in multiple different countries. Because different countries have different business environments, HR must adapt to these differences. Important differences that affect the business environment include differences in laws, cultures, political systems, religions, geographies, climate, labor standards, ethical frameworks, traditions, and risk, to name just a few![3] Thus, doing business globally requires firms to adapt their human resource practices, at least to some extent, to the country they are operating in. Second, firms doing business globally are multinational enterprises (MNEs), while those doing business in only one country are considered domestic firms. Because MNEs must now coordinate their practices among many countries, HR is considerably more complex in MNEs than in domestic firms. HR in MNEs is very different from domestic HR because there are more HR activities, a broader perspective is needed, there is more involvement in employees' personal lives, there is greater risk exposure, and there are considerably more external factors influencing HR decisions.[4] Therefore, managing HR is very different in global environments because it occurs in different country contexts and involves coordination across many countries. We now explore how this global nature affects each of the four functional HR areas, specifically looking at its impact on the key questions and how to maximize the functions' effectiveness.

Staffing in This Global Environment

An extremely global business environment completely changes how firms staff. As mentioned above, it requires firms to adapt their staffing practices to various different countries. It also complicates how firms staff jobs, because MNEs now must decide whether to staff foreign subsidiaries with employees from the home country, from the host country, or from some other country. Clearly, the global business environment changes the options of the four key staffing decisions.

Which labor market in the global context?

Similar to jobs in domestic firms, the labor market for jobs in MNEs tends to geographically resemble the availability of qualified labor. If qualified workers are available close to the work site, the local area tends to be the labor market the firm focuses on. When no qualified workers are locally available, firms must consider a regional or global labor market. For higher-level highly specialized, technical, or managerial jobs, the labor market is largely a function of the firm's staffing philosophy. The staffing philosophy largely drives the sources of higher-level labor MNEs use for foreign operations. Figure 3–5 shows the possible sources of labor for MNEs.

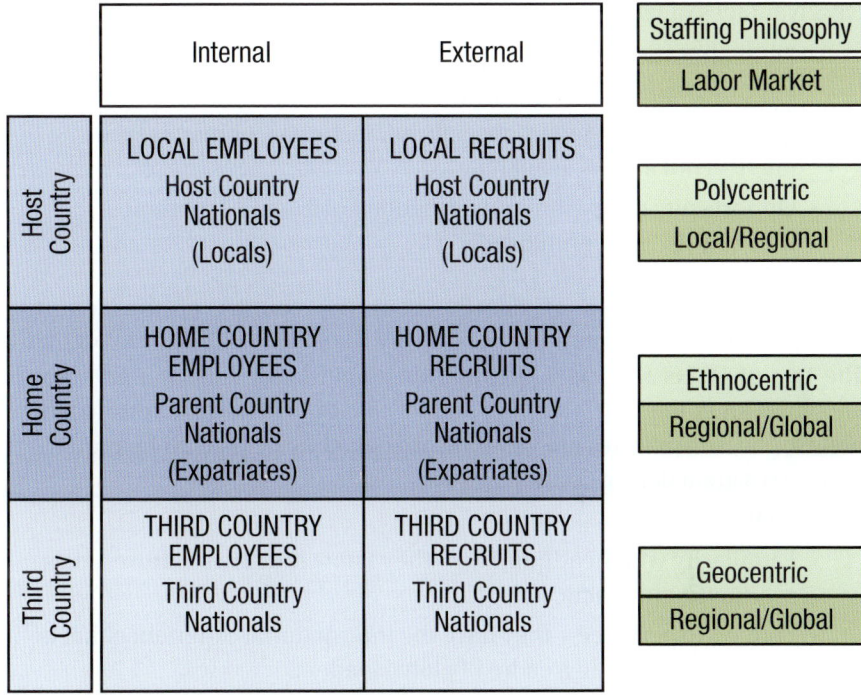

Fig. 3–5. Sources of labor for MNEs abroad

Locals. Employees or recruits from the country of the work site are known as *host-country nationals* (HCNs) or locals. The use of locals reflects a local or regional labor market and what is known as a *polycentric staffing philosophy*. In firms with a polycentric philosophy, subsidiaries are generally managed by locals with few transfers between subsidiaries and headquarters. There are many advantages to using local employees including:[5]

- No language or cultural barriers—employees are in their normal environment.
- Fewer legal complications—no work permits or local content requirement issues.
- Usually cheaper—compensation tends to be lower, and there are no transportation costs.
- Greater stability—generally there is no set assignment length.
- Motivational—promotions of locals create opportunities for lower levels.
- More familiarity with local business environment—the company's managers are more familiar with local markets, laws, and politics.

Disadvantages of using local employees include:

- Difficulties with headquarters—there are communication and cultural issues and less familiarity with headquarters' perspectives.
- Skill availability—highly specialized, technical, or managerial talent is not always available.
- Opportunity costs—limited availability of foreign experience for headquarters' employees.

Expatriates. Employees or recruits from the country of the company headquarters are known as *parent-country nationals* (PCNs) or expatriates. The use of PCNs reflects a regional or global labor market and what is known as an *ethnocentric staffing philosophy*. In firms with an ethnocentric philosophy, subsidiaries are generally managed by PCNs with headquarters that exert a great deal of control over all subsidiaries. The advantages of using PCNs include:

- Familiarity with headquarters—they know the firm's culture, goals, products, and practices.
- Foreign experience—they have the international experience that is important to create a cadre of global leaders.
- Greater control—they possess firm loyalty, good communication, and a history with headquarters.

Disadvantages of using PCNs include:

- Expensive—high salary, allowances, premiums, and training are usually required.
- Adaptation issues—cultural, language, and family adjustment problems are common.
- Unfamiliarity with local business environment—they probably do not know local markets, laws, or politics.
- Demotivating to locals—hiring PCNs stifles opportunities for advancement of locals.
- More legal complications—work permits and visas are needed, and there are frequently limits on how many PCNs can be employed.

An ethnocentric staffing philosophy runs counter to the strong commitment to diversity and inclusiveness the IOCs have made in their workforce in recent years. For example, among the IOCs, Shell has one of the most diverse top management teams with executives from Malaysia, The Netherlands,

Switzerland, United Kingdom, and United States. Shell's corporate website reports:

> We measure diversity and inclusion in part by the representation of women and local nationals in senior leadership positions. By the end of 2012, the proportion of women in senior leadership positions at Shell was 16.2%, down 0.4% from 2011 and up 0.9% from 2010. In 42% of countries where we operate, local nationals filled more than half the senior leadership positions—up 8% from 2011, and up 6% from 2010. In 2012, more than 90% of our employees worldwide were local nationals.

ExxonMobil is also working to leverage the diversity of its workforce and has said:

> Through a wide range of activities, activities and investments, Exxon Mobil strives to maintain a diverse, representative workforce, guided by the following objectives:
>
> - To be recognized as employer of choice by an increasingly diverse workforce;
> - To achieve a mix of employees with requisite skills culturally reflective of the locations where we operate;
> - To value diverse backgrounds, skills and abilities, recognizing individual contribution;
> - And to establish and improve programs, systems and practices that support diversity in our workforce.[6]

We will discuss inclusiveness and diversity in greater detail in chapter 6.

Third-country nationals. Employees or recruits from neither the country of the work site nor company headquarters are known as *third-country nationals* (TCNs). Hiring TCNs makes use of a regional or global labor market and reflects a *geocentric staffing philosophy*. In firms with a geocentric philosophy, subsidiaries are generally managed by TCNs, or whoever is deemed most qualified. The advantages of using TCNs include:

- Lower cost than PCNs—salary, benefits, and maintenance tend to be lower.
- TCNs generally adapt better than PCNs—they are usually highly experienced internationally.
- Truly global leadership—management has lots of international experience.
- Allows for hiring of best person for the job—ability is deemed more important than nationality.

Disadvantages of using TCNs include:
- Host-country difficulties—some nationalities may not be accepted by or adapt to the host country.
- Difficulties with headquarters—there may be communication and cultural issues.
- Demotivating to locals—hiring TCNs stifles opportunities for advancement of locals.
- More legal complications—work permits and visas are needed, and there are frequently legal limits on how many non-locals can be employed.

As figure 3.5 shows, another complication for MNEs is that internal recruiting may now be possible for any labor market because the company has employees worldwide. We discuss the issue of internal versus external staffing in the next section.

Which recruiting method should be used in the global context?

Just as the global context affects the labor market, it also affects our recruiting methods. Similar to domestic firms, the recruiting method is partially determined by a firm's staffing philosophy and preferences for internal or external candidates. Generally, PCNs and TCNs are internal candidates who then receive an international assignment; however, they could be external recruits. The overwhelming tendency of firms is to use internal candidates for international assignments (except in the rare cases where no qualified candidates are available or willing to take an international assignment). A recent study found that for most global positions, China, India, and the United States are the most attractive sources of talent in the oil and gas industry.[7] The availability of talent tends to be the most important factor in determining which source (or which staffing philosophy) firms use, with the default choice historically being internal home-country candidates for high level jobs.[8] However, as firms are becoming more global and local talent availability is increasing, the use of local candidates is increasing. Keep in mind that lack of talent in the long term can be anticipated and can be overcome through proactive staffing and training practices.

Another factor affecting the choice of internal versus exter*nal recruiting is the nature of the assignment. Foreign assignments tend to be learning driven* or *demand driven.*[9] Demand driven positions are assignments that are needed to fill open positions in foreign subsidiaries that cannot be filled by locals because of lacking technical, specialized, or managerial skills. Demand driven positions may also not be filled by locals to increase the transfer of knowledge between headquarters and the subsidiary. Thus, demand driven assignments are focused on the needs of the organization. Learning-driven positions are assignments that are given to employees for global competency development and the career enhancement of the employee. Although they are likely to benefit the firm in the long term, the focus is on the employee. Learning-driven assignments use internal candidates. Demand driven assignments where knowledge transfer is the reason for the assignment, also use internal candidates. When filling an open position is the reason for the assignment, internal candidates are preferred, but external candidates may be used if no internal candidates are available. However, firms must be aware that whether the candidate is external, internal, local, or an international assignee, will have a big impact on how and what knowledge gets transferred throughout the organization.[10]

Recruiting methods for internal candidates for higher-level MNE jobs include global talent management inventories, in-house global leadership programs, former/current PCNs, nominations from managers and firm leaders, internal job postings, and international succession-planning programs. Recruiting methods for external candidates for higher-level MNE jobs include employee referrals, job fairs, company Internet sites, executive search firms, professional associations, poaching, web-based job sites, global leadership programs in universities, and international volunteer organizations.[11] Recruiting success depends on a number of factors, but using more methods is generally better in attracting a bigger, more diverse applicant pool.

Recruiting success in general is related to job and organizational characteristics, recruiter behaviors, perceptions of the recruiting process, and hiring expectations.[12] So, in a foreign context it is important to adapt to local cultures and norms in both the recruiting practices and recruiter behaviors.[13] Using local recruiting or advertising firms can help in this. One of the firm characteristics that is important when recruiting foreign local talent is the company's reputation. However, as with most aspects of global business, its importance depends on the local culture.[14] The company's reputation is most important in cultures that have a group (rather than individual) mind-set (these are known as *collectivistic cultures*). See Current Challenge Box 3-1 for a more in-depth discussion of country cultures. The compensation, benefits, and other aspects of the job are more important in countries with an individual mind-set (*individualistic cultures*).

Current Challenge Box 3–1. Cultural Differences

The vast geographic differences across the global oil and gas industry are an indicator of vast cultural differences among countries in the industry. Cultural differences affect every aspect of doing business across countries and can negatively affect any cross-national interaction. Clearly, many things differ across countries, including laws, politics, economics, infrastructure, standards, geography, climate, and traditions. All of these affect culture, which is the overarching framework of a society's values, norms, and shared understanding. The most common way to assess culture is Hofstede's framework. This framework looks at national cultures in seven different dimensions:

1. *Collectivism versus individualism*—Do members of society tend to view everything relative to how it affects the group or themselves?
2. *Masculinity versus femininity*—What is seen as important: material goods and achievement or relationships and connections with others?
3. *Power distance*—Are people viewed as equals, or is it acceptable to view some people (or classes of people) as superior to others?
4. *Uncertainty avoidance*—How much of a concern is uncertainty and risk?
5. *Long term versus short term*—Do people generally view everything with a long-term or a short-term perspective?
6. *Pragmatic versus normative*—Do people believe that truth is not absolute and many things can't be explained, or do they have a strong desire to explain the world?
7. *Constraint versus indulgence*—Do people tend to control their desires and impulses or not?

How countries score on these dimensions is readily available (http://geert-hofstede.com/countries.html). Knowing how people in a country you are interacting with generally think using these scores can go a long way in helping to understand and communicate with different nationalities.

Certainly, some business practices readily fit or counter these different aspects of culture. For example, bonuses based on individual performance would not be as acceptable in a collectivistic culture or one with high uncertainty avoidance. So should firms adapt their practices to each country they are operating in? Historically, the pervasive view was that adapting to national cultures was not only desirable but also necessary for firm survival.

More recently, some scholars have argued that national culture scores just reflect averages in a society and that views vary in all societies. Therefore, firms would do better to have consistent, uniform practices globally that attract workers whose views are consistent with an organization's culture regardless of their national culture.[15] Currently, some suggest that firms are most successful when they are globally uniform with respect to practices that have strong strategic implications and adapt locally otherwise.[16]

The reputation of a company is also important in attracting internal workers to accept foreign assignments. Similar to attracting foreign local workers, attracting the appropriate employees for foreign assignments is easier for firms with a good global reputation and globally recognized brands.[17] However, an internal reputation of treating their PCNs and TCNs fairly while they are abroad (and upon their return) is also important in getting employees to accept foreign assignments. We now look at how selecting candidates from the applicant pool is affected by the global context.

Which selection method should be used in the global context?

Just as with recruiting, the global context changes selection in two ways. The first is that selection decision processes will occur in many different countries. The second is that we must select employees for international assignments. We start by looking at cross-national differences in selection methods. Because of cultural, legal, and historical differences, the use and acceptance of various selection tools varies among countries. Here are some examples:[18]

- Psychological assessments are frequently used by Belgian, Dutch, French, Irish, New Zealand, South African, Spanish, Swedish, and UK firms, but rarely used by German, Italian, and US firms.
- Cognitive ability tests are frequently used by Belgian, Dutch, New Zealand, Portuguese, South African, and Spanish firms, but rarely used by German, Hong Kong, and Italian firms.
- Only French and (French-speaking) Belgian firms like to use graphology (handwriting analysis).
- Panel interviews are frequently used in Australia, Ireland, Japan, and The Netherlands, but rarely used in Germany or Italy.
- Job trials are frequently used in Belgium, The Netherlands, and Portugal, but rarely used in Germany, Hong Kong, Italy, New Zealand, Sweden, and the UK.

- More than half of US and Dutch firms use peers as interviewers, while less than 10% of German, Greek, Italian, and Spanish firms do.

Recall from chapter 1 that validity, cost, and adverse impact are important factors in evaluating which tools to use for specific jobs. Clearly, these will all vary based on the national context. Selection methods such as assessment centers may be found to be highly valid for managerial jobs in the United States, but not so much in other countries where the process is unknown and performance varies a lot, because of the unfamiliarity with them. The wording in personality tests may be socially unacceptable in some countries, resulting in invalid responses. In some countries the effects of responding in a socially desirable way may be so strong as to render the test invalid. Expectations of interviewee behaviors may differ so strongly among countries that highly valid structured interviews designed in the United States are entirely invalid in other countries. Thus, it is important to realize that just because a test is valid in a US context, that does not mean it will be valid globally. The greater the cultural differences between the country it is being used in from the country it was validated in, the more questionable the use of the test.

The adverse impact of tests has been shown to vary across contexts, so it likely varies greatly across countries.[19] Different ethnicities, differences in gender roles, and differences in cultural norms associated with age or other criteria can all result in adverse impact. Further, the legal implications of adverse impact are also very different across different countries. For example, Canadian laws have found adverse impact illegal even for "genuine business reasons," a stricter interpretation than in the United States.[20] Finally, the costs of the selection methods may also vary considerably across countries and should be carefully considered when determining the most applicable methods.

The second difference of selection in a global context is the need to select employees for international assignments. For demand-driven positions, the temptation (and most common method) is to select employees in the same way as if the position were at headquarters: Hire the person most qualified to perform the tasks required of the job. Assignments determined in this way frequently end in failure because the demands of an international assignment go way beyond just doing the job. International assignment failures are more commonly due to the employees' (or their families') not being able to adjust rather than lack of work-related skills.[21] Depending on the location, a job in a foreign location is likely to be much more demanding on employees in many ways. These additional demands include cultural differences, language difficulties, tremendous uncertainty, greater risk, climate differences, differing gender roles, family difficulties, and safety and security concerns.

Skills not specifically related to the tasks of the job and other characteristics that will increase the probability of the success of international assignments are reported in table 3–1.[22] However, even when managers recognize the demands of foreign assignments and the necessity of non-task-related skills, the actual selection process is usually very informal and of highly questionable validity. The typical process is that a casual conversation between managers starts a short list and that the "selection" processes then just validate a decision that has essentially already been made.[23] This is consistent with the findings of a recent survey that showed that only 28% of firms had a formal candidate pool and used any candidate assessment tools.[24] Such an informal process may disregard non-task-related factors that are key to assignee success and is likely to lack validity, like the unstructured interview. As with other functions, the global context makes the selection process considerably more complex. In the case of international assignees, the selection process should now also include predictors of cultural adaptation as well as task performance. This is also true for learning-driven assignments.

Table 3–1. Non-task-related skills, competencies, and characteristics related to international assignee success

Skills	Other Characteristic
Conflict resolution skills	Flexibility
Leadership skills	Emotional stability
Communication skills	Resourcefulness
Language skills	Cultural adaptability
Social skills	Cultural sensitivity
Decision-making skills	Physical stamina
Negotiation skills	Openness to experience
Networking skills	Desire for foreign assignment
Relational skills	Social support
	Family support

So which predictors should be used? Similar to the process used for selecting personnel for domestic assignments, interviews, assessment centers, bio-data, standardized tests, work samples, and knowledge assessments can be used as selection methods for international assignments.[25] These selection methods should be tailored to the job as well as its more difficult international context. Some selection methods that are not generally used for domestic selection decisions are used for selecting international assignees. These include self-assessments, international tailored resume screening, foreign language assessments, personality assessments related to cultural adaptability, and family assessments. Although companies generally cannot predict international assignee success very well, success is improved if the process is formal, job-related, and related to the candidate's adaptability to the specific location. We can consider the selection as successful if the process is valid, seen as fair, and predictive of international assignee success.

Maximizing the effectiveness of staffing in the global context

Maintaining the legality and perceived fairness of the staffing process is much more complex in the global context. So are monitoring, evaluating, and adjusting the system. Unlike domestic firms, MNEs must understand the relevant labor laws of all countries they have operations in. Although this is more complex, it is relatively straightforward and can be dealt with to some extent by seeking the services of local legal experts. What is not as straightforward is when the laws of the host country compete with the laws of the home country. For example, one country's laws regarding discrimination based on national origin may contradict a foreign country's laws regarding percentage of its citizens that must be hired. Generally, the United States requires that foreign countries operating in the United States abide by US laws; however, trade treaties can preempt this principle.[26] Further, US MNEs operating abroad must abide by US laws if the employee is a US citizen and foreign laws are thereby not violated. Another legal constraint on MNEs consists of *local content laws*, which require that a certain percent of labor (or materials or parts) must be from the host country. Local content laws are seen to have a strong influence on MNE staffing in the oil and gas industry.[27] See Current Challenge Box 8-3, "Local Content Laws," in chapter 8 for a detailed discussion on this topic. Local content laws reflect nationalization programs that encourage, support, or mandate the employment of host nationals in preference to PCNs.[28] Similar to the case of talent shortages, long-term nationalization programs can be addressed through proactive staffing and long-term training practices focusing on youth.[29]

Perceived fairness is also more complex for firms operating in many countries, because perceptions are largely influenced by country culture (see Current Challenge Box 3-1). Because of differences in cultures, different practices (e.g., giving bonuses based on need rather than performance) will be perceived as unfair in some countries and fair in others. As discussed in Current Challenge Box 3-1, awareness of cultural differences is critical. When they are aware of the differences, firms can proactively determine whether to have globally consistent practices (where doing so does not violate local laws) or locally adaptive practices.

Monitoring, evaluating, and adjusting the system are also more complex for MNEs for a number of reasons. First, one way to measure the effectiveness of the staffing function is to evaluate the performance of the newly hired employees. Measuring performance is more difficult in MNEs because of the geographic distance, among other reasons (which will be discussed shortly in the performance management section). This is particularly true when the international assignee's manager is not in the same country and only long-term outcome data are available. Second, task performance may be at a satisfactory level, but assignees or their families may be having adjustment difficulties. Some organizational or national cultures may discourage assignees from communicating any difficulties for fear of appearing incompetent, uncooperative, or difficult. This prevents firms from getting honest feedback about issues that could perhaps be easily resolved for future assignees. Firms should be proactive in getting honest feedback about the entire process to be able to adjust the system with continual improvements. Finally, the evaluation process should include the assignee's family, since the family can have a large impact on the ultimate success of the assignment.

Training in the Global Environment

An extremely global business environment greatly changes the training function. As mentioned earlier, it requires firms to adapt their training practices to various different countries. It also complicates how and what firms train local versus international employees, because MNEs must have specialized training programs for both. Clearly, the global business environment changes the nature of the three key training decisions: who, what, and how.

Who gets trained in the global context?

Because companies in a very global context do business in many countries, there are categories of employees who need training that usually are not applicable in a purely domestic company. These include:[30]

- PCNs—the goal of PCN training is to prepare them their new role both in the workplace and out, this includes coping with the unexpected in the new culture.

- TCNs—the goal for TCNs is also preparation for their new culture, but they may additionally need training to be competent in their interactions with headquarters.
- Assignee's spouse and family—the goal is to prepare them for life in a foreign location, including coping with the unexpected in the new culture.
- Locals—beyond being competent at their specific jobs, the goal of local training is for them to be competent in their interactions with headquarters.
- Alternate international travelers—the goal for alternate international travelers is to maximize the returns from their unconventional assignments and minimize the costs, both to the company and the employee. Alternative international travelers include the following:
  - Virtual PCNs—managers of foreign locations who are not permanently located there
  - Short-term assignees—those going abroad for less than one year
  - International business travelers
- Potential global leaders—these include senior local nationals, or PCNs or TCNs who have been identified as having high potential as future leaders of the global firm.

These types of employees are unique to MNEs, and all require training. Yet not all companies provide it, and even if they do, it is not mandatory. For example, only 31% of all companies provide cross-cultural training for all assignments, and 17% do not provide any training. Further, for those that provide training, only 27% require it and only 46% provide it for the entire family. It is likely that cost prevents some organizations from providing training, yet for those that do, only 2% find it a poor value, while 85% find it a good or great value.[31]

What do we train in the global context?

When needed, all employees should be trained the skills they need to perform their jobs competently, and this is also true in the global context. However, beyond that, the global context requires many additional skills, depending on the category of employee, as discussed above. That is, what we train is largely dependent on who we are training.

PCNs. The most common training for PCNs is cross-cultural training, which provides PCNs the skills of cross-cultural awareness and adaptability. Ideally, such training should provide assignees with a framework for how cultures differ, country-specific knowledge, techniques to discover cultural differences, and an understanding of the adjustment process.[32] It can include training in host-country language, ethics, stress management, safety and security, risk management, and etiquette and protocol. Finally, training may also be needed in how to train locals, because part of assignees' job may be to transfer knowledge to the foreign subsidiary.

TCNs. TCNs may need all the training available to PCNs and, because of their possible unfamiliarity with headquarters and the headquarters' country, they may need basic organizational socialization, to teach them about the company's history, culture, policies, and practices. Country-specific knowledge, cultural training, and language training relevant to the headquarters' country may also be needed. [33]

Assignee spouse and family. This training may be the same as the cross-cultural training given to the assignees. Language training, stress management training, safety and security training, and etiquette and protocol training may also be provided to help the family adapt to the new context.

Locals. Of course, technical and skills-based training may be needed, as it would be in a domestic firm. Additionally, local employees should be socialized as company employees, learning the company's history, culture, policies, and practices. Language training may also be necessary if the local employee is expected to communicate with headquarters.

Alternate travelers. Alternate international travelers may benefit from the same training given to PCNs. However, because their interactions are of shorter duration or less immersive, the costs versus the benefits need to be considered more carefully.

Potential global leaders. Potential global leaders need to learn the skills that will allow them to become successful leaders for the global company. One of these is skills is cultural agility, the ability to accurately assess and respond effectively in situations where cultural context matters by adapting, integrating, or minimizing other cultures as appropriate.[34] Globally applicable leadership skills, global market knowledge, and global management skills are also frequently part of global leadership development programs.

As shown above, "what do we train?" largely depends on who is getting the training. Similarly, "how do we train?" is largely influenced by what gets trained as well as the cultural context.

How do we train in the global context?

Clearly, MNEs differ from domestic firms in who gets trained and what we train them. However, another important implication of the global context is that how people learn depends on their culture. Thus, how we train people needs to be adapted to the country culture. Adapting to the country culture goes way beyond just translating training materials. For training to be country appropriate, the training program must be created by someone who knows the local laws, local practices, the level of employee skills and knowledge, local learning styles, and the country culture.[35] For example, in most high-power-distance countries, education is expected to be very authoritarian, where the trainer is a respected expert who should not be questioned. Contrast that with the United States, where interactivity and student participation are encouraged.[36] Similarly, in countries with high uncertainty avoidance, having trainees publicly practice (with potential for failure) would be viewed negatively relative to more passive ways to learn.

Training methods also differ depending on the content of the training. The training of job-related skills will generally be of the types discussed in chapter 1, although they too should be adapted to the country culture. Other content uses different methods as follows.

Cross-cultural training. Cross-cultural training can use many different methods including area briefings, lectures and books, videos, role plays, using interpreters, case studies, simulations, preassignment visits, and web-based delivery methods. Web-based delivery methods are now used by half of all companies, a number that has been increasing steadily over the years.[37] Some popular providers of web-based cross-cultural training tools include Culture Wizard (www.rw-3.com), Culture Compass (www.living-abroad.com), Cultural Navigator (www.tmcorp.com), Country Navigator (www.tmaworld.com), and Global Road Warriors (www.worldtrade-press.com). Generally, it is better to expose assignees to less immersive methods until they gain a familiarity with the culture and then move to more immersive methods.

Socialization. Socialization generally uses briefings and multimedia to inform employees about the history, culture, practices, and policies of the company. Socialization ultimately teaches trainees the organizational culture (the values, norms, standards, and "way of thinking" in the company). Employees are not likely to become fully "socialized" into the organizational culture until they accept it and become immersed in it over time.

Global leadership development. One method of global leadership development is participation in global development centers. Activities in global development centers include team and individual exercises, work simulations, group activities, and possibly interviews and psychological assessments. Global development centers are much like managerial assessment centers with a global focus. Each participant receives a personalized assessment of strengths and weaknesses that can help form a development plan.[38] Another global leadership development method is mentoring. A mentorship program that pairs senior experienced global leaders with potential global leaders can help up-and-coming managers reach their potential by learning through the experiences of others. Other global leadership development methods tend to provide potential global leaders with opportunities to learn through global experiences. These include multinational team experiences, a variety of challenging international assignments, and virtual expatriation.[39]

Safety and security training, risk management training, and the like. These types of training tend to follow the standard training methods discussed in chapter 1. Again, the methods should be adapted to the culture they are being used in.

Which training method to use for global assignees is largely based on host location and type of assignment, but may also be based on characteristics of the person and the organization. For example, more training is likely to be needed for foreign locations that are considered to be the most culturally difficult to adjust to. Globally, these are China, Brazil, India, and the United States.[40] The most difficult global areas for US citizens to adjust to are Africa, the Middle East, and the Far East.[41] As shown in figures 3–6 and 3–7, many of these difficult locations are where the oil and gas resources for the major companies are located. As would be expected, the IOCs (e.g., BP, Chevron, ExxonMobil, and Shell) have the most diverse resource portfolios, requiring a greater global HR orientation. This is because the IOCs must search the world for oil and gas resources since they do not have the ownership advantages of the NOCs. In contrast, two large NOCs, CNOOC (Chinese National Offshore Oil Company) and Petrobras (Brazil), have most of their resources in their region of origin, allowing them to have more domestic-focused HR functions.

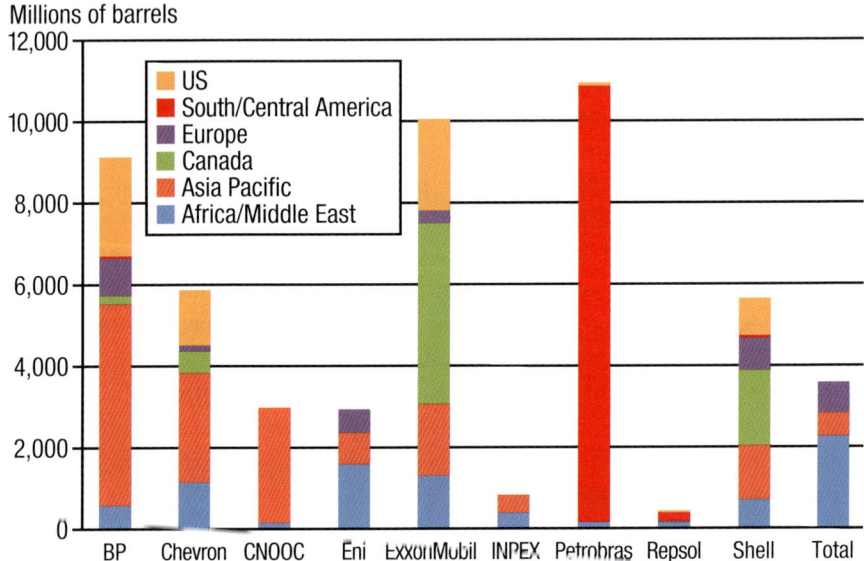

Fig. 3–6. Oil reserves by region for selected companies
Source: *Global Oil and Gas Reserves Study 2013*, Ernst & Young. Data from SEC and Annual Report filings; publicly traded companies only.

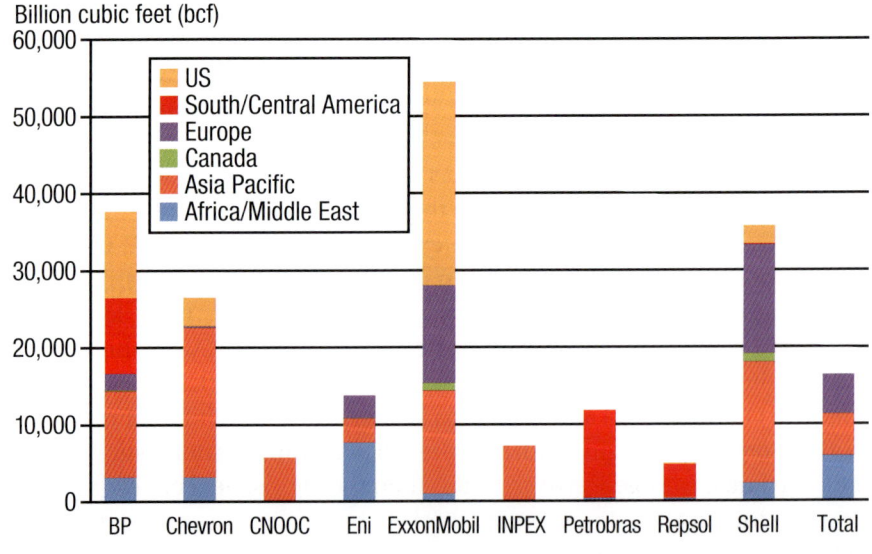

Fig. 3–7. Gas reserves by region for selected companies
Source: *Global Oil and Gas Reserves Study 2013*, Ernst & Young. Data from SEC and Annual Report filings; publicly traded companies only.

Maximizing the effectiveness of training in the global context

As with domestic companies, maximizing the effectiveness of training in the global context can be achieved by maximizing learning; facilitating training transfer, and monitoring; evaluating, and adjusting the system. This becomes more complex in a global environment. As we have already mentioned, adapting the training to specific cultures is necessary to maximize learning. This includes not only the training method but also the training setting, how the trainees are motivated to learn, the training style, and the training schedule. Of course, an organizational culture that values learning, not just at headquarters but throughout the subsidiaries, will also increase the effectiveness of training.

Transferring the learned material to the job (and in the case of international assignees to their daily lives) is also more complex in a global environment. This is because of the likely geographic distance from where the material was learned to where it will be applied. Monitoring the implementation of the learned material is much more difficult if it occurs in a place far away. Nevertheless, the technological advances in global communication and information systems make following up on training periodically much more affordable and feasible now than in the past. Also, web-based training sessions can be easily reinforced by continuing (or refreshing) training after departure.

Using technology to follow up on the training (or to get preliminary performance indicators) also applies to monitoring, evaluating, and adjusting the training system. The effectiveness of the training should be evaluated through a number of different means. First, the opinions of the trainees (usually through surveys) can provide some information about the effectiveness of the training (although these survey responses could be culturally biased). What aspects of the training did they find particularly useful? Or unhelpful? Assessments and tests of the trainees throughout the training session can also provide information about how much they have learned. Trainees can be monitored after the training to evaluate their level of proficiency of the learned material. Also, outcomes can be monitored to see if they improved after the training or if there are differences in the results of employees who underwent different training programs. Finally, professional success can be a helpful indicator of training effectiveness, although this measure is a longer-term indicator than the others. Because each way to measure training has strengths and weaknesses, collecting as many different measures of training effectiveness as possible can help determine which adjustments are necessary.

Performance Management in the Global Environment

As with staffing and training, a global business environment changes the performance management function in two basic ways. First, it requires firms to adapt their performance management practices to various different countries. Companies have to decide whether their performance system will be more centralized and standardized globally or more decentralized and adapted to each country's business environment and culture. Clearly, differences exist between countries in their typical performance management systems. Here are some examples:[42]

- The strength of labor laws is a key factor in the design of performance management systems in the United States, Germany, and Australia, but not so much in France.
- In Mexico, the culture makes it more difficult to base performance appraisals on individual rather than group measures.
- Goals tend to be set by managers in France, but include employee input in Germany.
- In Turkey, managers are likely to heavily weigh good organizational citizenship in their measure of performance.
- In India, performance appraisals are frequently just a form-filling exercise and few rewards are actually tied to appraisals.
- In China, the traditional practices of measuring moral behaviors are slowly moving toward also considering individual behaviors and their effect on organizational goals. No appraisal discussion and limited interpersonal feedback are common.
- In Japan, the focus of performance management systems tends to be on employee development consistent with their commitment to long-term employment.

Although firms may opt to disregard some of these cultural norms to have a global standard, they should do so knowing that going against cultural norms will likely meet with some resistance.

Second, performance management in MNEs is more complicated because of the differences in the job contexts of international assignees and local employees as compared to employees in domestic firms. These differences include difficulties in performance management because of geographic distances, the frequency of interactions with the employees, measurement issues, and role conflicts (assignees caught between performance expectations of headquarters versus local employees). Clearly, the global business environment changes the nature of the three key performance management decisions as we now discuss.

How is performance measured in the global context?

As with domestic employees, the measure of performance for MNE employees is largely driven by the job, and the thus the technical aspects of the job performance can largely be captured by standard appraisal methods (as described in chapter 1). For MNE employees, however, the choice of the measure may also be affected by the country context and culture as mentioned above. Also, the choice of measures is more complicated in an MNE for two reasons. First, the person who is usually responsible for determining and collecting the measures used to evaluate performance—the immediate supervisor—may be geographically far away and have only limited contact with the employee. Second, there may be difficulties in interpreting objective performance measures because of noncomparable data (because of differences or changes in accounting practices, exchange rates, labor laws, profit repatriation requirements, price controls, financing costs, depreciation allowances, inflation, etc.), different expectations, more volatile business environments, and arbitrary transfer pricing.

Given all these difficulties, how should performance be measured? The measure should be consistent with the goals of the parent company and business strategy, but should also carefully consider the context the employee was working in, including the country culture and how much of the resulting measure was in the employee's control.[43] Usually, multiple measures are used to capture all important aspects of performance. These measures may include hard measures such as objective targets of profits, sales, quality, production, return on investment, and market share, or softer (more subjective) measures such as knowledge transferred to the local team, communication skills, decision-making skills, interpersonal skills, leadership skills, technical skills, cross-cultural skills, relationships with government officials, and local public image of the firm. Although hard measures appear fairer because they are not subjective, they can also be problematic if they do not take the context into account, and they may motivate unethical behaviors where the ends justify the means. Therefore, multiple criteria of both hard and soft data are usually recommended.[44]

Who evaluates performance in the global context?

Who evaluates performance also depends on the country culture. For example, as mentioned in chapter 1, the direct supervisor of employees usually handles the performance appraisal and performance management function in the United States and most Western cultures. However, in China, self-evaluation and a consensus opinion of peers and subordinates are the most common sources of appraisal.[45] Again, although firms may opt to disregard cultural norms to have a global standard, they should do so knowing that going against cultural norms will meet with some resistance.

In MNEs, the direct supervisor may not be in the same country as the employee, adding another complication. As with domestic performance appraisals, getting more data points regarding performance is better than fewer. In fact, using more raters is likely to be more important in an international context because of rater misinformation or biases due to cross-cultural issues, greater geographic distances, or fewer interactions with the employees. Information sources that have international experience are likely to be better than those that do not, because they should have a better feel of the context in which the performance took place. One study found that firms use an average of three different raters for subjective measures of international assignee performance.[46] When using subjective measures, other useful sources of information regarding performance of MNE employees include cross-national colleagues (co-workers), the employees themselves, subordinates, customers, HR professionals inside the host country and sponsors, corporate HR professionals, and regional executives outside the host country.

Although using many different sources is likely to be a more valid and accurate measure of performance, doing so in the international context can be quite complicated give the different locations, cultures, and time zones of the raters. New technology is making this a lot easier. Web-based performance management systems are being increasingly used by MNEs and can help integrate performance data with other HR functions such as training, succession planning, and compensation.[47]

What format should be used in the global context?

Generally, the issues regarding format of the performance appraisal in the domestic context apply to the global context. The format usually largely depends on the type of measures used. Nevertheless, there are some differences for the global context. First, the issue of absolute versus relative comparisons now becomes more complicated because of the vastly different contexts employees are in. Clearly, comparative formats should be used carefully in MNEs, so that comparisons among employees are only done when the employees are in relatively similar contexts. Second, issues of language, translation, and interpretation can occur when adapting formats to different countries.

An additional question that we have not yet discussed regarding the format of the appraisals is: Should the format be standardized (for the same job category) or customized? Standardization helps reduce development costs, maintain baselines, and experience with the measures. However, in MNEs many jobs in the same job category have vastly different contexts, more volatile environments, and differing purposes of assignments. Thus, theoretically, customization makes more sense in the global context. Nevertheless, although most MNEs tend to use standardized forms for assignee appraisals, the increased use of performance management software makes customization easier and increasing in popularity.[48]

Maximizing the effectiveness of performance management in the global context

As with domestic firms, providing feedback; improving evaluation accuracy; including goals and rewards; and monitoring, evaluating, and adjusting can help maximize the effectiveness of performance management for MNEs. However, cultural differences and the different types of employees (PCNs, locals, TCNs, etc.) complicate these issues.

Different cultures have different expectations regarding feedback. Cultures with greater power distance would expect a one-way conversation with the supervisor informing the subordinate of the evaluation, while more egalitarian cultures would expect a dialogue between the supervisor and the subordinate. In some cultures, for example, in Japan, direct feedback itself, particularly negative, would likely be interpreted as a lack of trust and be insulting, resulting in strong negative reactions from the employee.[49] Rater biases are also likely to be much worse in cross-cultural situations because of national stereotypes and a lack of knowledge about the national context.[50]

The importance of goals and the value of external rewards also greatly vary among cultures. This will be discussed in detail in the next section, on compensation. Monitoring the system is more difficult because of cultural issues, geographical distances, and less direct interaction with employees in MNEs. Adjustments to the system become more important in a global setting because of the more volatile business environment, which can rapidly affect outcomes through no fault of the employees. In a global context, the performance criteria need to be more frequently reevaluated to make sure they reflect the current realities of the foreign location.[51] Involving foreign subsidiaries in the performance management of foreign employees can help overcome any culture issues, but will reduce the ability to have a globally standardized approach. In addition to considering cultural issues, the effectiveness of the performance management of international assignees can be improved by discussing the specifics of the performance management process prior to departure and putting specifics into the assignment plan. This will help set clear expectation for both the assignee and management.[52]

Compensation in the Global Environment

As with the other HR functions, a global business environment changes the compensation function in two basic ways. First, it requires firms to adapt their compensation practices to various different countries.[53] Companies have to decide whether their compensation system will be more centralized and standardized globally (within the constraints of legal requirements) or more decentralized and adapted to each country's business environment and culture.[54] Clearly, differences exist between countries in their typical compensation systems. Here are some examples:[55]

- In the United States, raises tend to be for performance or seniority. Pay for performance is common, usually through a merit system, bonuses, or profit sharing. Unlike in most countries, stock options are sometimes available for nonmanagerial employees.
- In Mexico, raises tend to be for seniority or across the board. Pay for performance is usually through attendance bonuses or profit sharing. Unlike in most countries, compensation practices that have been in place for two years are legally required to become permanent.
- In Japan, raises tend to be for seniority or skill improvement. Pay for performance is common, usually through group-based bonuses. Unlike in most countries, there is a prevalent use of "wage compositions," which include housing and commuting allowances.
- In Germany, raises tend to be for seniority, but sometimes for performance. Pay for performance is less common. Unlike in most countries, there is a strong harmonization of pay across organizations in the same industry.

Firms may opt to disregard some of these cultural norms (while still meeting legal requirements) in order to have a global standard consistent with their global strategy. They should do so knowing that countering cultural norms will meet with considerable resistance. For example, one petrochemical company ran into great difficulties resisting a common French practice of sharing regional profits with employees. However, the company was not run regionally, and providing such bonuses would have given their

French employees additional compensation that no other employees would have received.[56] In this case, the company had to weigh the costs of violating the country norms with the costs of creating internal inequities. In some countries, firms have very little freedom in determining wages (especially for lower level employees) because of centralized bargaining processes. For example, wages in Sweden, Italy, and Argentina are much more likely to be determined by centralized bargaining, compared to Canada, the UK, Japan, the United States, and the Czech Republic, where wages are generally up to each individual company.[57]

Second, compensation in MNEs is more complicated because of the differences in the compensation of international assignees compared to employees in domestic firms. These differences include determining the appropriate labor market (for pay comparison purposes) and the use of allowances and premiums. Clearly, the global business environment changes the nature of the four key compensation decisions, which we now discuss.

Determining pay level in the global context

The global environment does not directly affect a firm's pay-level strategy. That is, global firms can still decide to lag, match, or lead the market wages globally. However, determining pay level is more complicated for MNEs in two ways. First, firms can have a pay-level strategy that varies among countries. For example, for some countries where the qualified labor is plentiful and unemployment is high, they may choose a lag or match strategy to save on labor costs. In other countries, because of the difficulty in attracting qualified workers, or because of the strategic importance of the work being performed there, they may choose a lead policy. Nevertheless, the absolute pay level of workers greatly varies among countries, as shown in figure 3–8. For example, the average pay per hour (including direct pay, employer social insurance costs, and other labor taxes) in 2012 was $6.36 in Mexico, $35.67 in the United States, and $63.36 in Norway.[58] Big differences exist in the oil and gas industry, too. Figure 3–9 shows the average salaries of local employees in the oil and gas industry by country.[59]

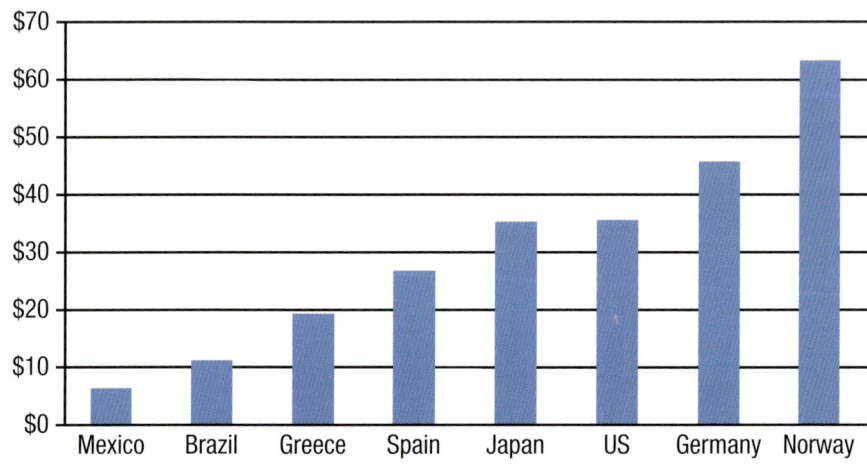

Fig. 3–8. Hourly compensation costs for production workers (2012)
Source: US Bureau of Labor Statistics, 2013.

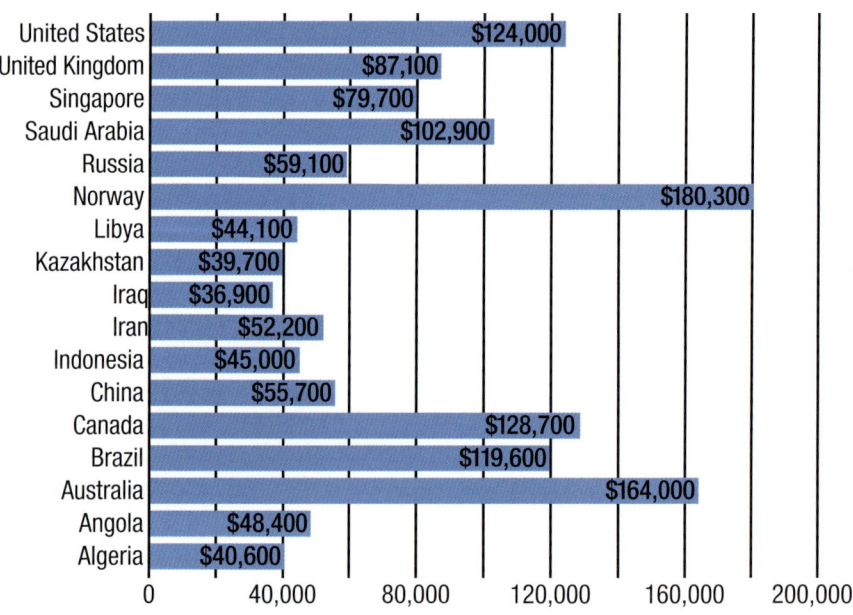

Fig. 3–9. Average oil and gas worker salaries, selected countries, 2012 local labor (US dollars)
Source: Constructed by authors based on data drawn from *Hays Oil & Gas Salary Guide, 2012*.

Second, as mentioned earlier, unlike domestic firms, MNEs must determine the appropriate labor market for pay determination purposes of international assignees. For example, if an American firm is employing a Pakistani national geophysicist to work at its subsidiary in the UAE, what is the appropriate labor market? Is it the going rate for the job in America, in Pakistan, or in the UAE? This is a strategic decision that firms tend to have a global philosophy about, but differences in assignments, regulations, laws, culture, and assignees may require some flexibility. Although country differences will affect pay levels, there are a number of different approaches to assignee pay (shown in table 3–2) that will have a strong effect on pay levels. Figure 3–10 shows the average salaries of foreign workers in the oil and gas industry by country.[60] Note that there are big differences between the pay of foreign and local workers. This is clearly illustrated in figure 3–11, which shows the ratio of the salaries of imported to local worker averages for various countries. These differences depend somewhat on which approach is used for foreign workers. The different approaches are:[61]

Table 3–2. Different types of international assignee compensation plans

Type of Plan	Salary Basis	Allowances	Premiums	Applicable Employees
Home-based plan	Home country salaries	Many: taxes, cost of living, housing, education, relocation, etc.	Hardship, foreign service, completion bonuses	Expatriates; TCNs (uses TCN's home country)
Host-based plan	Host country salaries	Minimal: tax protection	None	Localized expatriates; TCNs; Locals
Net-to-net	Host country salaries	Some: tax, housing, cost of living	None	Expatriates; TCNs
HQ-based	Home country salaries	Some: tax, housing, cost of living	Pay for performance	Expatriates; TCNs (uses HQ home country)

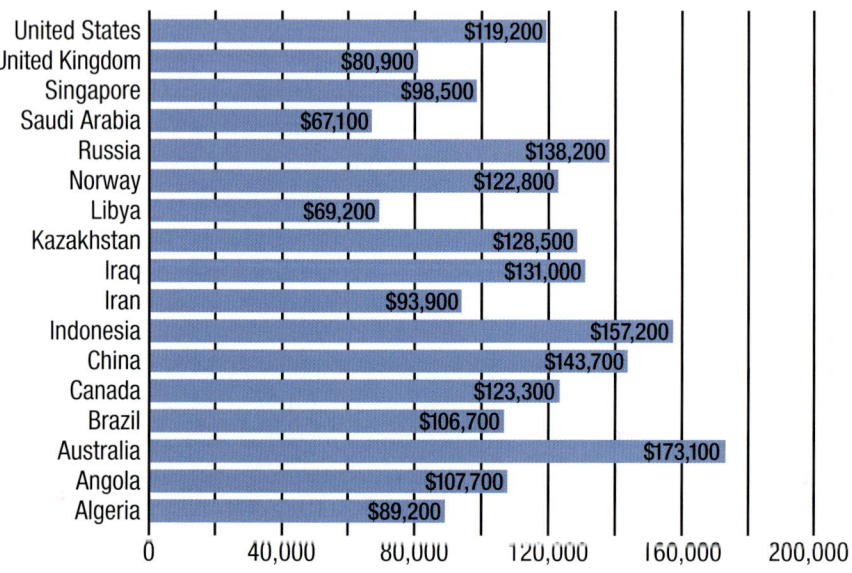

Fig. 3–10. Average oil and gas worker salaries, selected countries, 2012 imported labor (US dollars)

Source: Constructed by authors based on data drawn from *Hays Oil & Gas Salary Guide, 2012.*

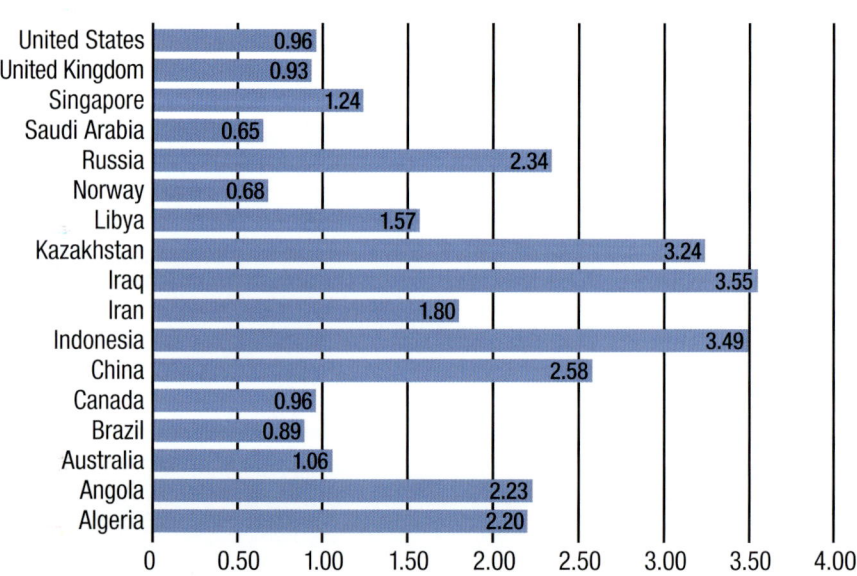

Fig. 3–11. Ratio of imported/local average oil and gas worker salaries for selected countries (2012)

Source: Constructed by authors based on data drawn from *Hays Oil & Gas Salary Guide, 2012.*

Home-based approach. The *home-based approach* starts with the assignees' existing salary, bonuses, and benefits at home. Usually, this approach then adds some premiums (bonuses to entice employees to accept the assignment), such as hardship premiums, foreign service premiums, or completion bonuses, and includes allowances (adjustments needed to make up for the differences in home- and host-country costs of living at a reasonably similar lifestyle), such as tax allowances, housing allowances, education allowances (for the assignee's children), inflation allowances, exchange rate allowances, cost-of-living allowances, relocation allowances, transportation allowances, medical allowances, utility allowances, furnishing allowances, home-leave allowances, and club membership allowances. The addition of premiums and allowances is referred to as the balance-sheet approach (with the home-based balance-sheet approach being the most popular). This approach tends be good at attracting international assignees and makes repatriation relatively easy, but it is very expensive. The home-based approach is the most popular approach, with more than 60% of companies using it.[62]

Host-based approach. In the *host-based approach*, international assignees are paid comparably to locals and receive only minimal allowances (e.g. tax allowances to prevent double taxation when applicable). This approach is more likely to be used for TCNs than PCNs, and it makes repatriation of the employee difficult. The host-based approach may also be used on PCNs who have overstayed their original contracts (see Current Challenge Box 3–2). This approach is used by less than 10% of MNEs.

Net-to-net approach. A *net-to-net approach* bases the salary on host-country levels, but provides allowances for cost of living, housing, taxes, and other expenses, which help assignees achieve a certain comfort level. This approach is similar to a host-based approach, but provides more allowances. This approach is used by less than 10% of MNEs.

HQ-based approach. The *HQ-based approach* uses a common global compensation package for certain job classes (usually higher level) worldwide that is determined by headquarters (HQ). The standard that is generally used is the rate paid at the firm's HQ (which will not be the home country of a TCN). Performance-based bonuses are usually a part of this package. This approach is used by less than 10% of MNEs.

Other approaches. Other approaches are also being used in the compensation of international assignees. These include the *ad hoc approach* (where firms negotiate independently with each assignee), *regional approaches* (where salary is based on regional norms), the *lump sum approach* (where assignees are given a lump sum of extra pay and can spend it on what they see fit), and the cafeteria approach (where assignees can choose from a menu of allowances up to a certain limit). About 15% of MNEs use approaches other than home-based, host-based, HQ-based, or net-to-net.

> ### Current Challenge Box 3–2. Expatriate Localization
>
> A current challenge for oil and gas companies is having to deal with PCNs staying in their foreign assignment location beyond their original assignment. This can be at the request of the PCN or MNE. These PCNs "go native" and frequently marry a local person and have a locally born family in the foreign assignment location. This is known as expatriate localization. On the positive side, expatriate localization indicates that the PCN has adjusted well and completely adapted to the new culture. On the negative side, these PCNs continue to receive their large allowances and premiums but live like locals (at frequently radically cheaper costs of living). One way to solve this issue is to have a policy that converts all PCNs to local compensation packages if they go beyond a contracted assignment period.[63] But, even in such cases expatriate localization can be a problem because localized employees may defeat the original purpose of the assignment if it was learning driven or its purpose was to facilitate transfer of knowledge back to headquarters. It is also a problem because once employees localize, they are usually no longer mobile. Companies must weigh the benefits and costs of keeping the employee happy and in the foreign location versus the benefits and costs of repatriating the employee back to headquarters and gaining the foreign knowledge the employee acquired abroad. Expatriate candidates should be informed of the localization policies at the selection stage of the assignment, so reasonable expectations are set and employees do not feel like they were treated unfairly and without warning later in their assignment.[64]

Determining pay mix in the global context

Again, as with pay level, we can expect mix differences based on country differences. Figure 3–12 shows the legally required and voluntary benefits costs as a percent of total compensation in different countries.[65] Clearly, legal differences and institutionalized factors have created big differences in the importance of benefits among countries. For example, in New Zealand, only

4% of total compensation consists of government-required benefits, while these benefits make up 30% of total compensation in France. Further, in New Zealand, total benefits account for 16% of total compensation costs. In France, benefits account for nearly half of total compensation costs. Clearly, country norms differ in the importance of benefits relative to other compensation. The same is true for performance-based pay, as will be discussed in the next section.

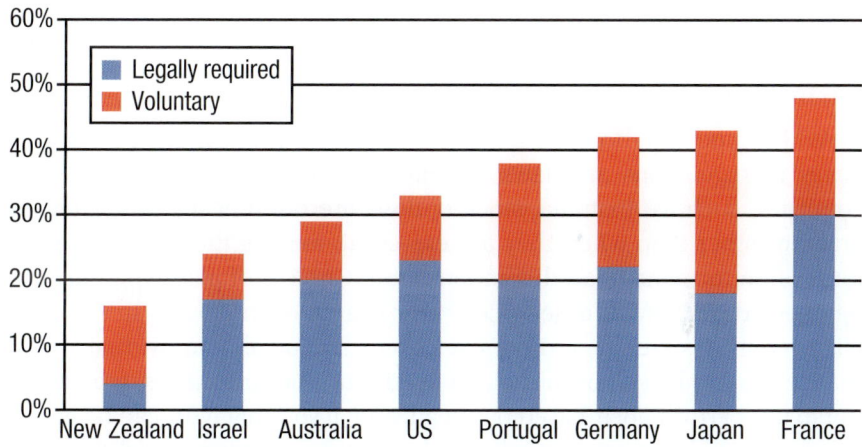

Fig. 3–12. Benefit costs as a percent of total compensation, 2012
Source: US Bureau of Labor Statistics, 2013.

In a global context, the pay mix is also affected by the existence of international assignees. The introduction of allowances and premiums into the pay mix occurs in MNEs, but not in domestic firms. For assignees, the pay mix will somewhat depend on assignment location factors, as described earlier, as well as the firm's compensation approach. For example, allowances and premiums are likely to be a far greater percentage of the pay mix in home-country-based plans than in host-country-based plans. Pay for performance is likely to be a greater percentage of the pay mix in HQ-based plans.

Determining the type of performance-based pay in a global context

Again, country differences can directly influence the compensation package. In the case of performance-based pay, historically, some countries were more open than others to having a large portion of pay being tied to performance (either individual or group-based). Some cultures are more accepting of risk, as can be seen in differences among countries in

Hofstede's uncertainty avoidance scores (See Current Challenge Box 3–1). Nevertheless, there has clearly been a convergence toward using variable pay: Worldwide almost 9 out of 10 companies offer some type of broad-based performance-based pay. However, differences still exist. For example, in Singapore, 96% of firms have performance-based pay plans, while only 70% of Swedish firms do.[66] However, just because most firms use some form of performance-based pay for some employees does not mean it is a prevalent proportion of pay for most employees. In fact, one study found that most managers in most countries found that pay for performance "should be" higher than it was, and that in many countries the amount of pay at risk was minimal.[67]

Which type of performance-based pay plan to use can be tied to differences in regulations and laws among countries. For example, in some countries, such as Australia, stock options are taxed at the time they are granted rather than when they are exercised, making them an unpopular form of performance-based compensation. Choices of plan type may also be affected by cultural differences.[68] For example, employees in individualistic countries are much more likely to be receptive to individual performance-based plans than those in collectivistic countries. Thus, whether or not to use performance-based pay at all, and, if so, what type, can be greatly affected by different country cultures, laws, and acceptance of risk.[69]

When compensating international assignees, the MNEs' general compensation approach is likely to also affect the type of performance-based pay. As previously mentioned, higher-level employees in HQ-based plans are likely to have a large component of their pay based on organization- or division-level performance. PCNs in home-based plans are more likely to receive bonuses based on successful completion of their assignments. Because performance-based pay should, when possible, be a part of the performance management of every international assignee, it too should be carefully communicated before departure.

Determining benefits in a global context

As shown in figure 3–12, the importance of benefits largely varies by country. Again, the prevailing norms, legal requirements, and tax implications are important factors in determining which benefits to include in compensation.[70] These differences can be large and may impact many different benefits, including health care, retirement income, unemployment insurance, vacations, holidays, accident and injury insurance, and leaves of absence. Because of these differences among countries, benefits are the most localized aspect of compensation.[71]

With respect to international assignees, premiums and allowances could be classified as benefits (although those that are cash-based could be arguably classified as performance-based pay). As mentioned earlier, the extensiveness of allowances and premiums is largely a function of the MNEs approach to compensating international assignees. For example, hardship premiums are much more likely to be part of a home-based plan than other plans. The US Department of State has a useful website that specifies hardship differentials, danger pay differentials and other cost-of-living estimates at the country and city level (http://aoprals.state.gov/Web920/location.asp?menu_id=95).

Allowances are also largely a function of plan type. International assignees are usually required to receive the legally mandated benefits specified by the host country, thus the benefits of international assignees are also heavily localized.

Because benefits are a big part of the cost of an international assignee, firms are doing more to make sure employees understand and appreciate their benefits. For example, Schlumberger holds benefit team meetings, organizes financial workshops, provides one-on-one consultations, gives online versions of workshops, and issues personalized total rewards statements to international assignees to help educate them on their benefits.[72]

Maximizing the effectiveness of compensation in the global context

As with domestic firms, gaining employee acceptance; being legally compliant; and regularly monitoring, evaluating, and adjusting the system are critical for maximizing the effectiveness of the compensation function in the global context. As with the other functions, doing so globally is much more complicated than in domestic firms.

As mentioned in chapter 1, fairness is largely based on employee perceptions. With respect to compensation, these perceptions are based on comparisons (in pay and benefits) between the employee and others. In domestic firms, these comparisons may be made between the employee and others in the same firm with the same or different jobs, or others in different firms with the same job. For example, petroleum engineers at BP may compare their pay with those of other petroleum engineers at BP, to their subordinates and supervisors at BP, and to petroleum engineers at Royal Dutch Shell, ExxonMobil, Chevron, and others. A global context greatly complicates these comparisons. What if the comparison BP petroleum engineers are in different countries, does that affect the perceptions of fairness? Do locals perceive their pay as fair compared to the much more highly paid

PCNs? Research shows that local managers do perceive their pay to be unfair relative to more highly paid PCNs, but that these negative feelings are largely offset if they are being paid more than other local managers at local competing firms.[73]

Addressing fairness perceptions in MNEs is also more complicated because of international assignees. International assignees are likely to compare their allowances and premiums to other assignees from their own company and from other companies. Showing that the processes used to determine allowances and premiums are reasonable and justifiable can go a long way in convincing assignees that they are being paid fairly.

Certainly, being legally compliant is much more complicated for MNEs because of cross-national differences and the use of international assignees. National labor laws will cover compensation issues such as minimum wage, overtime pay rate, night shift differentials, hazardous pay differentials (e.g., in Brazil, this is 30%), differentials for work that is dangerous to health, discrimination, vacation pay, Christmas bonuses, profit sharing requirements, payroll taxes, pension requirements, and health care requirements. However, legal issues that must be considered when dealing with international compensation, and benefits issues go beyond just labor laws. They also include tax laws, data privacy laws, immigration laws, finance laws, and accounting laws. The use of local experts can help MNEs navigate through the difficult terrain of evolving and varying country laws and conflicts between home-country and host-country laws.

Finally, monitoring, evaluating, and adjusting compensation are also complicated by the global context. Monitoring the system is more difficult because of cultural issues, geographical distances, and less direct interaction with employees in MNEs. For international assignees, difficulty in attracting assignees may show deficiencies in the compensation package. Because of the strong motivational effects of compensation, poor performance, increased absenteeism, or increased turnover may indicate issues with the compensation system. Adjustments to the system become more important in a global setting because of the more volatile business environment, which can rapidly affect performance results leading to unfairness in pay for performance systems. In a global context, the performance criteria for pay for performance plans need to be more frequently reevaluated to make sure they reflect the current realities of the foreign location.[74]

Conclusion

In this chapter we have shown that the oil and gas industry is probably the most global industry in the world. The industry is composed of companies, producers, and consumers from every country. The biggest companies in the industry are truly global companies with employees in more than a hundred countries. This global context affects all functions of HR and every key decision in each function. The global context greatly complicates those decisions. It does so because global companies must do business in many different countries with varying cultures, laws, standards, and so forth. Companies must balance the need to adapt to local contexts with the need for centralized standardization and control to maintain efficiency and strategic clarity. The global environment also complicates HR because of the different types of employees including PCNs, locals, and TCNs. Their selection, training, compensation, and performance management is very different from that of purely domestic employees.

References

1. Retrieved from http://corporate.exxonmobil.com/en/company/careers/employment-policies/employee-programs.
2. Werner, S. (2002). Recent developments in international management research: A review of 20 top management journals. *Journal of Management*, 28(3), 277–305.
3. Hofstede, G. H. (2001). *Culture's consequences: Comparing values, behaviors, institutions and organizations across nations.* Newbury Park, CA: Sage.
4. Dowling, P. J., Festing, M., & Engle, A. D., Sr. (2009). *International human resource management: Managing people in a multinational context.* Mason, OH: Thomson.
5. Briscoe, D., Schuler, R., & Tarique, I. (2012). *International human resource management* (4th ed.). New York, NY: Routledge.
   Dowling, P. J., Festing, M., & Engle, A. D., Sr. (2009). *International human resource management: Managing people in a multinational context.* Mason, OH: Thomson.
6. http://cdn.exxonmobil.com/~/media/Brochures/2007/news_pub_diversity.pdf
7. Gulati, R., & Carrera, L. (2012). An evidence-based approach to global talent-sourcing: Insights from the oil and gas industry. Towers Watson. Retrieved from http://www.towerswatson.com/en/Insights/IC-Types/Ad-hoc-Point-of-View/Perspectives/2013/Perspectives-An-evidence-based-approach-to-global-talent-sourcing.
8. Bonache, J., & Fernandez, Z. (1999). Strategic staffing in multinational companies. In C. Brewster & H. Harris (Eds.), *International HRM* (pp. 163–182). London, UK: Routledge.

9. Stahl, G. K., Chua, C. H., Caligiuri, P., Cerdin, J. L., & Taniguchi, M. (2009). Predictors of turnover intentions in learning-driven and demand-driven international assignments: The role of repatriation concerns, satisfaction with company support, and perceived career advancement opportunities. *Human Resource Management*, 48(1), 89–109.
10. Mäkelä, K., Björkman, I., & Ehrnrooth, M. (2009). MNC subsidiary staffing architecture: building human and social capital within the organisation. *The International Journal of Human Resource Management*, 20(6), 1273–1290.
11. Dowling, P. J., Festing, M., & Engle, A. D., Sr. (2009). *International human resource management: Managing people in a multinational context.* Mason, OH: Thomson.
    Caligiuri, P. (2013). *Cultural agility: Building a pipeline of successful global professionals.* Hoboken, NJ: Wiley.
12. Chapman, D. S., Uggerslev, K. L., Carroll, S. A., Piasentin, K. A., & Jones, D. A. (2005). Applicant attraction to organizations and job choice: A meta-analytic review of the correlates of recruiting outcomes. *Journal of applied psychology*, 90(5), 928.
13. Perkins, S. J., & Shortland, S. M. (2006). *Strategic international human resource management: Choices and consequences in multinational people management.* Philadelphia, PA: Kogan Page.
14. Caligiuri, P., Colakoglu, S., Cerdin, J. L., & Kim, M. S. (2010). Examining cross-cultural and individual differences in predicting employer reputation as a driver of employer attraction. *International Journal of Cross Cultural Management*, 10(2), 137–151.
15. Milkovich, G. T., & Bloom, M. (1998). Rethinking international compensation. *Compensation & Benefits Review*, 30(1), 15–23.
16. Yanadori, Y. (2011). Paying both globally and locally: An examination of the compensation management of a US multinational finance firm in the Asia Pacific Region. *International Journal of Human Resource Management*, 22(18), 3867–3887.
17. Caligiuri, P. (2013). *Cultural agility: Building a pipeline of successful global professionals.* San Francisco, CA: Wiley.
18. Ryan, A. N. N., McFarland, L., & Shl, H. B. (1999). An international look at selection practices: Nation and culture as explanations for variability in practice. *Personnel Psychology*, 52(2), 359–392.
    Shackleton, V., & Newell, S. (1994). European management selection methods: A comparison of five countries. *International Journal of Selection and Assessment*, 2(2), 91–102.
19. Hough, L. M., Oswald, F. L., & Ployhart, R. E. (2001). Determinants, detection and amelioration of adverse impact in personnel selection procedures: Issues, evidence and lessons learned. *International Journal of Selection and Assessment*, 9(1–2), 152–194.
20. Hunter, R. C., & Shoben, E. W. (1998). Disparate impact discrimination: American oddity or internationally accepted concept. *Berkeley Journal of Employment and Labor Law*, 19, 108.
21. Gregersen, H. B., Mendenhall, M. E., & Stroh, L. K. (1999). *Globalizing people through international assignments.* Reading, MA: Addison-Wesley.
22. Briscoe, D., Schuler, R., & Tarique, I. (2012). *International human resource management* (4th ed.). New York, NY: Routledge.
    Gregersen, H. B., Mendenhall, M. E., & Stroh, L. K. (1999). *Globalizing people through international assignments.* Reading, MA: Addison-Wesley.

Perkins, S. J., & Shortland, S. M. (2006). *Strategic international human resource management: Choices and consequences in multinational people management.* Philadelphia, PA: Kogan Page.
Caligiuri, P. (2013). *Cultural agility: Building a pipeline of successful global professionals.* San Francisco, CA: Wiley. Dowling, P. J., Festing, M., & Engle, A. D., Sr. (2009). *International human resource management: Managing people in a multinational context.* Mason, OH: Thomson.

23. Harris, H., & Brewster, C. (1999). The coffee-machine system: How international selection really works. *International Journal of Human Resource Management*, 10(3), 488–500.
24. Brookfield Global Relocation Services (2014). *Global relocation trends: 2013 survey report.* Retrieved from http://knowledge.brookfieldgrs.com/global_trends-Trends_and_Insights.
25. Briscoe, D., Schuler, R., & Tarique, I. (2012). *International human resource management* (4th ed.). New York, NY: Routledge.
    Black, J. S., Gregersen, H. B., Mendenhall, M. E., & Stroh, L. K. (1999). *Globalizing people through international assignments.* Reading, MA: Addison-Wesley.
    Perkins, S. J., & Shortland, S. M. (2006). *Strategic international human resource management: Choices and consequences in multinational people management.* Philadelphia, PA: Kogan Page.
    Caligiuri, P. (2013). *Cultural agility: Building a pipeline of successful global professionals.* San Francisco, CA: Wiley.
    Dowling, P. J., Festing, M., & Engle, A. D., Sr. (2009). *International human resource management: Managing people in a multinational context.* Mason, OH: Thomson.
26. Jackson, S. E., Schuler, R. S., & Werner, S. (2011). *Managing human resources.* CengageBrain.com.
    Posthuma, R. A., Roehling, M. V., & Campion, M. A. (2006). Applying US employment discrimination laws to international employers: Advice for scientists and practitioners. *Personnel Psychology*, 59(3), 705–739.
27. IPIECA. (2011). *Local content strategy: A guidance document for the oil and gas industry.* Retrieved from http://www.ipieca.org/sites/default/files/publications/Local_Content.pdf.
28. Rees, C. J., Mamman, A., & Braik, A. B. (2007). Emiratization as a strategic HRM change initiative: Case study evidence from a UAE petroleum company. *The International Journal of Human Resource Management*, 18(1), 33–53.
29. Darkwah, A. K. (2013). Keeping hope alive: An analysis of training opportunities for Ghanaian youth in the emerging oil and gas industry. International Development Planning Review, 35(2), 119–134.
30. Noe, R. A. (2013). *Employee training and development.* Boston, MA: McGraw-Hill/Irwin.
    Dowling, P. J., Festing, M., & Engle, A. D., Sr. (2009). *International human resource management: Managing people in a multinational context.* Mason, OH: Thomson.
31. Brookfield Global Relocation Services (2014). *Global relocation trends: 2013 survey report.* Retrieved from http://knowledge.brookfieldgrs.com/global_trends-Trends_and_Insights.
32. Caligiuri, P. (2013). *Cultural agility: Building a pipeline of successful global professionals.* San Francisco, CA: Wiley.

33. Briscoe, D., Schuler, R., & Tarique, I. (2012). *International human resource management* (4th ed.). New York, NY: Routledge.
34. Caligiuri, P. (2013). *Cultural agility: Building a Pipeline of successful global professionals*. San Francisco, CA: Wiley.
35. Perkins, S. J., & Shortland, S. M. (2006). *Strategic international human resource management: Choices and consequences in multinational people management*. Philadelphia, PA: Kogan Page.
36. Briscoe, D., Schuler, R., & Tarique, I. (2012). *International human resource management* (4th ed.). New York, NY: Routledge.
37. Dowling, P. J., Festing, M., & Engle, A. D., Sr. (2009). *International human resource management: Managing people in a multinational context*. Mason, OH: Thomson.
    Brookfield Global Relocation Services. (2014). *Global relocation trends: 2013 survey report*. Retrieved from http://knowledge.brookfieldgrs.com/global_trends-Trends_and_Insights.
38. Perkins, S. J., & Shortland, S. M. (2006). *Strategic international human resource management: Choices and consequences in multinational people management*. Philadelphia, PA: Kogan Page.
39. For more information on virtual expatriates and virtual HR, see Wood, T. (2007.) *The promise of the virtual oil company* (Cisco IBSG white paper). Retrieved from https://www.cisco.com/web/about/ac79/docs/wp/Virutal_Oil_Company_0629_FINAL.pdf.
    Heneman, R. L., & Greenberger, D. B.(Eds.). (2002). *Human resource management in virtual organizations*. Charlotte, NC: Information Age Publishing.
40. Brookfield Global Relocation Services. (2014). *Global relocation trends: 2013 survey report*. Retrieved from http://knowledge.brookfieldgrs.com/global_trends-Trends_and_Insights.
41. Black, J. S., Gregersen, H. B., Mendenhall, M. E., & Stroh, L. K. (1999). *Globalizing people through international assignments*. Reading, MA: Addison-Wesley.
42. Varma, A., Budhwar, P. S., & DeNisi, A. S. (Eds.). (2008). *Performance management systems: A global perspective*. New York, NY: Taylor & Francis.
43. Perkins, S. J., & Shortland, S. M. (2006). *Strategic international human resource management: Choices and consequences in multinational people management*. Philadelphia, PA: Kogan Page.
44. Dowling, P. J., Festing, M., & Engle, A. D., Sr. (2009). *International human resource management: Managing people in a multinational context*. Mason, OH: Thomson.
45. Lindhom, N., Tahvanainen,, M., and Bjorkman, I. (1999). Performance appraisal of host country employees: Western MNEs in China. In C. Brewster & H. Harris (Eds.), *International HRM* (pp. 144–158). London, UK: Routledge.
46. Black, J. S., Gregersen, H. B., Mendenhall, M. E., & Stroh, L. K. (1999). *Globalizing people through international assignments*. Reading, MA: Addison-Wesley.
47. Dowling, P. J., Festing, M., & Engle, A. D., Sr. (2009). *International human resource management: Managing people in a multinational context*. Mason, OH: Thomson.

48. Briscoe, D., Schuler, R., & Tarique, I. (2012). *International human resource management* (4th ed.). New York, NY: Routledge. Dowling, P. J., Festing, M., & Engle, A. D., Sr. (2009). *International human resource management: Managing people in a multinational context.* Mason, OH: Thomson.
49. Dowling, P. J., Festing, M., & Engle, A. D., Sr. (2009). *International human resource management: Managing people in a multinational context.* Mason, OH: Thomson.
50. Black, J. S., Gregersen, H. B., Mendenhall, M. E., & Stroh, L. K. (1999). *Globalizing people through international assignments.* Reading, MA: Addison-Wesley.
51. Black, J. S., Gregersen, H. B., Mendenhall, M. E., & Stroh, L. K. (1999). *Globalizing people through international assignments.* Reading, MA: Addison-Wesley.
52. Briscoe, D., Schuler, R., & Tarique, I. (2012). *International human resource management* (4th ed.). New York, NY: Routledge.
53. Marin, G. S., (2008). The influence of institutional and cultural factors on compensation practices around the world. In L. R. Gomez-Mejia & S. Werner (Eds.), *Global compensation: Foundations and perspectives* (pp. 3–17). New York, NY: Routledge.
    Marin, G. S. (2008). National differences in compensation: The influence of institutional *and cultural context. In L. R. Gomez-Mejia & S. Werner (Eds.), Global compensation: Foundations and perspectives* (pp. 18–28). New York, NY: Routledge.
54. Mearns, K., & Yule, S. (2009). The role of national culture in determining safety performance: Challenges for the global oil and gas industry. *Safety Science, 47*(6), 777–785.
55. Jackson, S. E., Schuler, R. S., & Werner, S. (2011). *Managing human resources.* CengageBrain.com.
56. Bradley, P., Hendry, C., and Perkins, S. (1999). Global or multi-local? The significance of international values in reward strategy. In C. Brewster & H. Harris (Eds.), *International HRM* (pp. 120–142). London, UK: Routledge.
57. Opute, J. (2010). Managing reward in developing economies: The challenge for multinational corporations. *Policy Futures in Education, 8*(1), 37–47.
    Milkovich, G. T., Newman, J. M., & Gerhart, B. (2014). *Compensation* (11th ed.). Burr Ridge, IL: Irwin/McGraw-Hill.
58. Bureau of Labor Statistics. (2013). *International comparisons of hourly compensation costs in manufacturing, 2012.* Retrieved from http://www.bls.gov/fls/ichcc.pdf.
59. Hays Oil and Gas. (2013). *The oil and gas global salary guide, 2013.* Retrieved from http://hays.clikpages.co.uk/Oil_and_Gas_Salary_Guide_2013/.
60. Hays Oil and Gas. (2013). *The oil and gas global salary guide, 2013.* Retrieved from http://hays.clikpages.co.uk/Oil_and_Gas_Salary_Guide_2013/.
61. Briscoe, D., Schuler, R., & Tarique, I. (2012). *International human resource management* (4th ed.). New York, NY: Routledge.
    Black, J. S., Gregersen, H. B., Mendenhall, M. E., & Stroh, L. K. (1999). *Globalizing people through international assignments.* Reading, MA: Addison-Wesley.
    Dowling, P. J., Festing, M., & Engle Sr., A. D., Sr. (2009). *International human resource management: Managing people in a multinational context.* Mason, OH: Thomson.

62. All usage statistics from Brookfield Global Relocation Services (2014). *Global relocation trends: 2013 survey report*. Retrieved from http://knowledge.brookfieldgrs.com/global_trends-Trends_and_Insights.
63. Briscoe, D., Schuler, R., & Tarique, I. (2012). *International human resource management* (4th ed.). New York, NY: Routledge. Hailey, J. (1999). Localization as an ethical response to internationalization. In C. Brewster & H. Harris (Eds.), *International HRM* (pp. 89–101). London, UK: Routledge.
64. Perkins, S. J., & Shortland, S. M. (2006). *Strategic international human resource management: Choices and consequences in multinational people management*. Philadelphia, PA: Kogan Page.
65. Bureau of Labor Statistics. (2013). *International comparisons of hourly compensation costs in manufacturing, 2012*. Retrieved from http://www.bls.gov/fls/ichcc.pdf.
66. Jackson, S. E., Schuler, R. S., & Werner, S. (2011). *Managing human resources*. CengageBrain.com.
67. Perkins, S. J., & Shortland, S. M. (2006). *Strategic international human resource management: Choices and consequences in multinational people management*. Philadelphia, PA: Kogan Page.
68. Franco-Santos, M. (2008). Performance measurement issues, incentive application, and globalization. In L. R. Gomez-Mejia & S. Werner (Eds.), *Global compensation: Foundations and perspectives* (pp. 41–56). New York, NY: Routledge.
69. Salimaki, A., & Heneman, R. L. (2008). Pay for performance for global employees. In L. R. Gomez-Mejia & S. Werner (Eds.), *Global compensation: Foundations and perspectives* (pp. 158–168). New York, NY: Routledge.
70. Martocchio, J. J., & Pandey, N. (2008). Employee benefits around the world. In L. R. Gomez-Mejia & S. Werner (Eds.), *Global compensation: Foundations and perspectives*. (pp. 179–191). New York, NY: Routledge.
71. Briscoe, D., Schuler, R., & Tarique, I. (2012). *International human resource management* (4th ed.). New York, NY: Routledge.
72. Lovewell, D. (2009). *Employer profile: Schlumberger's global challenge to deliver staff benefits*. Retrieved from Employee Benefits.com website: http://www.employeebenefits.co.uk/employer-profile-schlumbergers-global-challenge-to-deliver-staff-benefits/9352.article.
73. Chen, C. C., Kraemer, J., & Gathii, J. 2011. Understanding locals' compensation fairness vis-à-vis foreign expatriates: The role of perceived equity. *International Journal of Human Resource Management, 22*(17), 3582–3600.
74. Black, J. S., Gregersen, H. B., Mendenhall, M. E., & Stroh, L. K. (1999). *Globalizing people through international assignments*. Reading, MA: Addison-Wesley.

Case 1

Employment in an Upstream African Project: The Case of the Chad-Cameroon Petroleum Development Project

To illustrate a number of the issues described throughout the book, we now present a case showing the employment challenges borne by global oil and gas companies in the upstream: the Chad-Cameroon Petroleum Development Project. The project, one of the most transparent and highly publicized oil and gas developments of its time, was an investment undertaken by a consortium of three companies, ExxonMobil (Esso), Petronas, and Chevron, and the governments of Chad and Cameroon. Construction began in 2000. Even today, more than 15 years after production began, the project continues to employ more than 6,000 workers across two countries.

The project, a $3.7 billion investment, combined the development of an oil production area in southern Chad with a 1,000-kilometer pipeline extending from the production area to the southeast, running the length of Cameroon to the coastal city of Kribi on the Gulf of Guinea. The pipeline then extended underwater 11 kilometers out to a permanently stationed floating storage and offloading vessel (FSO). From the FSO, crude oil tankers loaded the crude oil for shipment to refineries around the world. As illustrated by figure C1-1, the project combined a multitude of technologies and facilities that had to be first constructed and then operated.

Project employment in an upstream development such as Chad-Cameroon ramps up rapidly in the early stages of construction. As illustrated in figure C1-2, employment in the project rose from only a few hundred people in 2000 to a peak of over 12,700 by the fourth quarter of 2002—an increase in employment in remote areas of Central and West Africa of 12,000 people in less than two years. Pipeline construction, most of which lies within the borders of Cameroon, was obviously the bulk of the early employment growth. Cameroon employment made up 7,000 of the 12,000 employees in 2002. The production area, the Kome region of southern Chad, was a slower ramp-up of employment as the power plant, primary petroleum processing facilities, and head of pipeline construction required staged construction in addition to field drilling.

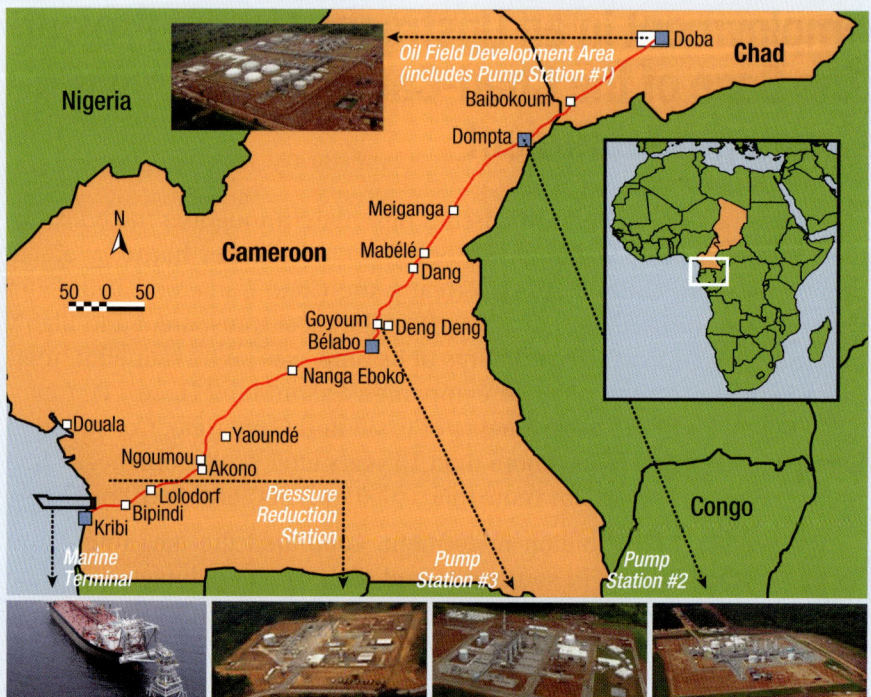

Fig. C1–1. The Chad-Cameroon Petroleum Development Project
Source: EssoChad.org.

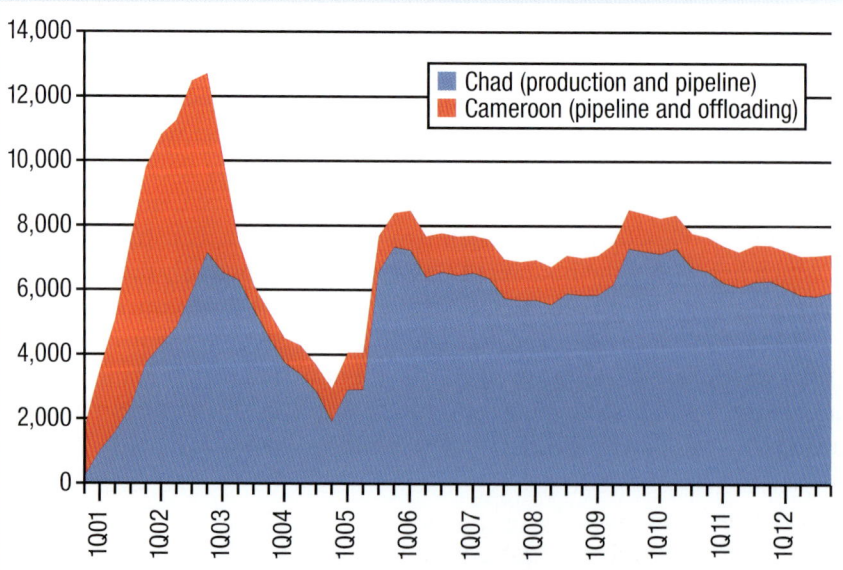

Fig. C1–2. Chad-Cameroon Petroleum Development Project employment by country
Data drawn by authors from *Chad Cameroon Petroleum Development Project Updates*, 1–33 EssoChad.

Construction lasted only two to three years, depending on the specific stage and structure. Just as quickly as pipeline construction employment grew, it fell as the pipeline was completed in the first quarter of 2003. At nearly the same time that most of the construction employment associated with the Chad production area was completed, total project employment fell from its peak of 12,700 to just below 3,000 in the fourth quarter of 2004. The first oil flowed through the pipeline in July 2003. Operations were underway.

Reinvestment

The Chad-Cameroon project, however, saw an employment rebirth in late 2005 with the initiation of a combination of the opening of several new oil fields in the main production area as well as a massive in-fill drilling and water injection effort in the existing fields. The Chadian production field proved to be more complex than expected, and a multitude of additional wells were needed in a closer matrix across the oil field to reach oil that proved difficult to access. The oil reservoir had also yielded a high water content, which the project reinjected into the reservoir to increase pressure and therefore support a higher petroleum yield. All of this investment required a massive buildup in employment in the Chadian production area, as seen in figure C1-2. Chadian employment then quickly increased from a low of 1,900 to over 7,000. Short-term strategic and operational changes required rapid recruiting and hiring responses.

These additional investments in the production area continued in the 2005–2014 period, and employment in both Chad's production area and Cameroon's pipeline and offloading operations and maintenance functions stabilized at these levels. Total project employment settled at a constant 7,000.

Employment in a project involving such complex activities in such a remote location requires a mix of both locals, called host-country nationals (HCNs), and expatriate employees, called parent-country nationals (PCNs). Figure C1-3 tracks the employment of these two major groups in both Chad and Cameroon over the construction and operation periods.

The oil company consortium made concerted efforts beginning prior to construction to recruit and train as many nationals as possible for employment. Even during the construction phases, the number of PCNs in both Chad and Cameroon were managed tightly. But an oil construction project like this requires a relatively larger number of skilled and semiskilled workers—skill levels that are relatively rare in countries like Cameroon and Chad—particularly Chad.

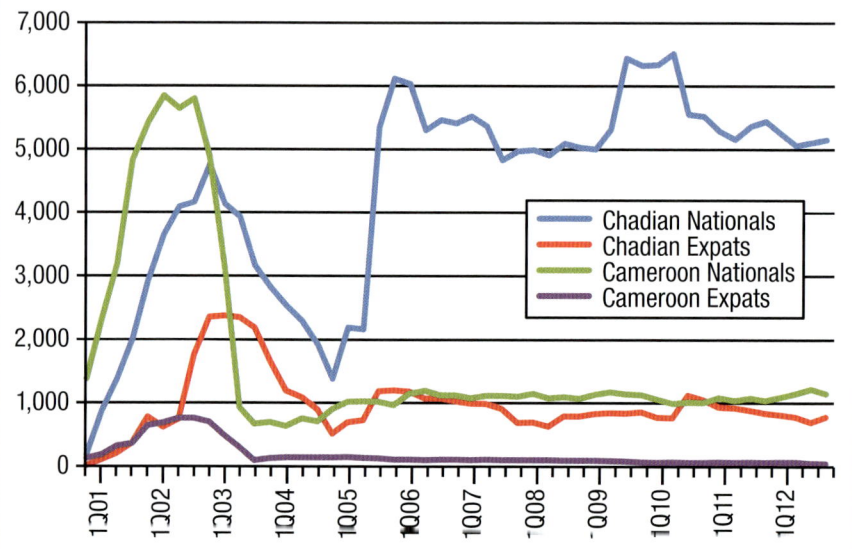

Fig. C1–3. HCNs and PCNs employed in the Chad-Cameroon Petroleum Development Project
Data drawn by authors from *Chad Cameroon Petroleum Development Project Updates*, 1–33 EssoChad.

The extreme difficulties of operating in Chad cannot be overstated. Consider the following:

- Chad at the time of construction was a country of about 10 million people. But with no official national census, this was only a rough estimate.
- Chad was consistently ranked at the bottom in global competitiveness rankings. Even by 2008 Chad was still ranked at 131 out of 131 countries on a global competitiveness basis.
- 5% of Chad's population was officially listed as HIV positive, but many authorities estimated that the true number could be over 20%.
- Chad's geographic size was enormous, being equal to three times that of California.
- More than 80% of the population relied on subsistence farming and livestock for survival. (Subsistence meant that no excess production was generated to constitute a cash crop.)
- Chad is landlocked (hence the pipeline across Cameroon), with few exports of substance. Exports were confined to some cotton, cattle, and cotton textiles.

- Chad was ranked number 172 out of 175 countries in terms of corruption.
- Chad had less than 1,000 kilometers of paved road, and within the entire country there was not a single kilometer of railroad.
- Chad borders on the east with the Darfur region of Sudan, which had been the center of war and famine for years.

In short, Chad was a country and a people of extreme poverty, completely lacking in the most elemental forms of infrastructure and economic development, and in severe need of new economic and social hope.

Upstream development and production in remote locations like central Africa present extreme personnel challenges for major oil and gas companies. Although they would clearly prefer to hire people as locally and domestically as possible, a country like Chad in 2000–2005 offered few human resources. Although the consortium commenced recruiting, training, and development of people immediately, much of the early technical staffing had to be through PCNs. Critical positions in mechanical, electrical, and petroleum engineering; geophysics; chemistry; and the like with both knowledge and experience in difficult and emerging project start-ups had to be staffed appropriately if the project was to succeed.

Structure of Employment

There were roughly three major categories of employees: (1) PCNs; (2) permanent HCNs, and (3) special-project nationals. Even the first category, PCNs, were a staffing and cost challenge. Many of the PCNs in Chad (fewer in Cameroon) were *rotators*. *Rotators* are PCNs who serve, in this case, four weeks on and four weeks off. They arrive from all parts of the globe and then work seven-day weeks, 12-hour days. At the end of four weeks, they return home for four weeks. This requires that most rotator positions be staffed back-to-back, such that the current worker is replaced with a second who has the same skills and same knowledge, and these efficient and effective handoffs occur every four weeks.

Like other PCNs around the world, their costs were high. Travel expenses were always significant to remote locations like Chad. In fact, there were few ways to reach Chad by air other than through Paris, regardless of where in the world you were traveling from. Although there were a few connections from Ethiopia and South Africa at the time, the vast majority of rotators flew from

Paris to the Chadian capital of N'djamena, where they were then flown by the oil consortium's charter airplane south to the production area in Kome.

PCNs also were paid country premiums, a variable component of compensation for accepting the supposed risks associated with working in a country with health and personal safety risks. Although crime was generally not an issue in Chad, all PCNs were provided with some degree of security. Health risks were probably higher, with major consortium-supported programs to reduce the threat of malaria undertaken from the very earliest of stages. Chad was part of the mid-African malaria belt, and as a result, medication and associated exposed skin precautions undertaken.

The second and third major categories of employees were both nationals, *permanent nationals* and *special-project nationals*, hired and trained in-country. Permanent nationals were employed on a day-to-day basis, working Monday through Friday shifts for shorter days (eight-hour days compared to the rotators) but on duty for most of the year with no rotation breaks. The third category, special project nationals, were temporary workers, often hired from villages near project facilities, under temporary contracts and working only a few weeks at a time to complete special projects (e.g., maintenance of the pipeline right of way).

As illustrated by figure C1-3, the vast majority of project employees, in both Chad and Cameroon, were national hires. PCN employment in Chad never rose above 2,500, with Cameroon never reaching 1,000. PCN employment in Chad has been relatively steady since 2005, reflecting the need for highly specialized reservoir engineering and production specialists. Cameroon, however, has been able—after continuous investment in training and development for more than a decade—to reduce PCN employment to near zero.

Project staffing fell into four broad categories: supervisory, skilled, semiskilled, and unskilled. Although the construction phase was characterized by a noticeably large portion of unskilled employees, the comparison between the construction and operating phases by skill category has been remarkably stable, as shown in Table C1-1. This stability is a clear indicator that even in the construction phases of the upstream industry, the need for skilled and semiskilled workers is critical for project execution.

Table C1–1. Proportion of labor skill by category

Project Phase	Supervisory	Skilled	Semi-skilled	Unskilled
Construction	6%	40%	20%	34%
Operating	8%	42%	29%	21%

In December 2012 the pipeline transportation companies announced that for the first time the pipeline system was staffed completely with Cameroonians. This was an impressive achievement, given that this pipeline system included two pumping stations along the 1,000-kilometer pipeline route, and a pressure reducing-station at the coastal city of Kribi, facilities that required highly technically skilled leadership and technical assistance. The effort to nationalize the project's workforce had been a stated objective of the project from the very beginning. This was seen as an important achievement because many of these positions were highly skilled and supervisory, and many had the highest wage classifications. The program for nationalization had combined intensive classroom training, on-the-job mentoring, and, in many cases, international assignments to broaden individual knowledge and experience.

Figure C1–4 illustrates the results of the Chad-Cameroon project's commitment to national human resource development. After transitioning from the construction phase to the operational phase, the project had consistently maintained national hires at 85% or more of total project employment

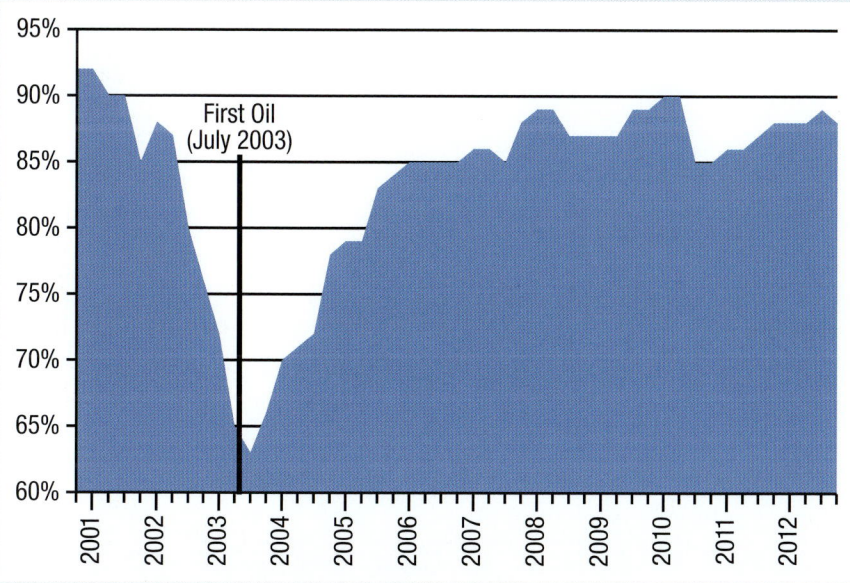

Fig. C1–4. HCNs' proportion of project employment
Data drawn by authors from *Chad Cameroon Petroleum Development Project Updates*, 1–33 EssoChad.

Conclusion

This case illustrates a number of the issues that arise in the international context of the oil and gas industry, including the increased complexities in managing human resources. These complexities are even greater when operating in countries that differ substantially from the company's home countries. Nevertheless, with careful planning and sound human resource management strategies these challenges can be managed.

4

The Importance of Health and Safety in the Oil and Gas Industry

Introduction

There is no more valuable resource or asset than the human resource. Maintaining and protecting the human resource is therefore of paramount importance, and that is particularly the case in the oil and gas industry. All firms in the oil and gas industry are concerned with the health, safety, and welfare of the people engaged in their global workplaces. Although labeled in a variety of ways—occupational safety and health, occupational health and safety, workplace health and safety—all focus on protecting and improving the health and welfare of all stakeholders in the business, including employees, family members, customers, suppliers, and communities.

Despite the efforts and initiatives put forward by governments and organizations of all kinds, occupational accidents and diseases still take a heavy toll globally every year. It is estimated that more than 2 million people die each year across the world from work-related cancers, circulatory and cerebrovascular diseases, and communicable diseases. An additional 270 million people are thought to suffer nonfatal injuries from occupational accidents, and another 160 million workers suffer from work-related diseases, two-thirds of which result in time away from work of four days or longer.[1]

There are many safety hazards across the oil and gas industry value chain, from the danger of falling objects on a drilling rig to a potential explosion at a refinery to malaria threats in Sub-Saharan Africa. Many industry products are volatile, toxic, and environmentally sensitive. Work sites such as refineries and oil production platforms have many complex pieces of equipment and must be carefully managed by a highly skilled workforce. Heavy equipment such as ships and gas turbines are built in some of the world's largest and most complex fabrication yards that may employ tens of thousands of workers. On the different job sites there is the potential for accidents of many types and levels of severity. According to the US Occupational Safety and Health Administration (OSHA), hazards in the well-drilling and servicing sector include being struck by various objects such as pipes, chains, whipping hoses, tools, and debris dropped from elevated locations; being caught between spinning chains and pipes; fires and explosions from well blowouts; releases of gas; poorly maintained electrical equipment; and aboveground detonation of perforating guns. Rigs can collapse, people can fall from elevated areas, and poisonous gases such hydrogen sulfide (H_2S) can be released. An additional hazard in the upstream in some countries is personal security, a reality of the industry's history and geographic scope.

The creation of a safe work environment and a culture of safety and security is a major leadership concern in human resource management. In the oil and gas industry, as in all industries, it is generally considered a moral obligation of the firm to assure a safe and healthy work environment, regardless of where on earth it occurs or how extremely difficult conditions may be. In many countries it is also a legal or common law obligation, and may be highly regulated and reported. In the global oil and gas industry, where companies vie for the rights and privileges of exploring, developing, and operating in other countries, a firm's health, safety, and environmental (HSE) record and reputation are fundamental to its license to operate.

For many years the major oil and gas firms have argued that safety of the workforce and their communities is their number-one priority. More recently, firms have shifted their safety perspective away from safety as a priority to safety as a core value, meaning that safety is part of firm's culture. Priorities can shift, but cultures are deeply rooted and difficult to change, so once a firm establishes safety as a core value, it should impact the entire organization and its workforce. Moreover, safety is good for business, because a safe operating environment creates a discipline that is transferable to all operational areas, including the office environment.

Occupational Safety and Health in the Oil and Gas Industry

The International Labour Organization and the World Health Organization share a common definition of occupational health.

> Occupational health should aim at: the promotion and maintenance of the highest degree of physical, mental and social well-being of workers in all occupations; the prevention amongst workers of departures from health caused by their working conditions; the protection of workers in their employment from risks resulting from factors adverse to health; the placing and maintenance of the worker in an occupational environment adapted to his physiological and psychological capabilities; and, to summarize, the adaptation of work to man and of each man to his job.[2]

Occupational safety and health programs and systems have three primary objectives:

1. Maintenance and promotion of workers' health and working capacity
2. Improvement of work and working environment to promote and preserve safety and health
3. Development of safe and healthy work organizations and working cultures

Safety and health are often further subdivided into (1) safety and injury hazards and (2) health and illness hazards. In recent years, the range of worker health concerns has expanded to encompass all potential environmental threats to worker health, both on and off the job, including air and water pollutants and a variety of disease threats like malaria, HIV/AIDS, and other diseases like Ebola in West Africa. Table 4-1 provides a brief introduction to the basic components of occupational safety and health, their meaning, and a variety of concerns and managerial solutions.

Table 4–1. Occupational safety, health, and environmental concerns

Safety	Health	Environment
Safe is a condition of being protected against physical, emotional, occupational, psychological, or other types of failure, damage, error, accidents, harm or any other even considered nondesirable.	*Health* is a state of complete physical, mental, and social well-being, not merely the absence of disease or infirmity.	*Environmental health* addresses all the physical, chemical, and biological factors external to a person, and all the related factors impacting behaviors.
A wide array of workplace *hazards* presents risks to the health and safety of people at work. These include chemicals, biological agents, physical factors, adverse ergonomic conditions, allergens, a complex network of safety risks, and a broad range of psychosocial risk factors.	*Occupational health* is a multi-disciplinary field of healthcare concerned with enabling an individual to undertake their occupation in a way that causes least harm to their health.	*Occupational environmental health* encompasses the assessment and control of those environmental factors as related to work or the workplace that can potentially affect health.
Personal safety or occupational safety focuses on preventing slips-trips-and-falls, driving safety, and other personal activities which could cause injury.	*Occupational health* refers to the identification and control of the risks arising from physical, chemical, and other workplace hazards including disease to maintain a safe and healthy working environment.	Growing recognition and accountability of the impact of the company and industry on the global environment, including particulate, carbon oxide, nitrogen oxide, and sulfur oxide emissions
Process safety focuses on the design for safety, hazard analysis, material verification, equipment maintenance, process upset reporting, by the unit or business.	*Working conditions* include working hours, workplace family policies, health promotion, and protection provisions.	Protection of biodiversity through research, improved cleaning technologies, and risk and impact management.
	An *occupational disease* is any disease contracted primarily as a result of an exposure to risk factors arising from work activity.	Increasing efforts to safeguard the environment by reducing emissions and discharges.

Figure 4–1 provides a basic safety comparison between the US oil and gas industry and other US private industry. Safety and health are particularly important to address in the oil and gas industry because of the materials and equipment used and the difficult locations and conditions of the work. Nevertheless, based on data presentations by the American Petroleum Institute, the oil and gas industry has a superior safety record compared to the overall US private sector (based on the total incident rate) for the entire decade 2003–2012. Figure 4–2 breaks down the rates by value chain components. Refining and the offshore upstream part of the US industry (exploration and production) have significantly improved in safety performance over the decade of data, with refining demonstrating the best performance. US upstream (exploration and production) is in the middle of the value chain performance, with retail and finally wholesale oil and gas marketing having the relatively highest incidence rates. It bears repeating, however, that the oil and gas industry—even by segments—shows superior performance to the US private industry sector as a whole.

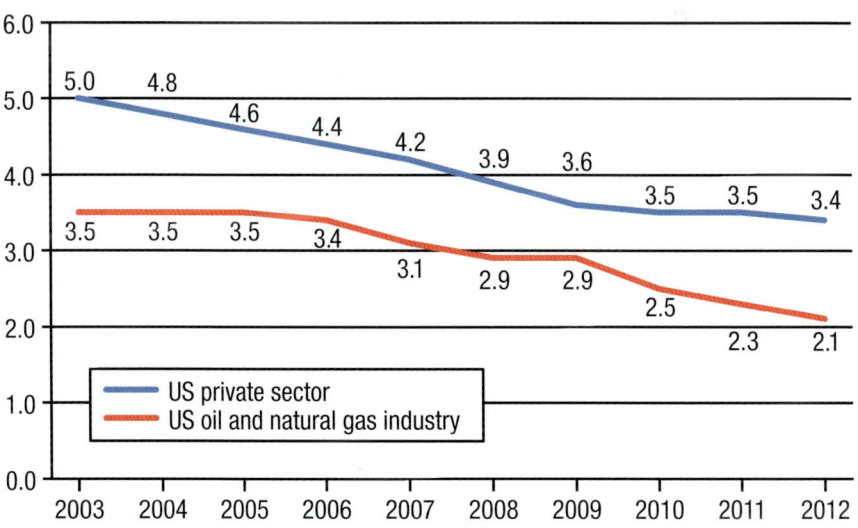

Fig. 4–1. Oil and gas industry safety and health versus private sector (nonfatal injury or illness rate per 100 workers)

Note: A nonfatal job-related injury or illness is an abnormal condition or disorder that results in days away from work, restricted work, or transfer to another job, medical treatment beyond first aid, or loss of consciousness. Injuries include cases such as, but not limited to, a cut, fracture, sprain, or amputation. Illnesses include both acute and chronic illnesses, such as, but not limited to, a skin disease, respiratory disorder, or poisoning.
Source: *Workplace Injuries and Illnesses Safety (WIIS) Report, 2003–2012*, American Petroleum Institution, 2013, p.1.

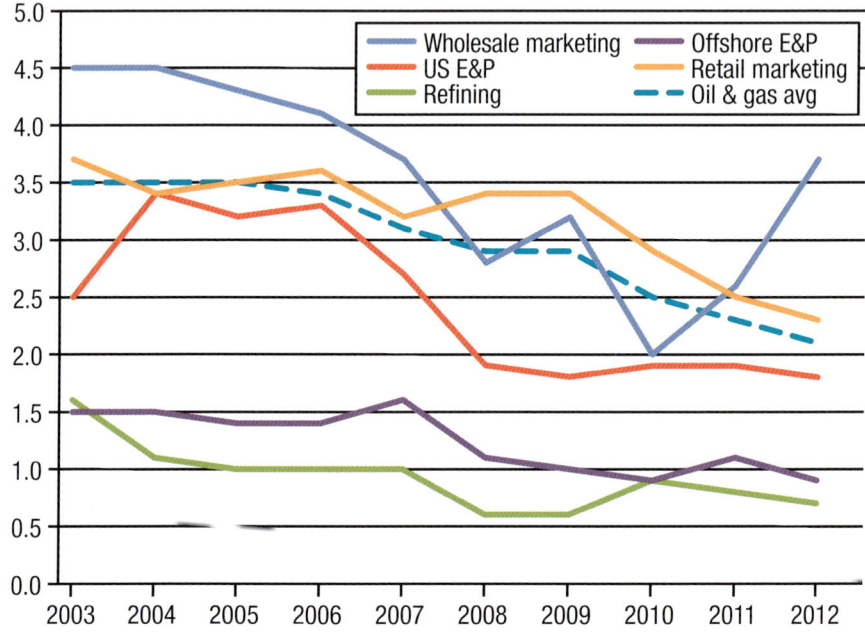

Fig. 4–2. Oil and gas safety and health rates by industry value chain (nonfatal injury or illness rate per 100 workers)
Source: *Workplace Injuries and Illnesses Safety (WIIS) Report, 2003–2012*, American Petroleum Institution, 2013, p.5.

Personal safety and process safety

One of the most important distinctions in safety across all industries is that between personal safety and process safety.

- Personal safety or occupational safety focuses on preventing slips, trips, and falls and encouraging driving safety as well as safety in other personal activities that could cause injury. The emphasis is on personal safety issues and personal or individual behavior.
- Process safety focuses on the design for safety, hazard analysis, material verification, equipment maintenance, and process upset reporting, by the unit or business. Process safety requires a system approach to safety, in order to analyze the interaction of all mechanical, chemical, biological, and human interactions in a business process.

This personal versus process safety distinction has been at the center of much change in the oil and gas industry in recent years as major accidents like that of the refinery explosion and fire at Texas City or the offshore explosion at the Macondo platform in the Gulf of Mexico have been studied in depth to understand what roles this distinction played. In general, the industry has concluded that although it has shown significant improvement in personal safety, the large incidences resulting in significant fatalities were largely the result of a lack of emphasis and investment on process safety. We now briefly describe each of these accidents to better illustrate the concepts of personal safety and process safety.

The 2005 Texas City Refinery explosion. The Texas City Refinery in Texas City, Texas, is one of the largest refineries in the United States. Texas City is a major industrial center, and in 1947, it suffered one of the worst industrial accidents in the United States when a ship full of fertilizer exploded, killing nearly 600 people. The Texas City Refinery had a capacity of 437,000 b/d. BP acquired the Texas City Refinery when it bought Amoco in 1998. A major explosion occurred in the Texas City Refinery in an isomerization unit at the site on March 23, 2005, killing 15 workers and injuring more than 170 others. The isomerization unit is used to convert raffinate, a low-octane blending feed, into higher-octane components for unleaded regular gasoline. The unit has four sections, including a splitter that takes raffinate and fractionates it into light and heavy components. The splitter consists of a surge drum, a fired heater reboiler, and a fractionating column 164 feet tall.

According to BP's initial internal report on the accident, operators overfilled and overheated a processing tower at the unit that housed hydrocarbon liquid and vapor. The liquid and vapor mix was overpressurized, flooded into an adjacent stack, and escaped into the atmosphere around the unit. The resulting vapor cloud was ignited, causing the explosion. The source of ignition is not known. The report said that supervisors did not verify that correct procedures were followed and they were absent from the unit at key times. "The failure of [the] managers to provide appropriate leadership and the failure of hourly workers to follow written procedures are among the root causes of this incident." [3]

A detailed investigation of BP, known as the *Baker Report*, reached the following conclusions:
- BP has not provided effective process safety leadership and has not adequately established process safety as a core value across all its five US refineries.
- BP measured safety performance through personal injury rate, rather than measuring process safety equipment performance.
- BP tended to have a short-term focus, and its decentralized management system and entrepreneurial culture have delegated

much discretion to US refinery plant managers without clearly defining process safety expectations, responsibilities, or accountabilities.
- BP has not demonstrated that it has effectively held executive management and refining line managers and supervisors, both at the corporate level and at the refinery level, accountable for process safety performance at its five US refineries.
- A corporate safety culture that may have tolerated serious and long-standing deviations from good safety practice.[4]

A key finding by the *Baker Report* was that personal safety had been given greater priority than process safety. The report made a number of recommendations for improving process safety and there was a strong emphasis on the need for a change in BP corporate culture associated with process safety. An investigation by the US Chemical Safety Board concluded that high-level decisions to defer overhauls, cut staff, and rein in costs at the Texas City plant contributed to the accident. BP's internal investigation identified several causes, including:

- Over the years, the working environment had eroded to one characterized by resistance to change, and lacking in trust, motivation, and a sense of purpose. Coupled with unclear expectations around supervisory and management behaviors, this meant that rules were not consistently followed, rigor was lacking, and individuals felt disempowered from suggesting or initiating improvements.
- Process safety, operations performance, and systematic risk reduction priorities had not been set and consistently reinforced by management.
- Many changes in a complex organization had led to the lack of clear accountabilities and poor communication, which together resulted in confusion in the workforce over roles and responsibilities.

The Chemical Safety Board had two findings that were disputed by BP senior management:

- The combination of cost cutting, production pressures, and failure to invest caused a progressive deterioration of safety at the refinery.
- BP's global management was aware of problems with maintenance, spending, and infrastructure well before March 2005. Unsafe and antiquated equipment designs were left in place, and unacceptable deficiencies in preventative maintenance were tolerated.

As a result of the incident, BP set aside $700 million to compensate victims of the explosion. BP also agreed to pay several fines totaling about $70 million and agreed to a number of corrective actions, including the hiring and placement of process safety and organizational experts at the refinery. In 2013 BP sold the refinery to Marathon Petroleum. The impact of the accident extended far beyond Texas City. All of the world's major refiners closely followed the results of the Texas City investigations and conducted their own internal analyses to ensure they were not at risk for similar process safety incidents.

The Macondo accident. On April 20, 2010, the drilling rig *Deepwater Horizon* caught fire and sank in the Gulf of Mexico. Eleven men were killed, and over the next 87 days, almost 5 million barrels of oil were discharged from the Macondo well into the Gulf of Mexico. The Joint Investigation Team of the Bureau of Ocean Energy Management, Regulation and Enforcement and the US Coast Guard identified a number of causes of the Macondo blowout. The investigation found that a central cause of the blowout was the failure of a cement barrier in the production casing string, a high-strength steel pipe set in a well to ensure well integrity and to allow future production. The failure of the cement barrier allowed hydrocarbons to flow up the wellbore, through the riser and onto the rig, resulting in the blowout.[5] The investigation identified BP practices that were contributing causes to the accident:

- The failure of the crew to stop work on the *Deepwater Horizon* after encountering multiple hazards and warnings.
- BP's failure to fully assess the risks associated with a number of operational decisions leading up to the blowout.
- BP's cost- or time-saving decisions without considering contingencies and mitigation.
- BP's failure to ensure that all the risks associated with operations on the *Deepwater Horizon* were as low as reasonably practicable.
- BP's failure to have full supervision and accountability over the activities associated with the *Deepwater Horizon*.
- BP's failure to document, evaluate, approve, and communicate changes associated with *Deepwater Horizon* personnel and operations.
- The failure of BP and Transocean [Transocean, the world's largest offshore drilling contractor, provided the rig and crew] to ensure they had a common, integrated approach to well control.
- The failure of the current Subpart O rule to identify (by definition) personnel who need to be trained in well control operations, specifically in kick detection.

The 2011 National Commission concluded that "Most of the mistakes and oversights at Macondo can be traced back to a single overarching failure—a failure of management. Better management by BP, Halliburton, and Transocean would almost certainly have prevented the blowout by improving the ability of individuals involved to identify the risks they faced, and to properly evaluate, communicate, and address them.... BP, Halliburton Co. and Transocean made breathtakingly inept and largely preventable missteps."[6]

As mentioned earlier, safety is equivalent to a license to operate. A firm that is viewed as unable to operate safely will see its reputation suffer, as was the case for BP after the Macondo well incident in 2010. The impact to BP's reputation can be seen in the fall in its share price. BP's share price peaked in January 2010 at $62.32. When the Macondo incident occurred, the share price was $60.48. On June 25, BP's share price was $27.02, a 57% drop in market capitalization. BP was forced to cancel three of its dividend payments in 2010 and had to create a $20 billion fund to pay for damages after the spill. BP was banned from participating in Gulf of Mexico lease auctions; the ban was lifted in 2011. Internally, BP made various changes to improve both its safety culture and its reputation, including a reorganization of the company's exploration, development, and production units; a new global safety division with direct reporting to the CEO; and an asset divestiture plan of $25–$30 billion to provide funds for damages and legal settlements.

Security

Security has long been a very high priority for the oil and gas industry. The industry has been a target for various forms of special interest or terrorist organizations for many years for at least three reasons: (1) the high value of products and projects of the industry, (2) the importance of the oil and gas industrial and commercial sector to many host countries, and (3) the potential for malicious release or use of the physical and chemical products and materials used in facilities.

The security risks are substantial, and the lessons learned have, in some cases, been very hard. For example, Statoil, the national oil company of Norway, suffered the single worst terrorist attack in the history of the oil and gas industry in 2013 when its facility in In Amenas, Algeria, was attacked by terrorists. Statoil's own assessment of the attack was presented in the company's *Sustainability Report for 2013* (p. 3):[7]

> Forty innocent people were killed, including five Statoil employees, in the brutal terrorist attack on In Amenas 16 January. Following the attack, on 26 February 2013 the Board of Directors commissioned

an investigation into the attack. The main objectives of the investigation were to clarify the chain of events and to facilitate learning and further improvements within risk assessment, security and emergency preparedness. The investigation team submitted its report on 9 September 2013. The report was discussed by the Board of Directors on 11 September and made public the following day. The main conclusions of the investigation were:

On the attack:
- The sum of outer and inner security measures failed to protect the people at the site from the attack on In Amenas on 16 January. The Algerian military were not able to detect or prevent the attackers from reaching the site. Security measures at the site were not constructed to withstand or delay an attack of this scale, and relied on military protection working effectively.
- Neither Statoil nor the joint venture could have prevented the attack, but there is reason to question the extent of their reliance on Algerian military protection. Neither of them conceived of a scenario where a large force of armed attackers reached the facility.
- The joint venture incident management team led the civilian crisis response, supported by Sonatrach and many other agencies on the ground. Statoil's contribution to the overall emergency response was effective and professional. The investigation team has not identified areas where a different response by Statoil could have changed the outcome.

On security in Statoil:
- Statoil has established a security risk management system, but the company's overall capabilities and culture must be strengthened to respond to the security risks associated with operations in volatile and complex environments.

The report made 19 recommendations within the following areas: security at In Amenas and other facilities in Algeria; organisation and capabilities; security risk management systems; emergency preparedness and response; and cooperation and networks. A new corporate unit for Security and emergency preparedness and a security improvement programme with a dedicated programme manager has been established in order to strengthen the security area and implement the report's recommendations. The company will now ensure that the recommendations are prioritized and integrated into the security improvement programme that has been initiated. The Board of Directors has endorsed the improvement programme, and will continuously monitor the implementation and consider the need for further measures.

Companies in the oil and gas industry obviously must take and do take the security of their personnel, their families, facilities, and materials very seriously. As with all safety and security efforts, however, it is an area in which the investment of time and money can seemingly never be enough. See Current Challenge Box 4–1 for further elaboration of this topic.

Current Challenge Box 4–1. Terrorism, Piracy, and Kidnapping

On January 16, 2013, a group of 32 terrorists armed with missiles, mortars, and bombs attacked a gas plant near In Amenas, Algeria. The plant, jointly operated by BP, Norway's Statoil, and Algeria's national oil company Sonatrach, was taken over by the terrorists. A day later, Algerian special forces stormed the plant, rescuing more than 750 workers. Three of the terrorists were captured, and the rest were killed. However, 40 employees were also killed including three American, five British, 10 Japanese, five Norwegian, and eight Filipino workers.[8] As the biggest assault of an energy facility ever, it showed the susceptibility of energy plants to large-scale attacks. But small-scale attacks through piracy and kidnapping are the more common types of attacks that threaten oil and gas workers. In recent years, piracy has been mainly occurring in offshore Africa. For example, there were 31 incidents of piracy or armed robbery in the Gulf of Guinea region during the first six months of 2013.[9] Kidnapping of oil and gas workers has also been on the rise, most recently in Nigeria, Yemen, Columbia, and Mexico. What can be done to increase security for oil and gas workers? In response to the Algerian assault, Statoil created a new corporate-level Security and Emergency Preparedness Unit as well as a number of other security improvement initiatives. To combat piracy, some companies have hired private armed personnel, but armed security personnel must frequently relinquish their weapons to local law enforcement depending on the location. To help keep PCNs secure, many companies provide security training or personal safety training. Security training can include awareness training, hands-on defense, evacuation training, traveling tips, and first-aid training.[10] Some companies are providing their employees with specialized insurance geared toward PCNs. This insurance can include medical evacuation services and may even include kidnap and ransom insurance. The bottom line is that oil and gas companies must always be mindful of security issues for their employees, particularly when they are located in countries that pose higher security risks. They should have detailed standard security operating procedures, training protocols, and crisis management plans tailored for each operating environment and regularly reviewed and updated.[11]

Occupational disease

An occupational disease is any disease contracted primarily as a result of an exposure to risk factors arising from work activity. Those exposures may then be categorized as arising from the work activity and the work environment. The risks and exposures borne by oil and gas workers globally are immense. Here are some examples:

- Oil and gas development activities are often conducted in compounds isolated from the general population. Workers in the compounds are transported to the site and confined to that site during the duration of their work stay. This intensifies worker interactions and the potential for illness or disease transmission. An additional complication is that as a result of the time and cost of some remote transportation of workers, once on the site they may work long shifts weeks at a time, raising concerns over general health and safety.
- Development projects are usually subject to significant local hiring requirements, where local employees may work in close proximity to foreign workers. Differences in social demographic backgrounds mean very different health histories, immunization records, and general health and sanitation practices. These differences, combined with high-wage jobs for local residents, can result in worker compound or camp disease issues. Many oil and gas developments in Sub-Saharan Africa have extremely high incidence rates of tuberculosis and HIV/AIDS.
- Onshore exploration, development, and production activities may occur in highly remote locations, sometimes in and around local native populations who have had limited exposure to the outside world. This has occurred everywhere from the Amazon to Papua New Guinea to Sudan. Risks accrue both to the workers and the communities as a result of this interaction.
- Offshore activities sometimes require long boat or air transport of workers to isolated sites in the Caspian Sea, the Gulf of Guinea

(West Africa), or the Gulf of Mexico. Offshore facilities involve confined spaces that intensify health, safety, and disease risks.

So far in this chapter, we have tried to show the importance safety and health in the oil and gas industry. Personal safety, process safety, security, and occupational diseases are all particularly important to address in the oil and gas industry because of the materials and equipment used and the difficult locations and conditions of the work. We have shown how accidents in this industry can be catastrophic. We now show how human resource management can help address these issues.

Safety through Human Resource Management

We have shown that safety is critical in the oil and gas industry. (We implicitly refer to personal safety, process safety, employee security, health, and well-being with the term "safety"). So how can the management of human resources (HR) help firms address the critical issue of safety? In the rest of this chapter we show how each of the HR functions can be used to foster a work climate that is safe. When all four HR functions create an overarching picture of what the organization expects, values, and considers acceptable this strongly contributes to the organization's culture. In chapter 3 (in the context of socializing local employees), we described how an organizational culture captures the values, norms, standards, and "way of thinking" in a company. Thus, given the importance of safety in the oil and gas industry, companies should strive to create a safety culture in their organization. In firms with a strong safety culture, the following would be commonplace:[12]

- Safety-conscious behavior is the norm.
- Accidents and injury rates are low, and zero incidents are achievable.
- Safety issues receive a great deal of attention.
- Employees share strong ideas and beliefs about risk and accidents.
- Employees all have a strong commitment to safety.
- The organization has extensive health and safety programs.
- Safety is a priority at all levels of the organization.
- Everyone in the company is willing to learn from safety-related errors, incidents, and accidents.
- The firm's HR practices strongly support safe behaviors.

A strong safety culture is important because it has a very powerful influence on the behaviors and decision making of employees. In companies with a strong safety culture, employees do not think twice about turning off a valve, stopping a machine, or intervening with a colleague who is doing something unsafe. In fact, safety culture has such a strong effect that it can even overcome the strength of a national culture. Research has shown that although national cultures differ in their attitudes, perceptions, and beliefs regarding safety, firm factors such as safety culture and management's commitment to safety have a stronger effect on safety behaviors and accident rates than national culture.[13]

So how does a firm go about creating a strong safety culture so that the features listed above describes them? First, practices have to be put in place that are consistent with a strong safety culture. These include extensive health and safety programs and HR practices that strongly support safe behaviors and safety-focused decisions. Second, management's attitudes and behaviors must show support for a strong safety culture. Interestingly, the introduction of extensive health and safety programs and HR practices that support safety lead employees to conclude that management has a strong commitment toward safety.[14] Finally, safety officers and safety committees should be given high status through rewards, recognition, and access to resources.[15] Although it is only one of the many ways to foster a strong safety culture, appropriate HR practices are critical in the development and maintenance of a safety culture. We now look at each of the four HR functions and their key decisions as mechanisms to build a safety culture. Staffing and training are currently particularly important with respect to safety because the current growth (and upcoming retirements) in the industry means there will be an influx of new employees who will need to understand the importance of safety.[16]

Safety through Staffing

How is a safety culture related to staffing? Staffing is related to safety in a number of ways.

First, if one includes scheduling as part as staffing, then scheduling appears to be a significant factor in accidents. The greatest cause of fatalities in the oil and gas industry in the United States over the last decade has been in highway crashes, and many of these are attributed to workers falling asleep at the wheel after long shifts. Fatigue can also cause accidents and greater exposure to hazards on the job. Thus, those scheduling work should be aware that even if regulations allow longer hours, working beyond 12-hour shifts can make workers more susceptible to errors or fatigue that can have health and safety consequences.[17]

Second, staffing appears to be related to safety based on the decision of whether or not to have work done in house or by contractors. Using contract workers increases staffing flexibility and the specialty knowledge of employees, but may increase threats to safety.[18] Figure 4–3 shows that in the upstream sector worldwide and the downstream sector in Europe, contract workers are more likely to have fatal accidents than company employees.[19] Exactly why the rate is so much higher for contract employees is unclear, but lack of control over staffing and training are likely to be factors. Companies that want to fully control the staffing, training, performance management, and compensation of workers at their work site will rely less on contractors.

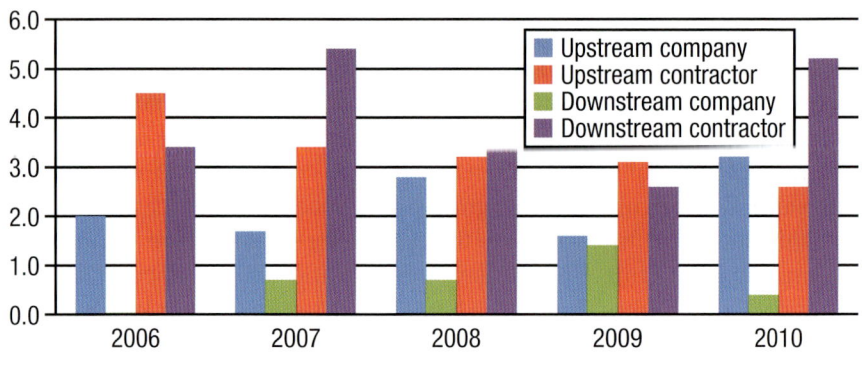

Fig. 4–3. Fatalities in oil and gas industry by worker type (per 100 million hours worked)
Source: ILO Issues Paper.

However, the most important way that staffing is related to safety is by recruiting and selecting employees that value safety and health. One way to do this is to focus on the notion of risk. There is an underlying characteristic of people related to risk: Some people have a general propensity toward high risk taking, while others have a general tendency toward risk aversion.[20] Although some people are generally more willing to take risks than others, it is likely that this trait is not necessarily stable and pervasive across contexts.[21] Risk propensity may apply to many different aspects including financial, social, recreational, health, ethical, and gambling-related risk taking. Although some people may have a general tendency to be risk takers across all of these areas, there are clearly some who are very high or very low propensity for risk taking within each specific type of risk. Companies that want a strong safety culture should avoid hiring employees who have a tendency to take health- and safety-related risks. We now discuss how this can be achieved through the framework of the key decisions.

Recruit applicants who tend to not take safety-related risks

Recruiting can be helpful in fostering a safety culture by generating a pool of applicants who tend to not take health and safety risks. This could be done by making sure that the job announcements and advertisements highlight the importance of safety on this job. For example, pictures of workers should show them in full safety gear. The description of the ideal applicant may include that they are "highly conscientious with respect to safety." Highlighting the importance of safety may help safety-focused applicants self-select into the applicant pool, and it will signal to all applicants that safety is important to your company. The recruiting material may also focus on risk aversion, in order to discourage those that have a high propensity for risk taking from applying. However, it is likely to be problematic to focus on general risk-taking tendencies, because in some jobs good performance may depend on being willing to take risks. For example, one study found that entrepreneurs have a higher risk propensity than managers.[22] So if you screen out those with higher risk propensity, you may be screening out many of the employees who will act entrepreneurial. Thus, focusing on risk taking that specifically applies to health and safety—such as health-related risk taking and recreational risk taking—is likely to be more beneficial in the long term. Once we have an applicant pool that is likely to include applicants with varying levels of safety consciousness, how do we then determine whom to hire?

Select applicants who tend to not take safety-related risks

Given that you want to select employees who don't take safety-related risks, how do we determine who would take these risks? Which selection tools can help us identify this characteristic? We know that past behaviors predict future behaviors. This is consistent with the findings of one study, which showed that selecting employees based on their safety records in their previous jobs was one of the strongest predictors of having low employee-injury rates.[23] Thus, one selection criterion could be identifying indicators of past safety- and health-related risk taking. This could include driving records, work records, and so forth.

There are other ways of identifying past or future safety behaviors. These could include questionnaires related to such behaviors, safety knowledge tests (assuming that safety knowledge is an indicator of safer subsequent behaviors), and interview questions focusing on safety issues. For example, using a behavioral descriptive interview question such as "Tell me about a time when you encountered a situation that appeared to be unsafe to you" can help provide information about applicants' probable responses to safety incidents. Interview questions regarding safety issues in general may also help identify past safety behaviors and will signal to applicants that safety is a priority in this company.

Another way of assessing future risk taking is through personality tests. Studies have shown that personality is somewhat related to risk taking in general and to safety-related risks. The most common personality measures are called the "Big 5"—these are five traits are that have been shown to capture most of the differences in personality across people. These five traits are extraversion, neuroticism, openness to experience, agreeableness, and conscientiousness. All five have been related to risk taking in general, with those low in neuroticism, agreeableness, and conscientiousness and high in extraversion and openness to experience being the most likely to take risks. The same pattern was found among these traits and safety-related risk taking, with the exception of openness to experience not being related to taking safety-related risks.[24] Thus, using personality traits can help predict future safety risk taking. Focusing on conscientiousness may be the most helpful since conscientiousness has also been found to be positively related to good job performance across a broad range of jobs.[25]

Finally, we should also note that studies have consistently found that women take considerably fewer risks than men in all facets of risk taking. Thus, when safety is an important factor, women should be given particular consideration (although hiring based on gender should always be done judiciously since it may violate antidiscrimination laws).[26] Chapter 6 will describe this issue in more detail.

Use a multiple-hurdles approach on safety-related selection tools

If a company wants to firmly establish a safety culture, and safety is a priority, then selection tools focusing on safety should not be disregarded in any circumstances. However, if a company uses a compensatory approach to different selection tools, that is what could happen. In a compensatory system, bad scores on one tool can be offset by good scores on others. However, using a compensatory system does not make any sense when one is using tools that the organization deems essential to the hiring decision. Thus, for companies who want to make safety a priority, using safety-related selection tools and using a minimal cutoff (multiple-hurdles) approach on those tools makes the most sense.

Maximize the effectiveness of safety-related staffing

As with staffing in general, legal compliance, treating applicants fairly, and monitoring, evaluating, and adjusting the system all can help maximize the effectiveness of safety-related staffing. Because research has found demographic differences in those who are risk taking and those who are risk averse (most notably gender), women will score lower than men on most measures of risk taking. Although, this will benefit women, who are greatly underrepresented in the oil and gas industry (see chap. 6), the validity of those measures should be established before using them.

Safety through Training

Training is a critical component of a safety culture. Training appears to greatly improve safety knowledge.[27] Safety knowledge has been found to be strongly related to safety compliance (e.g., following procedures, using protective equipment, and practicing risk reduction), safety participation (e.g., initiating safety-related changes and whistle-blowing), and, to a lesser extent, accident and injury rates.[28] Because safety knowledge leads to more safe behaviors, it is not surprising that safety training also improves safety performance, and thus safety and health outcomes. However, the relationship between safety training and safety performance is not as strong as the relationship between safety training and safety knowledge. This is consistent with the idea that having safety knowledge is necessary for safe performance, but employees could be knowledgeable about safety and still perform unsafely. This suggests that employees must be given the ability to perform safely (through safety training), and they must be motivated to perform safely (through performance management and compensation, as will be discussed shortly).

Who should receive safety training?

As in the previous chapter, who receives the training is largely dependent on what is being trained. Clearly, there are some aspects of safety and health that everybody should be trained on, including safety-related policies, safety-related procedures, and general safety information that helps establish and reinforce the organization's safety culture. Nevertheless, in the safety context, there are some clear factors that help determine who gets trained. One factor is geographic location: Employees in specific locations should receive security, safety, and health training specific to that geographic location. For example, all employees on a deepwater rig need to be given deepwater survival training. Employees in a country where kidnapping of foreign workers is prevalent need to be trained on techniques and strategies to maximize employee security (see Current Challenge Box 4–1). Another factor that can determine who gets training is the job itself. Job-specific safety instructions in the oil and gas sector may include hotwork safety training, electrical safety training, drilling simulations, and blowout retention training. This may also include training related to exposure to hazardous materials and machinery, as applicable to the job. Training can also be designed for specific employee groups, where the training is not directly applicable to their location or jobs. Training for new hires or training for parent-country nationals (PCNs)—expatriates—is an example of this type of training.

An interesting issue about who gets trained is the case of contract workers. Contract workers are technically not employees of the company (we will discuss them in greater detail in chaps. 5 and 6). In the United States, the legal liabilities of the costs of accidents penalize companies that exercise direct control over contract employees. This has led to plant managers generally giving primary responsibility for safety training of contract employees to the contractors. Unfortunately, research indicates that plants offer more effective safety training than contractors do, which may explain the lower safety rates of contract workers, as described earlier.[29] Thus, employers should at least verify that their contract employees have had satisfactory training or should conduct the safety training themselves, accepting that this may increase their legal risk.

The content of safety and health training

Beyond the obvious—safety and health training is related to safety and health—there are many different aspects of safety and health that warrant training in the oil and gas sector. Table 4–2 shows the various domains of training in the oil and gas sector and relates them to different employee groups.[30] As discussed earlier, general topics that are applicable to all employees include safety policies, safety procedures, first-aid training, personal safety training, incident reporting training, emergency response training, fire prevention training, accident prevention training, and emergency preparedness training.

Other types of training are geared toward the location of the work. This could include the specific geographic location of the work (e.g., Nigeria), location environmental factors (e.g., extremely cold or hot locations), and location based working conditions (e.g., confined space, exposure to H_2S). Location-related topics include security training, survival training, offshore training, hazardous environment training, cold water survival training, H_2S awareness training, and helicopter underwater rescue training. For example, BP Norway requires an offshore health certificate as well as a number of different training courses for any employee before going offshore, even if it is just as a visitor.[31] These include a one-day course on HSE issues and BP Norway's safety culture, an in-house safety training verification course, and a safe job analyses and work permits course. Additionally, employees who go offshore on a regular basis must take a basic offshore safety and emergency preparedness course that includes first aid, firefighting, helicopter evacuation, and marine rescue operations.

Table 4–2. Examples of different types of safety training plans

All Employees	Location Specific	Job Specific	Applicable Employees
First aid training	Security training	Drilling simulators	Expatriate security training
Emergency response training	Survival training	Blowout prevention training	Fire crew training
Fire prevention training	Offshore training	Incident command skills training	Fast rescue craft training
General safety rules	Marine first aid	Hazardous materials training	Crew resource management training
Housekeeping	Confined space training	Electrical safety training	Protective equipment training
Safety roles and responsibilities	Hazard communication	Hotwork safety training	Risk assessment training
Stress management training	Hazardous environment training	Lockout/tagout training	Chemical safety training
Personal safety training	Heat stress training	Equipment safety training	New employee orientation
Safety reporting training	Cold water survival training	Safe use of hand tools	Safe driver training
Accident prevention training	H_2S awareness training	Transportation safety training	Process safety training
Incident reporting training	Emergency evacuation training	Trenching safety training	Oil spill emergency response training
Back injury prevention training	Shipping safety training	Safety regulation compliance training	Prevention of workplace violence
Emergency preparedness training	Helicopter underwater escape training	Pipeline safety training	Ionizing radiation safety training

The content of many training courses is job specific, for example, drilling simulators, blowout prevention training, hazardous materials training, electrical safety training, hotwork safety training, lockout/tagout training, equipment safety training, and trenching safety training. Finally, some training is designed for specific employee groups and is not directly applicable to their location or jobs. This could be based on their level in the organization, their exposure to certain hazards, their use of certain equipment, or their participation in special assignments or task forces. Examples of training based on these factors include protective equipment training, PCN security training, fire crew training, fast rescue craft training, risk assessment training, new employee orientation, process safety training, oil spill emergency response training, and crew resource management training. *Crew resource management training* is geared toward teams of employees and focuses on enhancing their understanding of the people-related aspects of effective teamwork to reduce operational errors and enhance the skills needed to deal with emergencies.[32] Crew resource management training may focus on decision making, communication, assertiveness, and stress.

Safety and health training methods

Similar to general training, many different methods to teach safety and health are available, including coaching, workshops, conferences, videos, mentoring, teleconferencing, simulations, and interactive multimedia (computer-based learning). Safety training through computer-based methods is increasing in popularity because of its low cost and customizability. However, research has shown that computer-based safety training appears to be less effective on older employees, and some types of computer-based safety training may be less effective than others.[33] Specifically, text-only computer-based training seems to be the least effective computer-based method, while more interactive methods including pictures and narration are more effective for all ages of employees. This is consistent with many other studies on safety training that find that more engaging methods are more effective.[34]

Level of engagement has been found to increase the effects of safety training on safety knowledge (particularly regarding high-hazard situations), safety performance, and safety and health outcomes (accident rates, injury rates, etc.). Figure 4–4 shows the effects of safety training on safety knowledge, safety performance, and safety and health outcomes for low, moderate, and high levels of engagement across many studies.[35] In these studies, lectures, films, and video-based training are classified as low in engagement; computer-based instruction is classified as moderately engaging; and behavioral modeling, simulation, and hands-on training is classified as high in engagement. Highly engaging methods of safety training were consistently found to have much better results. But even methods such as lectures can be made more engaging by letting participants interact with each other, by sharing what they have learned on the job, telling stories about their experiences, and sharing their "insider" knowledge.[36] Clearly, engaging methods of training safety and health topics should be preferred by companies that want to have a strong safety culture. Less engaging methods should only be used when more engaging methods are not feasible.

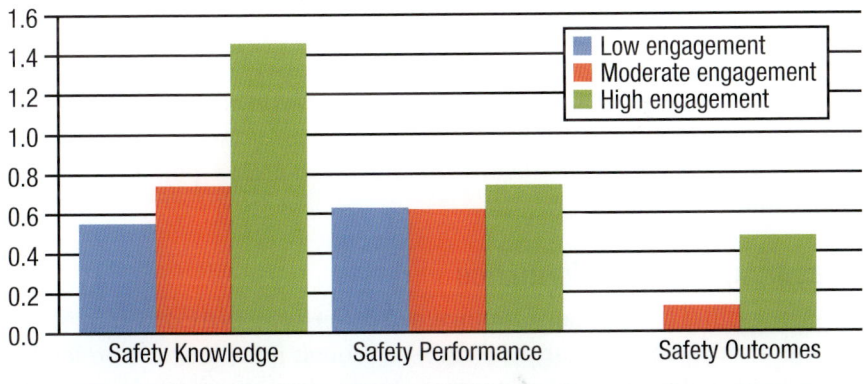

Fig. 4–4. Improvement from safety training (in standard deviations)
Source: Burke et al. (2006).

Maximizing the effectiveness of safety and health training

As with training in general, maximizing the effectiveness of safety training can be achieved by maximizing learning; facilitating training transfer; and monitoring, evaluating, and adjusting the system. Training improves safety knowledge more with higher engagement and feedback. However, higher safety knowledge does not necessary result in better safety performance or better safety outcomes unless that knowledge is transferred to the job.

Facilitating transfer of safety training can be increased by incorporating the training content into performance measures. This is clearly illustrated by a study that found that one of the best predictors of low injury rates in companies was the answer to the question: To what extent does the training program perform assessments following instruction to verify that the safe work practices are being carried out in the work areas?[37] Thus, feedback during and after training by monitoring and evaluating trainee safety performance can help maximize the effectiveness of the training.[38] Further, goal setting (another component of performance management) of the safety training content has also been found to improve safety performance.[39] We now further discuss the role of performance management in creating a safety culture.

Safety through Performance Management

Fostering a safety culture through performance management is, theoretically at least, very straightforward: Make safety an important part of performance. In doing so, assuming the firm has a good performance management system, goals will be set for safety performance, safety performance will be measured, feedback will be given on it, employees will be developed relative to it, and meaningful rewards will be tied to it. We now discuss safety through performance management relative to the key decision points in performance management.

Measuring safety performance

As with other types of performance, safety performance can be measured through traits, behaviors, or outcomes—although here, too, it is preferable to measure behaviors and outcomes. Evaluating outcomes is generally preferable because it is a more objective measure, and outcomes are ultimately what are important. The most common measures of safety performance regularly reported and used in the oil and gas industry are the *total recordable incident rate* (TRIR), the lost time incident rate (LTIR), and the *days away from work case rate* (DAWC).

Total recordable incident rate (TRIR). The most common measure of safety used by organizations of all kinds, including those operating in the oil and gas industry, is the TRIR, sometimes also referred to as the total recordable injury rate or the total recordable case rate, which calculates the number of recordable injuries experienced by a project, firm, or industry. According to OSHA, a recordable injury is any event resulting in fatalities, lost workdays, restricted workdays, or medical treatment cases. The TRIR, as it is required to be reported by OSHA, is a standardized value based on 100 workers working 50 weeks a year at an average of 40 hours a week, or 200,000 total work hours ($100 \times 50 \times 40$).

Recall the Chad-Cameroon Petroleum Development Project described in Case 1. It is an extremely transparent project from a safety performance standpoint, due to the rigorous disclosure provided by the project sponsors. Reporting in this case is comprehensive, as it includes both the production area in southern Chad as well as the operations of the 1,000-kilometer pipeline extending from southern Chad across Cameroon to the Gulf of Guinea. Figure 4–5 illustrates the TRIR for the project from its initiation in 2001. Although the individual values are significant, the time series or trend analysis is normally the subject of much managerial attention as well in attempt to continually improve performance and improve the development and execution of a safety culture. Note that the TRIR is declining over time—the trend is downward. Also, because it is quarterly data on an individual project, and therefore a relatively small statistical sample (e.g., compared to an industry), it shows more volatility.

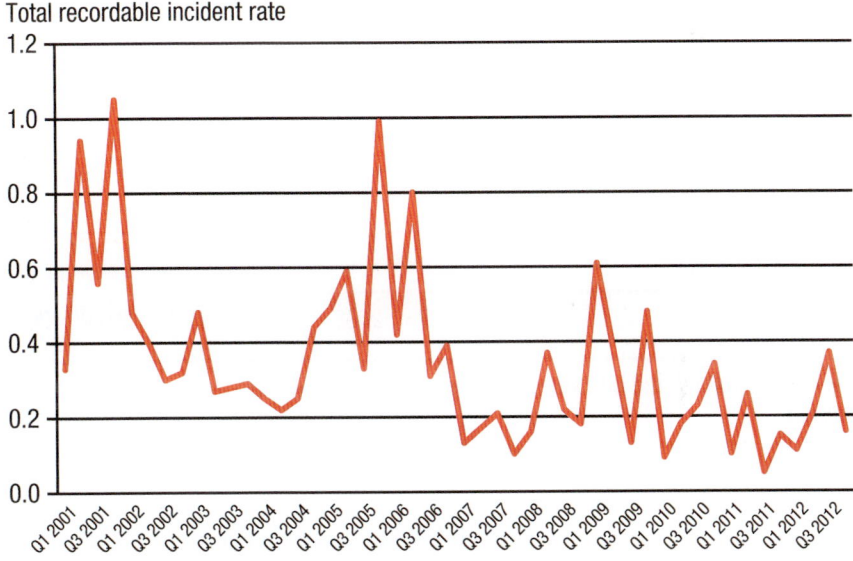

Fig. 4–5. Chad-Cameroon TRIR safety performance
Source: Chad Camerron Project Update series, 1–33.

Lost time incident rate (LTIR). The LTIR follows the same basic formulation as the TRIR, but includes only fatalities and lost time events in the reported incidents, excluding restricted work and medical treatment cases. Figure 4-6 shows the average scores of these two basic safety metrics for the global oil and gas industry in general. These data, from the International Association of Oil and Gas Producers (IOGP), show consistent improvement in both TRIR and LTIR metrics for the 2003–2012 period.

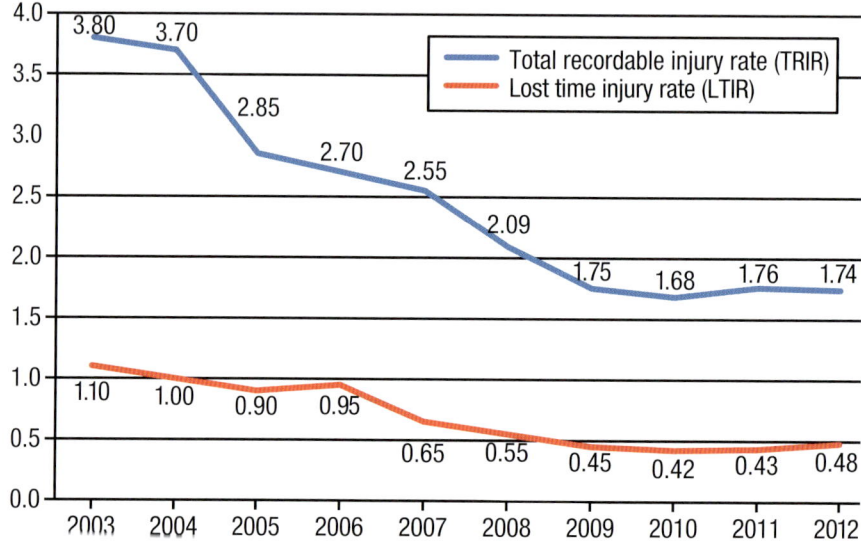

Fig. 4–6. Global oil and gas safety performance (incident rate per million hours worked)
Source: *OGP Safety Performance Indicators, 2012 Data*, Report No. 212s, International Association of Oil and Gas Producers, June 2013, vii.

Days away from work case rate (DAWC). The DAWC measure the rate of occupational injury or illness cases that result in an employee being unable to work a full assigned work shift; that is, the employee is off from work, resulting in a lost workday. As defined by OSHA, a fatality is not considered a lost time case. Restricted cases are defined as any occupational injury or illness that results in a limitation in their job (i.e., no lifting, climbing, etc.) or transferred to another job (restricted days).

Other measures. Oil and gas industry safety performance data are regularly collected and published by a number of organizations, including the US Bureau of Labor Statistics, the IOGP, and the American Petroleum Institute (API). In the United States, OSHA requires firms to report accident rates on a regular basis. According to OSHA, recordable injuries and illnesses include all work-related deaths, illnesses, and injuries that result in a loss of consciousness, restriction of work or motion, or permanent transfer to another job within the company, or that require some type of medical treatment other than first aid. Although different countries and companies use a variety of slightly different definitions, most measures differentiate on fatalities, lost time from work, restricted activities at work, and medical treatment and first aid at work. Some other common outcome measures include fatalities, injuries requiring hospitalization, days away on restricted duties, medical treatment injury frequency rate, first-aid injury frequency rate, occupational illnesses, number of days lost to occupational illnesses, leaks or spillage of workplace substances, and collapse or partial collapse of any building or structure.[40] Nevertheless, measuring safety outcomes for performance purposes can be problematic for a number of reasons, including:

- Most outcome measures are at the group or organization level and not at the individual level.
- Relatedly, it is difficult to find outcomes that can be directly attributable to one person.
- Good outcomes can result even if bad safety behaviors have occurred.
- Bad outcomes can result even if good safety behaviors have occurred, through no fault of the employee.
- Outcome measures can be manipulated.
- Negative outcomes may not arise until years after the causal actions occurred.

Thus, to better capture the entire aspect of safety performance, behaviors generally should also be evaluated. Safety-related behaviors may include complying with all safety policies and procedures, reporting potential safety hazards immediately, reporting all injuries promptly, recognizing and addressing unsafe working conditions or practices, cooperating with safety investigations, maintaining required licenses and certifications, operating machinery only with authority, maintaining proper speed in vehicles, using equipment properly, never working under the influence of drugs or alcohol, participating in safety meetings, and always wearing protective equipment.[41]

There are many measures (or metrics) of performance that are used by firms with a strong safety culture. These measures tend to be at the group, division, plant, or organization level. Therefore, these measures may be used to evaluate individual performance if one can reasonably make the argument that this particular employee had some impact on the measure. That is, a plant-level measure would be a better performance indicator for a plant manager than for a lower-level plant employee. Safety metrics can be categorized as leading or lagging.[42] *Leading indicators* are those that measure aspects of the safety system. *Lagging indicators* tend to be outcomes such as injuries and fatalities. Safety metrics can also be categorized as measuring process safety or personal safety. As we discussed earlier in the chapter, process safety concerns hazards arising from the processing activities of a plant, such as the release of flammable material. Process safety hazards have the potential to damage the plant and generate multiple fatalities. Personal safety concerns the safety of individual workers, regarding injuries resulting from hazards such as falls, crushing, electrocution, trips, slips, vehicle accidents, and the like. Personal safety hazards may not have anything to do with the processing activity of the plant. Table 4–3 reports some common safety metrics categorized as leading or lagging and as process or personal safety measures.[43] Note that many of these indicators are rates rather than absolute numbers, to control for the number of employees. Following the guidelines of the IOGP, these rates are usually specified as per 1 million hours worked for injury or illness rates, and per 100 million hours worked for fatality rates. In the United States, the OSHA rate guidelines are per 200,000 hours worked for injury or illness rates and per 1 million hours worked for fatality rates.[44]

Table 4–3. Safety metrics in the oil and gas industry

Type	Leading Safety Metrics	Lagging Safety Metrics
Personal Safety	Number of Inspections	Lost Time Incident Rate
	Number of Safety Meetings	Total Recordable Injury Rate
	Safety Meeting Attendance	Fatal Accident Rate
	Time from incidents to investigations	Days Away from Work Cases Rate
	Percentage of at risk employees under health surveillance	Severity of Lost Work Day Cases
	Percentage of worksites at which health and safety concerns of employees are represented	Exposure Hours
	Safety/Health Risk Assessment Percentage	Number of Fines
	Employee percentage that have completed safety/health training programs.	Number of Prosecutions
	Frequency of Medical Emergency Drills	Number of Claims
	Response Time of Medical Emergency Drills	Number of Complaints
Process Safety	Inspection of Mechanical Integrity Rate	Process Safety Incident Rate
	Percentage of procedures reviewed/updated	Process Safety Incident Severity Rate
	Safety Action Processing Rate	Number of Near Misses
	Percent of Safe Start-ups Following Changes	Number of incidents of any release of a hazardous substance
	Percentage of Employees Trained on Process Safety	Number of deviations or excursions of normal processes
	Percentage of employees successfully completing process safety training on first try	Number of inspections or testing results outside acceptable limits
	Percentage of employees educated on fatigue risk	Number of incidents where demands were put on safety systems
	Percentage of Overtime/Extended Shifts	Mechanical Failed State Rate

The IOGP and the International Petroleum Industry Environmental Conservation Association (IPIECA) also recommend that companies track metrics on eight elements of health-related performance: health risk assessment and planning, industrial hygiene and control of exposures, medical emergency management, management of ill health in the workplace, fitness for task assessments, health impact assessments, health reporting, and promotion of good health.[45] Measures of psychosocial health (psychological and social well-being related to how work is organized, designed, and managed) may also be useful, given that psychosocial health has been related to the health and safety of workers (and their quality of work life) in the oil and gas industry.[46] These measures tend to use surveys to capture perceptions of employees' job demands, role clarity, relationships, job control, and support.

Another aspect of safety-related performance that may be measured is safety training. Safety training in itself may be considered an aspect of performance. Measures of safety-training performance are generally objective measures such as attendance, tardiness, participation, and safety knowledge tests scores during and after training; however, they may also include more subjective behavioral measures such as attentiveness and cooperation.

Who should be involved in measuring safety performance?

As with all measures of performance, involving everyone who has accurate information regarding the employee's safety performance generally provides the most accurate assessment of an employee's performance. Thus, as with regular performance, involving the supervisor, subordinates, the employee, and coworkers when looking at safety outcomes and behaviors is likely to provide the best overall assessment of safety performance. Anonymity for all assessors other than the supervisor usually leads to more accurate evaluations. In the case of safety performance, others who may have useful information include safety trainers (when evaluating performance in safety training) and HSE officers who are familiar with the safety performance of many employees.

The format for measuring safety performance

Again, the format for safety performance follows the format of performance in general. For measuring safety behaviors, behavioral observation scales are generally preferable because they deal with frequencies, (e.g., how frequently the employee followed safe driving protocols), which captures the essence of what you are trying to measure: Did the employee always behave safely? Behaviorally anchored rating scales (BARS) may be used to capture more egregious safety violations, where even one incident would be unacceptable.

Maximizing the effectiveness of safety performance management

As with other aspects of performance, providing feedback, accurate evaluations, goals, and rewards is necessary to maximize the effectiveness of safety performance management. Of course, the safety performance system should be monitored and evaluated, and adjustments should be made when necessary. Feedback has been shown to be critically important in helping to achieve high levels of safety performance. In this case, feedback appears to have a strong effect when it goes beyond the feedback of the annual performance evaluation. That is, feedback should be regular, such as weekly updates on safety performance, and immediate, as soon as when safety violations are discovered. Studies have shown that training alone has little effect or only slightly improves safety performance; however, training with regular feedback greatly improves safety performance and helps maintain it.[47]

Although feedback has been shown to improve safety performance after training, safety performance is even higher when feedback is coupled with goal setting.[48] Safety goals should be specific and challenging, but perceived to be obtainable. And as mentioned with goal setting in general and consistent with the discussion above, regular feedback should be given to inform employees how they are doing relative to achieving the goals. Finally, to maximize the effects of feedback and goal setting, tying rewards to safety performance, as with performance in general, will further help motivate employees to achieve safety goals and maximize safety performance.

Safety through Compensation

Rewards are a powerful motivator to influence safety-related behavior, both positively and negatively. You tend to get what you reward. If you give big bonuses for safe behavior, you get safe behavior. If you reward productivity, you get productivity, perhaps at the expense of safe behaviors. Bonuses appear to have the greatest influence on safety performance. Nevertheless, all aspects of compensation may have some impact on safety performance, as we will now discuss, through the key decisions discussed in chapter 1.

Pay level and safety performance

Pay level is likely to have only an indirect effect on safety performance, and then only in certain cases. Pay level affects the attraction, retention, and motivation of employees. With respect to attraction, it is a strong factor that can help attract workers to your organization. Assuming that your selection system does a good job of determining who will be the more safety-conscious workers, paying more than average will help attract them to become your employees. The same is true for retention: Once you have good, safety-conscious workers, paying above the average will retain them more than paying at or below average. Pay level can also effect motivation, particularly if a good performance management system is in place. Employees will be motivated to be good performers to keep their high-paying jobs. Thus, assuming that safety performance is an important part of performance, and a good performance management system is in place, paying above the average will motivate employees to maintain good levels of safety performance.

Pay mix and safety performance

As previously mentioned, bonuses can have a strong influence on safety performance. Thus, a pay mix with a large pay-for-performance component is likely to have the greatest impact on safety performance. Of course, the specifics of that pay-for-performance component are critical in actually motivating the behaviors that the organization wants. What is the measure of performance? What is the reward? What is the link between the two? Nevertheless, the greater the pay-for-performance component is relative to salary, the more motivational it will be. An exception to this is if the pay to performance component is largely based on recognition awards. Recognition awards tend to be quite motivational, and yet their low cost makes them a small component of total rewards. We now discuss how the different types of performance-based pay can affect safety performance.

Types of performance-based pay and safety performance

At least three types of performance-based pay can impact safety performance. The first is bonuses based on safety performance. Bonuses based on both individual and group measures of have been shown to lead to better safety outcomes, such as number of employee injuries.[49] Group measures seem to be more effective than individual measures because (1) good group measures are more readily available than individual measures; (2) group measures foster a social component to the goal attainment where other employees will help and correct unsafe behaviors so that their bonuses will not be negatively affected; and (3) an organizational culture of safety is more consistent with a group mentality. Bonuses tied to measures almost always result in improvements in what is measured. For example, Schlumberger found improved driving performance and lowered accident rates when it implemented a driving improvement plan in Qatar based on training and rewards for improvement on those and other measures related to motor vehicle safety.[50] However, as noted earlier in the discussion of performance management, group performance measures of safety also have their downsides.

Because of the powerful effect of bonuses, some caution regarding the use of safety bonuses must be noted. One is that the safety performance measure must be carefully chosen. Because, with respect to getting the bonus, the measure is performance, employees will focus on the measure. Thus, if a bonus is based on the number of reported incidents, then employees will be less likely to report incidents. If the bonus is based on the number of injuries requiring medical treatment rather than first aid, employees will most likely choose to receive first aid rather than medical treatment, even when medical treatment is warranted. Thus, the measure must be chosen carefully. Another related caution is that safety bonuses can cause unintended consequences, such as those mentioned above. To receive their bonuses, employees may find loopholes, game the system, or just ignore other aspects of performance that do not lead to bonuses. These unintended consequences can be due to motivating "cheating," such as purposely underreporting failure rates (see Current Challenge Box 4–2). Unintended consequences can also be due to rational decisions made in good conscience.[51] Take, for example, the case of a training manager who receives a bonus based on the number of employees who have undergone a specific type of safety training. The manager may reduce the number of days of training per employee to be able train greater numbers. Thus, quality could be sacrificed for quantity. This decision is not necessarily dishonest, but is merely adjusting performance to focus on the aspects the organization has indicated as important. A final caution about group measures of safety is to be aware of the downsides discussed earlier. Outcomes, the typical group measure, can occur years after the behaviors occurred due to systemic effects. Also, a clear causal link does not always exist between individual safety efforts and outcomes (particularly with respect to process safety). Finally, outcome measures can sometimes be manipulated.[52] Bonus systems must be carefully designed and, after implementation, carefully monitored, evaluated, and adjusted to properly focus their powerful motivational effects and reduce possible unintended consequences.

Current Challenge Box 4–2. Underreporting of Safety Incidents

The underreporting of safety incidents can be detrimental to companies in the oil and gas industry for a number of reasons. First, not reporting certain safety incidents can violate regulations (e.g., in the United States, OSHA regulations regarding reporting and documentation requirements). Second, not reporting violations can be an opportunity cost for companies since there is much that can be learned from each incident that can prevent future incidents. Even near misses should be recorded because next time it may not be a near miss. Third, if underreporting incidents becomes the norm, an organization's safety culture may be undermined. In a strong safety culture, safety should always be openly discussed, analyzed, and acknowledged, rather than buried when the news is not good. Thus, underreporting tends to undermine safety in the long term. However, research estimates that between 68% and 78% of accidents and safety incidents go unreported.[53]

So why is there so much underreporting? There are several reasons why individuals do not report accidents and safety incidents. These include fear of reprisal, fear of a loss of benefits, fear of social retaliation, belief that injuries are a normal consequence of work, lack of management responsiveness after prior reporting, misinformation about reporting requirements, and a desire to meet company goals.

So what can companies do to reduce underreporting? First, they can address some of the factors that lead to employees not reporting accidents or incidents. This includes being responsive when someone reports an accident or incident and training employees about reporting requirements. It also includes removing the punishments associated with reporting and instead rewarding those who report. Thus, promptly reporting accidents, incidents, and injuries should be a part of every employee's performance criteria. Valuable incentives tied to low reported numbers also appear to increase underreporting. Large bonuses attached to low reporting may be more likely to affect reporting behaviors than safety behaviors, while recognition awards are less likely to have adverse effects. Any safety incentive program should also discourage underreporting. Second, studies have shown that a strong safety culture and having supervisors who enforce safety policies not only results in far fewer accidents, but also far less underreporting.[54] Thus, enforcing safety policies should be a clear part of every supervisor's performance criteria. Finally, some of the negative effects of underreporting can be somewhat overcome by having anonymous worker surveys and symptom reports that can help identify safety issues that need to be addressed. Underreporting appears to be commonplace in most organizations, but it does not have to be.

A second type of performance-based pay that can impact safety performance consists of bonuses based on productivity, quantity, profits, sales, or any other measures that safety may reduce. Clearly, taking greater precautions, undergoing more training, stopping machinery, and so forth, increase costs and slow productivity, particularly in the short term. Productivity-based bonuses have been shown to increase risk-taking behaviors.[55] For example, driving faster usually gets you there quicker. Making bonuses largely tied to factors that counter safe behaviors in the short term will cause many people to forgo those safety behaviors. An example of this was shown in the *Baker Report* of the BP Texas City Refinery explosion, mentioned earlier in the chapter. The report showed that 50% of the bonuses that most of the employees received were based on cost cutting. Only 10% of the bonuses were based on safety (the number of OSHA recordable injuries). For senior managers, less than 20% of their bonuses were based on HSE measures. Further, measures of process safety (hazards arising from the processing activities of a plant such as the release of flammable material) were completely missing from the incentive system.[56]

A third type of performance-based pay that can impact safety performance consists of recognition awards based on safety performance. Although they may not be quite as motivational as big cash bonuses, they do not have a lot of the downsides that come with safety bonuses. Employees are not as likely to underreport accidents and incidents or focus exclusively on safety for a plaque, certificate, or T-shirt as they would for a large bonus. Further, recognition awards signal the importance of safety, are low cost, and yet are still quite motivational.

Finally, we should briefly mention the use of pay-for-performance for health-related behaviors of employees. This type of compensation provides incentives for participation and achievements in wellness programs. Wellness programs, which may include on-site exercise programs, stress management training, free health-screening clinics, nutritional training, weight-loss management training, and smoking-cessation training, appear to be a cost-effective way to improve employee overall health. Further, incentives such as small rewards (e.g., gift cards) for achieving health-specific goals seem to be effective in increasing participation. Figure 4–7 shows which types of incentives HR professionals found very effective in encouraging participation in their workplace wellness programs.[57] However, in the United States, companies should implement such programs with caution, as they may discriminate against employees with disabilities. The Patient Protection and Affordable Care Act of 2010 clarifies the conditions under which wellness incentives are legal in the United States.[58]

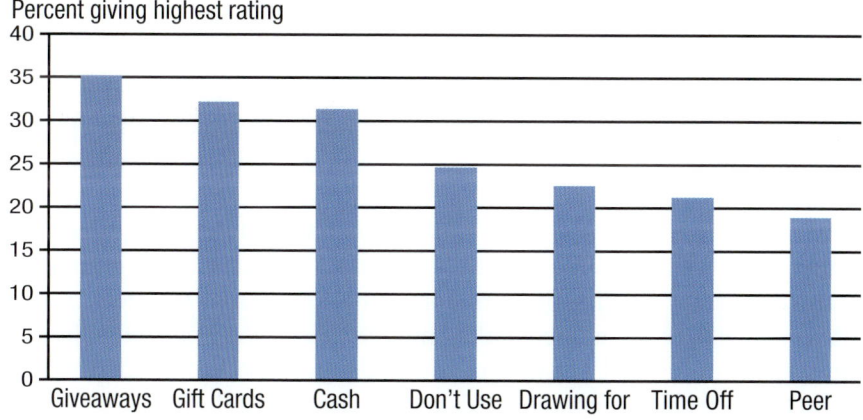

Fig. 4–7. Most effective incentives for encouraging participation in wellness program
Source: HR.com Wellness Survey Study.

Benefits and safety performance

As with pay level, benefits can indirectly affect safety performance. Benefits can be related to safety (and health) issues in a number of ways. First, as mentioned above, wellness programs (which are usually considered a health-care-related benefit) can positively influence the health of employees. Second, similarly, employee assistance programs, which can help employees deal with substance dependence issues, legal problems, financial problems, and mental health issues, tend to positively influence the health and general well-being of employees. Third, health insurance is an important benefit that can help employees maintain or regain good health. This is especially applicable to international assignees, who, without it, may have to face suboptimal levels of health care. Health insurance for international assignees usually also includes evacuation and repatriation coverage if an employee needs to be flown home for emergency medical care. Fourth, companies can provide employees with safety gear and other safety-related materials that can improve job safety. Providing equipment and materials beyond what is legally required helps signal the importance of safety to organization, helps ensure a consistent level of protection of employees (because some employees may buy inferior equipment to save money), and helps ensure that employees will have the needed equipment.

Maximizing the effectiveness of compensation for safety performance

As with compensation in general, compensation used to foster safety performance will be more effective if employees accept the system. Providing employee bonuses for measures over which they have no control, for achieving goals that they do not believe they can possibly achieve, or for achieving goals that they perceive as unfair relative to the goals of others will detract from the effectiveness of those systems. Of course, legal compliance is also necessary in all aspects of compensation related to safety performance. Finally, as mentioned earlier, monitoring, evaluating, and adjusting the system are critical because of the many possible unintended consequences when changing the rewards the employees receive.

Conclusion

We have shown that the safety is critical in the oil and gas industry. A strong safety culture is important because it has a very powerful influence on the behaviors and decision making of employees. Appropriate HR practices are critical in the development and maintenance of a safety culture. Staffing is related to safety through recruiting and selecting employees that value safety and health, through scheduling, and through the use of contract labor. Companies that want a strong safety culture should avoid hiring employees who have a tendency to take health- and safety-related risks. Employees must be given the ability to perform safely (through safety training), and they must be motivated to perform safely (through performance management and compensation). Firms should make safety an important part of performance by setting goals for safety performance, measuring safety performance, providing feedback on it, developing employees relative to it, and tying meaningful rewards to it. Although pay level and benefits can be indirectly related to safety performance, bonuses and recognition for safety performance are likely to have the greatest motivational effect on employees.

References

1. Alli, B. (2008). *Fundamental principles of occupational health and safety* (2nd ed.). Geneva, Switzerland: International Labour Office.
2. Coppee, G.H. (2011). ILO Encyclopaedia of Occupational Health & Safety. http://www.ilo.org/iloenc/part-ii/occupational-health-services/item/155-occupational-health-services-and-practice.

3. Cummins, C., & Herrick, T. (200, May 5). BP takes blame for lethal blast, citing mistakes by its employees, *Wall Street Journal*, p. A3.
4. Baker, J., Leveson, N., Bowman, F., Priest, S., Erwin, G., Rosenthal, I., ... & Wilson, L. D. (2007). The report of the BP US refineries independent safety review panel. BP US Refineries Independent Safety Review Panel.
5. The Bureau of Ocean Energy Management, Regulation and Enforcement. (2011). *Regarding the causes of the April 20, 2010 Macondo Well blowout*. http://docs.lib.noaa.gov/noaa_documents/DWH_IR/reports/dwhfinal.pdf.
6. National Commission on the Deepwater Horizon Oil Spill and Offshore Drilling. (2011). *Deepwater: The gulf oil disaster and the future of offshore drilling* (Report to the President). http://www.gpo.gov/fdsys/pkg/GPO-OILCOMMISSION/pdf/GPO-OILCOMMISSION.pdf.
7. Statoil ASA. (2013). *The In Amenas attack: Report of the investigation into the terrorist attack on In Amenas*. Retrieved from http://www.statoil.com/en/NewsAndMedia/News/2013/Downloads/In%20Amenas%20report.pdf.
8. Statoil ASA (2013). *The In Amenas attack: Report of the investigation into the terrorist attack on In Amenas*. Retrieved from http://www.statoil.com/en/NewsAndMedia/News/2013/Downloads/In%20Amenas%20report.pdf
    Walt, V. (2013, February 11). We did not predict this: Top Algerian minister discusses January hostage crisis. *Time*. Retrieved from http://world.time.com/2013/02/11/we-did-not-predict-this-top-algerian-minister-discusses-january-hostage-crisis/.
    Ouali, A., & Kebir, K. (2013, January 21). 2 Canadians among militants in Algerian gas-plant attack. *Denver Post*. Retrieved from http://www.denverpost.com/ci_22421592/2-canadians-among-militants-aglerian-gas-plant-attack
9. Mainwaring, J. (2013). *Private protection: Combating the piracy threat*. Retrieved from Rigzone website: http://www.rigzone.com/news/oil_gas/a/129963/Private_Protection_Combating_the_Piracy_Threat
10. Sawyer, M. (2006). Personal safety for expatriates. *Workspan, 49*(8), 37–38.
11. Baker, T., & Collard, S. (2013, April). The risk of kidnapping: What oil companies need to do. *World Oil*, 52, 175–179. Retrieved from http://www.nyainternational.com/downloads/world_oil_article.pdf.
12. Cooper, M. D. (2000). Towards a model of safety culture. *Safety Science, 36*(2), 111–136.
    Glendon, A. I., & Stanton, N. A. (2000). Perspectives on safety culture. *Safety Science, 34*(1), 193–214.
    Zhang, H., Wiegmann, D. A., von Thaden, T. L., Sharma, G., & Mitchell, A. A. (2002, September). Safety culture: A concept in chaos? *Proceedings of the Human Factors and Ergonomics Society Annual Meeting, 46*(15) 1404–1408.
13. Mearns, K., & Yule, S. (2009). The role of national culture in determining safety performance: Challenges for the global oil and gas industry. *Safety Science, 47*(6), 777–785.
14. Murugeson, S., & Chelliah, S. (2013). Participatory program and employees' perception towards management commitment in the safety management system. In *Proceedings of International Conference on Business Management & IS*. Available at: http://www.ijacp.org/ojs/index.php/ICBMIS/article/view/101.
15. Clarke, S. (1999). Perceptions of organizational safety: Implications for the development of safety culture. *Journal of Organizational Behavior, 20*(2), 185–198.
16. Williams, B. (2007). Oil & gas UK accelerating staffing initiatives for UK offshore oil and gas industry. *Offshore, 3*, 4–9.

17. Urbina, I. (2012, May 14). Deadliest danger isn't at the rig, but on the road. *New York Times.* http://www.nytimes.com/2012/05/15/us/for-oil-workers-deadliest-danger-is-driving.html?_r=0.
18. Kochan, T. A., Smith, M., Wells, J. C., & Rebitzer, J. B. (1994). Human resource strategies and contingent workers: The case of safety and health in the petrochemical industry. *Human Resource Management, 33*(1), 55–77.
19. International Labor Organization. (2012). *Current and future skills, human resources development and safety training for contractors in the oil and gas industry* (Issues paper). Geneva, Switzerland. Retrieved from http://www.ilo.org/wcmsp5/groups/public/---ed_dialogue/---sector/documents/meetingdocument/wcms_190707.pdf.
20. Dahlbäck, O. (1990). Personality and risk-taking. *Personality and Individual Differences, 11*(12), 1235–1242.
21. Hanoch, Y., Johnson, J. G., & Wilke, A. (2006). Domain specificity in experimental measures and participant recruitment an application to risk-taking behavior. *Psychological Science, 17*(4), 300–304.
    Wang, X. T., Kruger, D. J., & Wilke, A. (2009). Life history variables and risk-taking propensity. *Evolution and Human Behavior, 30*(2), 77–84.
22. Stewart, W. H., Jr, & Roth, P. L. (2001). Risk propensity differences between entrepreneurs and managers: A meta-analytic review. *Journal of Applied Psychology, 86*(1), 145.
23. Vredenburgh, A. G. (2002). Organizational safety: Which management practices are most effective in reducing employee injury rates? *Journal of safety Research, 33*(2), 259–276.
24. Nicholson, N., Soane, E., Fenton-O'Creevy, M., & Willman, P. (2005). Personality and domain-specific risk taking. *Journal of Risk Research, 8*(2), 157–176.
25. Hurtz, G. M., & Donovan, J. J. (2000). Personality and job performance: The big five revisited. *Journal of applied psychology, 85*(6), 869.
    Judge, T. A., & Ilies, R. (2002). Relationship of personality to performance motivation: A meta-analytic review. *Journal of Applied Psychology, 87*(4), 797.
26. Byrnes, J. P., Miller, D. C., & Schafer, W. D. (1999). Gender differences in risk taking: A meta-analysis. *Psychological bulletin, 125*(3), 367.
27. Burke, M. J., Sarpy, S. A., Smith-Crowe, K., Chan-Serafin, S., Salvador, R. O., & Islam, G. (2006). Relative effectiveness of worker safety and health training methods. *American Journal of Public Health, 96*(2), 315.
    Burke, M. J., Salvador, R. O., Smith-Crowe, K., Chan-Serafin, S., Smith, A., & Sonesh, S. (2011). The dread factor: How hazards and safety training influence learning and performance. *Journal of Applied Psychology, 96*(1), 46.
28. Christian, M. S., Bradley, J. C., Wallace, J. C., & Burke, M. J. (2009). Workplace safety: A meta-analysis of the roles of person and situation factors. *Journal of Applied Psychology, 94*(5), 1103.
    Vinodkumar, M. N., & Bhasi, M. (2010). Safety management practices and safety behaviour: Assessing the mediating role of safety knowledge and motivation. *Accident Analysis & Prevention, 42*(6), 2082–2093.
29. Rebitzer, J. B. (1995). Job safety and contract workers in the petrochemical industry. *Industrial Relations: A Journal of Economy and Society, 34*(1), 40–57.
30. Anderson, A. (1984). Training for the offshore petroleum industry in Atlantic Canada. *Journal of Canadian Petroleum,* July–August, 70–72.
    API. (2007). Recommended practice for occupational safety for onshore oil and gas production operation (API Recommended

Practice 74). Retrieved from http://www.4cornerssafety.com/uploads/ LYK1F56aLIQMHCmslmTPu6NgegDaYGlw.pdf.
31. BP Norway. (2014). *Required health and safety training before going offshore*. Retrieved from http://www.bp.com/sectiongenericarticle.do?categoryId=9004953&contentId=7009071.
32. Flin, R. H. (1995). Crew resource management for teams in the offshore oil industry. *Journal of European Industrial Training, 19*(9), 23–27.
33. Wallen, E. S., & Mulloy, K. B. (2006). Computer-based training for safety: Comparing methods with older and younger workers. *Journal of Safety Research, 37*(5), 461–467.
34. Burke, M. J., Sarpy, S. A., Smith-Crowe, K., Chan-Serafin, S., Salvador, R. O., & Islam, G. (2006). Relative effectiveness of worker safety and health training methods. *American Journal of Public Health, 96*(2), 315.
    Christian, M. S., Bradley, J. C., Wallace, J. C., & Burke, M. J. (2009). Workplace safety: A meta-analysis of the roles of person and situation factors. *Journal of Applied Psychology, 94*(5), 1103.
35. Burke, M. J., Sarpy, S. A., Smith-Crowe, K., Chan-Serafin, S., Salvador, R. O., & Islam, G. (2006). Relative effectiveness of worker safety and health training methods. *American Journal of Public Health, 96*(2), 315.
36. Cullen, E. T. (2011). Effective training: A case study from the oil and gas industry. *Professional Safety*, March, 2011, 40–47.
37. Vredenburgh, A. G. (2002). Organizational safety: Which management practices are most effective in reducing employee injury rates? *Journal of Safety Research, 33*(2), 259–276.
38. Komaki, J., Heinzmann, A. T., & Lawson, L. (1980). Effect of training and feedback: Component analysis of a behavioral safety program. *Journal of applied psychology, 65*(3), 261.
39. Ray, P. S., Bishop, P. A., & Wang, M. Q. (1997). Efficacy of the components of a behavioral safety program. *International Journal of Industrial Ergonomics, 19*(1), 19–29.
40. Herbertson, W. (2009). *The practical safety guide to zero harm: How to effectively manage safety in the workplace*. Salamander Bay, Australia: The Value Organisation Party Ltd.
41. Gordon, R. P. (1998). The contribution of human factors to accidents in the offshore oil industry. *Reliability Engineering & System Safety, 61*(1), 95–108.
42. Hopkins, A. (2009). Thinking about process safety indicators. *Safety Science, 47*(4), 460–465.
43. IPIECA. (2007). *Health performance indicators: A guide for the oil and gas industry*. Retrieved from www.ipieca.org/system/files/publications/HPI.pdf.
    Carson, P. A., & Snowden, D. (2010). Health, safety and environment metrics in loss prevention—Part 1. *Loss Prevention Bulletin, 212*, 11–15.
    Carson, P. A., & Snowden, D. (2011). Health, safety and environment metrics in loss prevention—Part 2. *Loss Prevention Bulletin, 221*, 12–17.
    CCPS. (2010). *Process safety leading and lagging metrics*. Retrieved from http://www.aiche.org/sites/default/files/docs/pages/CCPS_ProcessSafety_Lagging_2011_2-24.pdf.
44. IPIECA. (2010). *Oil and gas industry guidance on voluntary sustainability reporting*. Retrieved from http://www.api.org/environment-health-and-safety/~/media/files/ehs/environmental_performance/voluntary_sustainability_reporting_guidance_2010.ashx.

45. IPIECA. (2013, June). *Health leading performance indicators* (Report 2012h). Retrieved from http://www.ogp.org.uk/pubs/2012h.pdf.
46. Vestly Bergh, L. I., Hinna, S., Leka, S., & Jain, A. (2014). Developing a performance indicator for psychosocial risk in the oil and gas industry. *Safety science, 62*, 98–106.

    Al Muftah, H., & Lafi, H. (2011). Impact of QWL on employee satisfaction case of oil and gas industry in Qatar. *Advances in Management and Applied Economics, 1*(2), 107–134.
47. Komaki, J., Heinzmann, A. T., & Lawson, L. (1980). Effect of training and feedback: component analysis of a behavioral safety program. *Journal of applied psychology, 65*(3), 261.

    Ray, P. S., Bishop, P. A., & Wang, M. Q. (1997). Efficacy of the components of a behavioral safety program. *International Journal of Industrial Ergonomics, 19*(1), 19–29. Laitinen, H., & Ruohomäki, I. (1996). The effects of feedback and goal setting on safety performance at two construction sites. *Safety Science, 24*(1), 61–73.
48. Ray, P. S., Bishop, P. A., & Wang, M. Q. (1997). Efficacy of the components of a behavioral safety program. *International Journal of Industrial Ergonomics, 19*(1), 19–29.

    Laitinen, H., & Ruohomäki, I. (1996). The effects of feedback and goal setting on safety performance at two construction sites. *Safety Science, 24*(1), 61–73.
49. Lauver, K. J. L. (2007). Human resource safety practices and employee injuries. *Journal of Managerial Issues, 19*(3): 397–413.
50. NG Oil and Gas. (2013). Safety first! Retrieved from Http://www.ngoilgas.com/article/Safety-first/
51. Hopkins, A. (2009). Thinking about process safety indicators. *Safety Science, 47*(4), 460–465.
52. Osmundsen, P., Aven, T., & Erik Vinnem, J. (2008). Safety, economic incentives and insurance in the Norwegian petroleum industry. *Reliability Engineering & System Safety, 93*(1), 137–143.
53. Langford, D., Rowlinson, S., & Sawacha, E. (2000). Safety behaviour and safety management: Its influence on the attitudes of workers in the UK construction industry. *Engineering Construction and Architectural Management, 7*(2), 133–140.
54. Pransky, G., Snyder, T., Dembe, A., & Himmelstein, J. (1999). Under-reporting of work-related disorders in the workplace: A case study and review of the literature. *Ergonomics, 42*(1), 171–182.

    Probst, T. M., & Estrada, A. X. (2010). Accident under-reporting among employees: Testing the moderating influence of psychological safety climate and supervisor enforcement of safety practices. *Accident Analysis & Prevention, 42*(5), 1438–1444.
55. Probst, T. M., & Estrada, A. X. (2010). Accident under-reporting among employees: Testing the moderating influence of psychological safety climate and supervisor enforcement of safety practices. *Accident Analysis & Prevention, 42*(5), 1438–1444.
56. Hopkins, A. (2009). Thinking about process safety indicators. Safety Science, 47(4), 460–465.
57. Reasen, R. (2012). *HR.Com Wellness Survey study*. Retrieved from http://www.hr.com/system/app/media/rs/2013/2/1/hcnz6ja7/og.pdf
58. Jackson, S. E., Schuler, R. S., & Werner, S. (2011). *Managing human resources*. CengageBrain.com.

5

Projects in the Oil and Gas Industry

Introduction

The oil and gas industry is very capital intensive, with investment decisions made in all activities across the value chain—from upstream exploration and development to midstream shipping and pipelines to downstream refining and chemicals. The investments lead to the construction and development of facilities such as onshore oil and gas production facilities, offshore production facilities, onshore petroleum storage facilities, offshore storage facilities, oil tankers, liquefied natural gas (LNG) liquefaction facilities, pipelines, refineries, terminals, petrochemical plants, lubricant blending units, gas stations, and carbon capture and storage facilities. There can be major differences and subareas within these categories. For example, onshore production facilities can include conventional oil wells and unconventional shale gas and oil developments. The type of crude oil will determine whether it must be upgraded or processed before shipment. In remote regions, oil production sites may include infrastructure such as roads, housing, ports, power plants, and office facilities.

The projects in the oil and gas industry are now bigger, more complex, and riskier than ever before and thereby create increasingly difficult challenges including technical, managerial, and human resource (HR) challenges.[1] (See Current Challenge Box 5-1 for more details on this topic.) Table 5-1 describes the five most expensive energy projects currently in the world, starting with the Kashagan complex in Kazakhstan, at an estimated cost of $116 billion. This project is a great example of the cross-national cooperation prevalent in the oil and gas industry as its contributors include KazMunayGas, Eni, Shell, Exxon, Total, ConocoPhillips, and INPEX.[2]

Current Challenge Box 5–1. The Increased Complexity of Projects

Complexity is a much used word in the oil and gas industry for good reason. Many projects undertaken today, from the Arctic to the Amazon, have increased in sheer technological and construction complexity, cost (which often exceeds the financial size and resources of any individual firm), and consortium of relationships—sometimes the ownership is made up of several major firms (e.g., Kashagan in Kazakhstan) and literally hundreds of different suppliers of services, materials, and technology.

Size alone has become an enormous source of risk and complexity. Many of today's oil and gas developments across the world are considered megaprojects—projects of such size and scope that they require the strategic, managerial, capital, technological, and organizational capabilities of several major firms combined. The capital expense (capex) of a number of these projects is summarized by new oil and gas–producing countries in figure 5–1. Obviously, in many cases, the capex alone equals, or in some cases exceeds, the gross domestic product (GDP) of the subject countries.

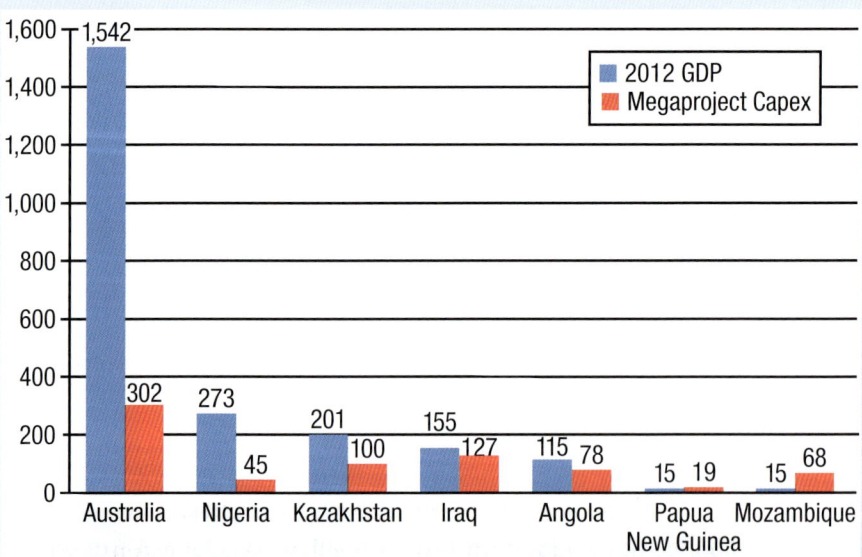

Fig. 5–1. Megaprojects capex versus 2012 GDP (billions of US dollars)
Source: "The Challenge of Renaissance: Managing an Unprecedented Wave of Oil and Gas Capital Projects," Deloitte Center for Energy Solutions, 2013, p. 3.

When a mega-project is developed in more remote locations, such as Papua New Guinea, Siberia, or interior Africa, logistics become critical. In remote locations, there may be no roads, airports, telecommunications, potable water, electricity, housing, labor force, or medical care. For some of these projects, the technical challenges of the oil or gas field are dwarfed by the challenge of getting equipment, materials, and people to the job site.

Complex projects create many challenges for staffing and HR support. The project management skills required must be organized in new and innovative ways—including the ability to work for one firm while operating effectively across several firms. A recent report by Bain & Company, *Large Project Management in Oil and Gas*,[3] listed the following essential skills for successful project management:
- Project management
- Local content and other stakeholders
- Costs and schedule
- HR and support functions
- Production operations
- Engineering
- Procurement
- Contracting
- Risk management
- Quality and health, safety, and environment

A study of mega oil sands projects in northern Alberta found that many projects suffered from huge cost overruns.[4] These projects require multi-billion-dollar investments and may require scheduling work for 7,000 to 8,000 people daily, with a personnel turnover as high as 300% annually. The reasons for the cost overruns include HR issues, such as:
- Underestimating the cost to attract and maintain the labor (craft) workforce
- Underestimation of the labor productivity loss associated with working in cold weather climates and locations with severely shorter daylight hours in northern regions
- Shortages of skilled labor and lower-than-anticipated labor productivity due to mismanagement of the construction phase
- High labor turnover mainly due to the harsh working environment and competition between employers attracting labor

These issues are likely to become even more important in the future as the complexity and remoteness of locations continue to increase.

Table 5–1. The five most expensive oil and gas projects in the world

Project, Location: Cost	Companies Involved	Description
1. Kashagan, Kazakhstan: $116 billion	KazMunayGas, Eni, Shell, Exxon, Total, INPEX, ConocoPhillips	This project moves oil out of Kashagan, a remote corner of central Asia through a series of interconnecting rail lines, pipelines, and ships using a series of artificial islands.
2. Gorgon, Australia, $57 billion	Chevron, Exxon, Shell	This natural gas project includes a liquefaction plant on Barrow Island to handle the estimated 40 trillion cubic feet of natural gas located off Australia's northwest coast.
3. Ichthys, Australia: $43 billion	INPEX, Total	Another natural gas project located off the coast of northwest Australia. This floating facility separates the natural gas and liquids brought to the surface.
4. Bovanenkovskoye, Russia: $41 billion	Gazprom	This facility is located on a peninsula of the Kara Sea, in a challenging remote arctic region.
5. Australia Pacific Australia: $37 billion	Origin, Sinopec ConocoPhillips	This facility liquefies natural gas from coal seams in the northeastern part of Australia.

Source: Money.CNN.com.

Regardless of the type of project that must be developed, each approved investment decision creates a new project that must be managed from conception to implementation. As a result, project management is one of the most critical business areas in the oil and gas industry. Whether the project involves a $30 billion oil field development or a $10 million storage tank at a fuel terminal, the decision-making and project-development processes that occur are conceptually similar: Project sponsors need to evaluate qualitative and quantitative risks; assess project economics; identify the source of capital; engage project development teams; create contractor supply chains; manage the project design, construction, and execution; hand the project over to the operators; and integrate the project into a larger organization.

Project managers in the oil and gas industry and the people who work in the various activities that support the project management function have a unique set of technical, management, and HR management challenges. This unique set of management and human resource management (HRM) challenges are the focus of this chapter.

Project Development in the Oil and Gas Industry

Although there is no one type of project or project-development process in the oil and gas industry, major projects of large scale, cost, and complexity will often follow the basic development path described in figure 5–2. In the upstream, which is largely the focus of this chapter, the typical project-development path generically is a very traditional three-step process of exploration-development-production. A defined development process path requires recognition of the activities summarized in figure 5–2 and the various agreements, contracts, organizational players, and specifically the skilled people to execute each of the phases.

Prior to entering the first phase of project feasibility, a long period of discussion, study, and negotiations between various parties must take place. For example, if the project were an oil-production project, the major oil company or companies wishing to develop the asset would most likely have spent years in discussion with the host country and government and among their own subunits. Along the way, the various parties use a variety of different agreements—letters of intent, memorandums of understanding, and heads of agreement—to increasingly provide definition and commitment to the various organizations participating.

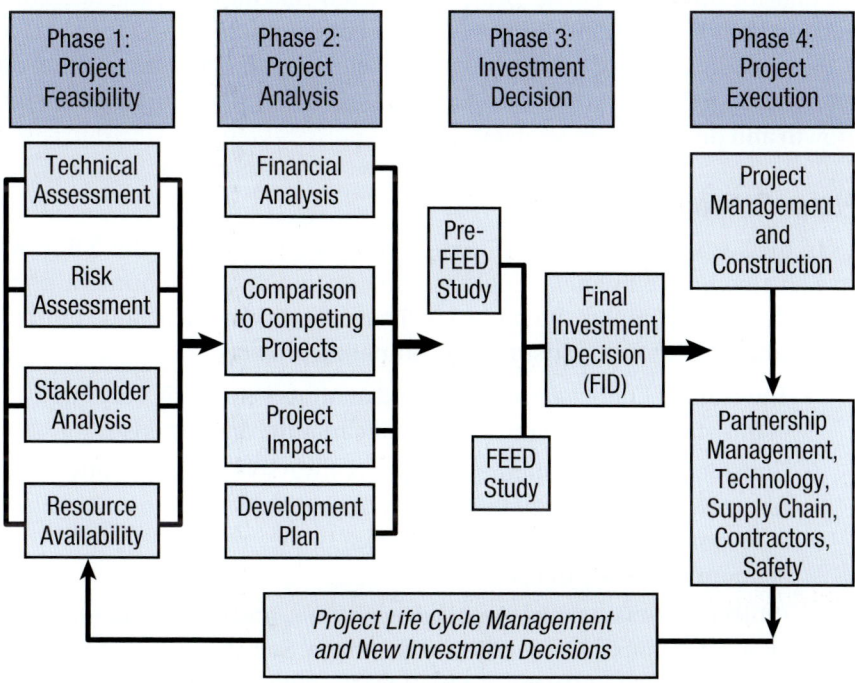

Fig. 5–2. Project development phases in the oil and gas industry

Phase 1: Project feasibility

Each task in the project feasibility evaluation phase requires experts from a variety of disciplines.

- The technical assessment will need input from scientists and engineers with expertise in the development of oil and gas projects.
- The risk assessment requires experts in finance, risk modeling, science and engineering, and business.
- The stakeholder analysis requires expertise in business relationship management.
- The analysis of resource availability can require input from logistics and supply chain experts, construction managers, and procurement.

In most cases, an individual or small team will drive the feasibility phase and have overall responsibility for ensuring that all of the tasks in the phase are completed on time and on budget. This individual or team will collect data and information from experts in all areas of the business and then prepare the data for submission to the project's decision-making body. The decision-making body is generally composed of executives and managers who have a deep understanding of the various business, financial, and technical risks associated with project development, as well as the company's overall growth and development strategy.

Phase 2: Project analysis

In the second phase, a development project goes through the following stages: a complete financial analysis, the creation of a development plan, an analysis of the impact the project will have on stakeholders and communities, and finally a comparison with competing projects.

- The team of financial specialists and analysts is typically very large at this stage and often requires significant time and expense.
- Consideration of competing projects is always important in any resource industry, given that a flood of new products into a market in response to price increases can often undermine the future economics of the project.
- Project impacts—social, political, economic, and environmental— all need serious consideration at this stage. Social and environmental impacts carry more and more weight in project analysis (discussed in detail in chapter 7).
- Development planning begins in parallel with financial and impact assessments, requiring the same detail and time-path considerations.

Phase 3: Investment decision

If the project makes it past the analysis phase, it is ready to move to Phase 3, when the final investment decision (FID) will be made. The FID is the final "go, no go" decision, and it is based on a series of detailed front-end engineering and design studies—pre and final. Companies often announce upcoming FIDs through press releases.

The significance of an FID cannot be overstated. As one recent article noted, "Reaching what the industry calls a "final investment decision"— the golden milestone signaling that steel can be ordered, workers hired and construction can begin—takes time, money and tight coordination between multiple parties working on different parts of a project."[5] More than 90% of project costs occur only after the FID. If the FID results in a "go" decision, the project will move into Phase 4.

Phase 4: Project execution

The last phase of the oil and gas development process is the actual construction of the production facilities. For most oil and gas projects, this is an enormously complex undertaking that includes a myriad of business tasks such as project management and construction, partner management, procurement, and supply chain management, technology acquisition and implementation, safety, and preparation of the project for handover to the project's operators.

While all oil and gas development projects are uniquely complex, they share some common business elements that span the entire life of the project from the point in time the FID gives the project a green light to the point at which the development team hands the project off to the production operators.

The common attributes include project management, partnership management, contractor management, supply chain and procurement management, risk management, and safety management.

- **Project management.** Oil and gas development projects generally consist of the primary project timeline plus the subsidiary projects that, when combined, complete the primary project.
- **Partnership management.** Partners are critical to the success of any oil and gas project. Sponsor firms rarely if ever complete an oil and gas development project entirely on their own.
- **Contractor management.** All major projects have an array of contractors that participate in or supply the design, engineering, and construction of the project.

- **Supply chain and procurement management.** Projects in oil and gas today are technology based and require many different technological and technical hardware and service providers. This supply chain and procurement management stretches across the entire life of the development project.
- **Risk management.** As projects move from feasibility assessment to execution, assessing and mitigating risks become part of the constants in the project management process.
- **Safety management.** For all of the major oil and gas companies, safety is the number-one management priority in project development. The focus on safety extends to all contractors and across the supply chain.

Completion of the development project is the last phase in a development project. Once the development project is complete, production of oil and gas can begin.

So far in this chapter we have briefly shown the prevalence of projects in the oil and gas industry, particularly in the upstream sector. (See Badiru & Osisanya, 2013, for an in depth analysis of project management in the oil and gas industry.[6]) We have discussed the common phases of projects and the different ways they are contracted. Project-based work differs substantially from the way work is typically organized. We now show how HRM can help address these differences.

Managing Human Resources in Project-Oriented Companies

We have shown that projects tend to be an important component of the work in many companies in the oil and gas industry. The differences between HRM in typical companies and in project-oriented companies are just starting to be considered by companies and in research. Historically, top management has generally only considered the technical and financial aspects of the project (such as economic viability, detailed financial analysis, project specifications, budgets, technical feasibility, technical support, and risk evaluations) when evaluating projects, selecting projects, and ultimately determining project success. However, with the general increase in the number of project-oriented and project-intensive firms, and the greater focus on the strategic implications of HRM, the role of HRM in project-oriented firms is being more closely examined.[7] Further, the increased physical, technical, environmental, and political challenges associated with increasingly large and complex projects, particularly in the oil and gas industry (recall table 5-1), make HR an important area of concern. A survey of oil and gas companies found that HR was considered to be more important to project success than knowledge management and supply chain issues, and equally important as engineering issues.[8] For suppliers, only project planning, risk management and project selection were seen as more important to project success than HR. Further, both oil and gas companies and suppliers were less satisfied with HR than most other aspects. So what makes HR different in project-oriented companies?

Several aspects of projects, as discussed earlier in this chapter, substantially relate to HR in project-oriented companies, for example:[9]

- The project as the organizing component of work
- The temporary nature of projects
- Greater uncertainty because of the strong rate of change
- Multirole demand on employees
- Process- and team-oriented work

These aspects generate a number of management problems, such as the separation of authority and responsibility, jurisdictional conflicts resulting from difficult assignment decisions, the frequent integration of new teams, and greater stress on employees. Greater stress on employees occurs in project-based firms for two basic reasons: First, there is greater frustration with multiple supervisors, "make work" at the end of projects, lack of formal procedures, lack of role definitions, multiple levels of management, and fluctuating workloads. Second, greater anxiety occurs because of uncertain requirements, career setbacks due to project failures, lack of anyone being concerned about their personal development, greater conflict, and more uncertainty regarding one's long-term security.[10]

These problems all have HR implications because they negatively affect employee well-being. However, companies do not seem to be doing a good job of providing employees with support through their HR functions in response to these issues.[11]

Beyond the introduction of complexities based on the dynamic nature and temporary nature of projects, there are other fundamental differences in the HR of project-oriented organizations. Three basic differences are (1) the widespread use of contractors, (2) the application of HR functions specifically to the job of project manager, and (3) the application of HR functions within the project in addition to within the organization. For example, HR functions within a project include determining assignments to the project, training for a specific project, performance management of the project, and compensation related to the project. In this chapter we focus on the HR functions as applied to project-oriented organizations, the projects themselves, and the job of project manager (for discussion of the widespread use of contractors, see chapter 6). We now look at how firms can improve the effectiveness and success of projects and improve the performance of the project-oriented firm through different HR functions, focusing on the key questions within each.

Staffing Projects

Clearly, staffing quality affects project performance.[12] In this section we discuss three specific aspects of staffing in project-oriented companies. First, we look staffing at the organizational level for project-oriented firms. That is, what types of employees are needed to succeed in the project-oriented environment? Second, we look at the staffing of projects. That is, how are assignments to specific projects made? Third, we look at the recruitment and selection of project managers. Because the job of project manager is a critical job in project-oriented companies, we explore how firms can attract and select those most qualified. We look at these three issues using the framework of the key questions in staffing consistent with past chapters.

Labor markets in project-oriented companies

The choices of labor market, recruitment practices, and selection practices in project-oriented companies depend on a number of factors. First, are we looking for people for a specific project or to join the organization long term? If we are looking for people for a specific project, companies will generally begin with the internal labor market, that is, assign (or recruit) current employees to the project. This is particularly true when hiring project managers. However, if no internal candidates are available, then external hires can be considered. External hires for project managers have the benefit of not having any strong internal ties and thus can be more impartial in their decisions. If we are looking for people to join the organization for the long term (rather than for just one project), then the company recruits from the external market. However, even in cases where the intention is to have the new employees join the company in the long-term, project-oriented firms usually reconsider the continued employment of the workers on a project-by-project basis.

A second key issue in project-oriented companies is: Are the knowledge and skills that are needed core to the company or are they peripheral and easily found in the external labor market?[13] In many aspects of projects, such as construction, security, and some services, skills are relatively uniform and acquired in a relatively short time frame. In such cases, workers can be hired on a contract basis or the work can be outsourced to specialist service providers. These workers are likely to be in the same geographic labor market as the work site. When the skills are core to the organization, not easily learned, and difficult to replace, employees with these skills provide the core competencies of the firm and add value. In such cases, employees with longer tenures and considerable relevant experience are needed. Thus, they are much more likely to be internal candidates, and because of the limited number of internal candidates, a broader geographic net must be considered. Projects managers are a good example of this.

Recruiting for project-oriented companies

As previously mentioned, candidates may be recruited for a job in the company in general or for a specific project. We discuss recruiting for a specific project and recruiting project managers in the next sections. When recruiting for employees in general for a project-oriented company, the focus should be on the candidate's broader abilities that may apply to numerous different projects, rather than the specific abilities that apply to only one project. Further, the recruitment process for employees intended to be retained in the long term is somewhat different in project-oriented firms than in other companies. Recruiting for long-term employees with core skills frequently involves the use of headhunters, personal contacts, and the grapevine, as well as relationships with professors and administrators at universities and technical schools.[14] Thus, the recruiting process tends to be more interpersonal and less formal.

Recruiting for specific projects. Selection for projects may also be thought of as assignments to projects depending on how much freedom is given to employees to accept or decline working on specific projects. This freedom can range from being able to decline a project without explanation to having no choice in declining the project without serious employment repercussions. If assignments are voluntary (and thus more accurately labeled as selections rather than assignments), then recruiting plays a more important role. In cases where you are recruiting for specific projects, of course the focus should be on the project. Project factors important to the recruitment and selection process include the project's content, geographic location, complexity, and, of course, the skills needed. These factors then should drive the recruitment process. Key questions to ask to help recruiting include the following:

- Who has expertise in this area?
- Who has worked with these particular clients?
- Who has worked with key contractors?
- Who has experience with this location?
- Who is available?

Project recruitment should also consider long-term strategic implications, including the career development of employees, the importance of developing functional staff through job rotation, and the company's ability to retain employees. Thus, the organization should also consider the employee's perspective.

Looking at the assignment from the employee's perspective raises different issues. The employee's desire to be on the project and the specific developmental opportunities of the project should play a role in the assignment decision. In cases where the assignment is not voluntary, these aspects should still play into the assignment decision. Recruiting for specific projects can be done through internal advertising or personal interactions for internal labor markets, or it can follow conventional recruiting methods for external markets. Again, contract work is a popular option for less-value-added jobs.

A further consideration is that many projects in the oil and gas industry are medium to long term in duration. Construction projects may take anywhere from two to five years, and in some cases longer. The more senior and experienced the project roles are, the more challenging it may be for individuals to take on these job roles. As professionals grow in experience and age, they often take on more personal and family responsibilities. This can, in many cases, make them less mobile and possibly make moving them more difficult and costly to the firm itself. Long-term projects pose significant recruiting and retention challenges for the oil and gas industry; as one professional noted, "It's like getting a new life every five years."

Recruiting project managers. A brief discussion of the title "project manager" is needed here before we discuss their recruitment. We use "project manager" throughout this chapter for simplicity, but in the oil and gas industry, the responsibilities, duties, tasks, and requirements can vary widely within the broad scope of this title. Depending on the phase of the project, size of the project, nature of contracts, and so on, other commonly used titles such as project developer, project superintendent, general contractor, or construction manager may better specify the nature of the job.[15]

Project managers are important to the success of projects. Poorly performing project managers will increase the time, cost, and quality of the project, possibly damaging the reputation of the firm.[16] To recruit effective project managers, we need to have an idea of which skills are needed to do the job. Table 5-2 shows the results of several studies that have looked at the skills possessed by effective project managers.[17] These skills tend to be categorized as interpersonal, managerial, and technical skills. Interpersonal skills include communication, motivation, leadership, and negotiation. Managerial skills include planning, organizing, problem solving, goal setting, and analyzing. Technical skills include expertise in the content area, the technology used, computer technology, and project processes. Interestingly, surveys tend to show that interpersonal skills are believed to be most important, followed by managerial and technical skills (although all are seen as important).[18]

Table 5–2. Skills related to project manager effectiveness

Interpersonal Skills	Managerial Skills	Technical Skills
Motivating skills	Delegation skills	Project knowledge
Communication skills	Planning skills	Process knowledge
Coping skills	Organizing skills	Computer skills
Conflict resolution skills	Problem solving skills	Required technology skills
Political sensitivity	Goal setting skills	Tool skills
Self-confidence	Analytical skills	Content expertise
Team-building skills	Entrepreneurial skills	Software skills
Leadership skills	Administrative skills	
Negotiation skills	Resource allocation skills	
Relationship skills	Critical thinking skills	
Emotional intelligence		
Cultural awareness		

The importance of these skills changes depending on the phase the project is in. For example, content and technical expertise is considered more important during the early phases of a project (the feasibility and analysis phases) than in later phases. Managerial expertise becomes more important in later phases. Thus, companies should be focusing recruitment on potential candidates with the skills needed in all phases. However, the skill requirements may be so different across phases of a particular project that a planned change in project managers may need to occur in these different phases.[19]

Project managers may be recruited as employees who are expected to move from project to project, or may be recruited with one specific project in mind. We discuss some of the differences when selecting for specific projects later in this chapter.

Selecting in project-oriented companies

Again, selection in project-oriented companies can be for the company in general or for a specific project. We discuss selecting for specific projects and selecting project managers in the next sections. Selecting for the company in general implies long-term intent and thus is more likely to occur for jobs that require skills and knowledge that are core to the firm. For project-oriented organizations, the hiring of such employees tends to be less mechanical and prescribed than in other firms. Specifically, project-oriented firms are more likely to hire exclusively based on work experience and with the mind-set that the first project could be considered a trial period. Further, because specific skills can be learned, they are not as critical as being adaptable to different projects, clients, or locations. Most project-based firms do not use specific selection tests, but rather observe employees over time in different situations to determine long-term viability.[20] Nevertheless, based on the criteria specified above, one could easily argue that one of the Big 5 personality traits, openness to experience, would appear to a viable predictor of long-term success in project-oriented companies.

Selecting for projects. Clearly, the primary focus of the project assignments should be the project. Although assignment decisions are frequently made by functional managers, project managers should be involved in the selection of their teams. This will help assure buy-in and accountability for the performance of the teams as well as allow the project managers to provide insight into team compatibility. Research has shown that greater stability in the membership of project teams across projects increases trust among the project partners. This, in turn, increases the acquisition of new knowledge by the team members, which subsequently increases innovation.[21] Thus, the possible gains from using stable project teams across projects has to be weighed against the gains from introducing new members to the teams.

Relatedly, another consideration for each project is the how a new recruit will blend in with the rest of the team.[22] Because projects are a team effort, consideration must be given to how the applicant (or assignee) will work with the other team members. This would suggest that applicants with a strong team orientation would be preferable. It also suggests that assignments should be avoided where personality conflicts between team members have occurred in the past. The importance of teamwork for projects also suggests that, depending on the people and the project, the project manager's team-building skills may be more important on some projects than on others. Team cohesion appears to be particularly important for projects that are more complex, have more strategic importance, are more visible to the marketplace or internally, introduce new technology, have higher risk, and are long term.[23] Note that most of these points describe the typical project in the oil and gas industry. That said, in recent years projects have grown in size and complexity for the major oil and gas firms. Thus, firms must ensure that project teams understand and can mitigate the associated risk of the added complexity. How well the employee will fit in with the team depends not only on interpersonal fit, but also on how well the employee's technical skills, experience, and professional attitudes and behaviors mesh with the rest of the team.

In organizations where multiple simultaneous and successive projects form a large portion of the business, allocating appropriate personnel is a huge issue. Mathematical modeling (incorporating parameters such as the window size of projects, the skills needed and available, and employee workloads) can also be used in multiproject organizations to help clarify possible alternatives.[24] However, modeling or scheduling software is not always the solution because it cannot incorporate all issues that should be considered, the parameters frequently change (some projects go long, some do not), and many companies tend to overcommit.[25]

Selecting project managers. Once a pool of candidates has been recruited to apply to the company as project managers (ideally with the needed skills), those who are hiring must decide what tools to use to make the final selection decisions. As mentioned earlier, project managers need a wide range of skills and knowledge to be successful. However, other ways to assess the competence of project managers include looking at their personality characteristics or demonstrable project management performance.[26] Standard assessments of the skills and knowledge of project managers and of their demonstrable project management performance are available from a number of professional associations or standard setting bodies. Specifically, the following are examples of certifications that are available:

- In the United States and globally, the Project Management Institute (PMI) offers nine different certifications including Project Management Professional (PMP), PMI Risk Management Professional (PMI-RMP), and PMI Scheduling Professional (PMI-SP).
- In the UK, the Association for Project Management (APM) has six different certifications including an APM Introductory Certificate, a APM Risk Certificate, and Practitioner Qualification Certificate.

- In Japan, the Project Management Association of Japan has four different certifications including Project Management Coordinator (PMC), Project Management Specialist (PMS), Project Manager Registered (PMR), and Project Management Architect (PMA).
- In Australia, the Australian Institute of Project Management has four different certifications based on meeting differing levels of Professional Competency Standards for Project Management that have been established, which include Certified Practising Project Practitioner, Certified Practising Project Manager, Certified Practising Project Director, and Certified Practising Portfolio Executive.
- In South Africa, the South African Qualifications Authority has a national certification in project management specified as National Diploma: Project Management.
- In Europe, the International Project Management Association has four levels of certification (many which are have been aligned with national associations), including Certified Project Management Associate, Certified Project Manager, Certified Senior Project Manager, and Certified Projects Director.

Certification can provide prospective employers with some degree of confidence that the applicant has a certain level of competence regarding project management skills and knowledge. Scores on project management knowledge tests are related to higher performance, as judged by senior managers.[27] Although we can find no research that definitively shows that certified project managers outperform those without certifications, logically certification appears to be a reasonable selection criteria, other things being equal.

Interviews are also generally used for project managers. As mentioned in chapter 1 they should be structured and based on a job analysis. If the hiring is for a specific project, the questions should be tailored to that project. However, final selection decisions appear to be usually made based on past experience.

Combining tools in project-oriented companies

How tools should be combined in project-oriented firms depends on the job, the project specificity of the hiring, and the project. In general, the principle for combining the tools in project-oriented firms is the same as in other firms, use the multiple hurdles approach for selection criteria that are critical for success on the job. These may be technical, managerial, or interpersonal skills, and are likely to be estimated by using experience and background information rather than direct tests. For the job of project manager, it may be that all three types of skills are critical. However, companies should be aware that as the number and difficulty of the hurdles (e.g., cutoff scores on the tests or number of years of applicable experience) increase, it will be less and less likely that even one applicant survives the multiple hurdles. Thus, cutoffs should be set that reflect the minimal level of skill or knowledge needed, rather than the optimal level. For example, if requiring a high level of technical skill for a project manager leads to the elimination of all the applicants with higher levels of managerial and interpersonal skills, then it may be better to require only the minimally needed level of technical skills leaving a larger pool of candidates with high levels of the other needed skills.

When hiring or assigning workers for specific projects, multiple hurdles should be used for predictors that are deemed critical for the success of that project. However, a compensatory approach may be better for employees who are being assigned for developmental purposes or when the project is seen as a short-term assignment. The bottom line is that a compensatory approach is more likely to provide a larger pool of applicants to choose from, but a multiple-hurdles approach will ensure that all those remaining in the pool have at least the minimum level of proficiency of all the skills needed. Although it is tempting to use a compensatory approach so that someone can be hired, there is a substantial risk in hiring someone without even the minimum skill proficiency needed for some aspect of the job. This risk may be great enough to warrant continuing recruitment rather than settling for "what's available."

Maximizing the effectiveness of staffing in project-oriented companies

The general principles of being legally compliant; treating applicants fairly; and monitoring, evaluating, and adjusting the system all apply to project-oriented firms, but with a few differences. First, they all apply at the project level as well as at the organizational level. For example, with respect to legality in the United States, project assignments are considered a condition of employment and are thus subject to the same federal discrimination laws as when hiring employees. With respect to fairness, assignments to projects may be considered unfair if assignees have little input and employees perceive that they are being assigned to undesirable projects more than others.

Second, the shorter-term aspect of projects provides firms with more frequent selection decisions, particularly when there are multiple short-term projects. Thus, monitoring, evaluating, and adjusting become easier because there are many more opportunities to do so. Thus, there are more opportunities for firms to learn from their mistakes and correct the project-level staffing issues so that they do not reoccur.

Finally, the dynamic rate of change in project-oriented companies makes more frequent monitoring and adjusting more important. This includes more frequent monitoring of not only the staffing system but also the hiring and assignment decisions. Changes may need to be made during the project if personnel are unable to adapt to changes in or different phases of the project.

Training for Projects

Training and development in project-oriented firms differs in three substantial ways from other firms. First, training may occur for specific projects. Second, certain project assignments may be made for largely developmental reasons. Third, the development of project managers differs from that of other jobs. We look at these three issues using the framework of the key questions in training, consistent with previous chapters.

Who gets trained in project-oriented companies?

As in most companies, everyone in the company is a candidate for some type of training. In project-oriented firms, who gets trained will vary depending on the reason for the training (e.g., is it project specific or not?), the training needs of employees, and the employees' roles in the project. We now discuss who gets trained in project-specific training, who gets trained when projects are used as developmental assignments, and who gets trained in project manager training.

Project-specific training. Training of the project team appears to be an important factor in the success of projects.[28] Training has been shown to be related to better performance, reduced projection costs, increased satisfaction of team members, and increased motivation of team members. Training is also instrumental in helping project employees attain their full potential.[29] For some projects, all team members may need specialized training before they start a large, new project. This is determined by a front-end analysis of the project team's training needs. A training needs assessment should include all members of the project team as well as vendors, contractors, and subsequent operators and maintenance personnel.

Of course, the ultimate success of any project lies in not only on how well the project was designed or built but also on how well it is subsequently maintained and operated. Thus, training for those running and maintaining the processing facility, refinery, pipeline, or platform is critical to a company's overall success.[30] Insufficient training of these operators, maintenance personnel, and managers can lead to long problematic start-ups, increased accidents, damaged equipment, and serious injuries.

Finally, vendors or contractors may also need to be trained. Although the temptation may be to assume that they are responsible for their own training, ensuring that all participants have the necessary information and skills needed to achieve their goals on the project can save many headaches later. Thus, vendors and contractors should be included in any training needs assessments, although they may be asked to conduct their own training.[31]

Projects as training. The location, scope, and types of projects can vary dramatically in project-oriented companies. Project assignments can be developmental in nature, when the assignment will provide desirable on-the-job training for the employee. Employees can gain valuable experience through assignments in different geographic locations, through assignments on projects with greater value or complexity, and through assignments on projects of different content. Further, assignments to projects that are similar to employees' past experience can be developmental if they are given different roles on the project team than they have had in the past.

Any team member may be assigned to a project largely for developmental reasons, particularly if the assignment is to a project that differs from the employee's past experiences (e.g., a larger project or a different geographic location) but is in the same role. Assigning employees to similar roles to projects in new geographic locations or with increased project scope can develop employees by exposing them to different experiences within their same roles. When the assignment involves a new role, employees with high potential to be project managers are the most likely to be assigned to projects for largely developmental reasons. However, greater knowledge about project management processes in general can help all team members improve the efficiency of projects. Further, many team members value developmental opportunities, and thus assignments positioned as such are likely to be received favorably by most team members. Projects that are assigned as training vehicles are also career development vehicles for employees.

Training project managers. High-potential employees who are seen to possess or are likely to be able to possess many of the skills listed in Table 5–1 are the most obvious candidates for receiving training as project managers.

However, current project managers are also viable candidates not only because this training will help them improve their skills in areas where there are currently shortcomings, but it will also help them keep up with the current body of knowledge needed to be an effective project manager. Thus, training programs can be designed specifically for entry-level, standard, and advanced project managers.

What do we train for in project-oriented companies?

Training topics depend on whether the training is for specific projects, applies to multiple projects, or applies only to project managers. Based on a survey in 100 firms, basic courses that project-oriented organizations like to see employees in general take include introductory project management, planning and control, accounting and finance, organizational behavior, systems management, law, information systems, management policy, Program Evaluation and Review Technique/Critical Path Method (PERT/CPM), computers, management science, managerial economics, production management, statistics, marketing, international trade, and quality control.[32] In the oil and gas industry, content would include political risk management, molecule management, upstream and downstream asset integration, and oil and gas demand/supply economics. Clearly, projects and project management cover a wide range of content areas. Organizations appear to prefer generalist training through courses and later more specific focused training on the job, although this may vary based on the type of project.

Project-specific training. When a front-end analysis shows that a project team needs specialized training for a particular project, the training is usually customized to focus on those needs. This training may also be targeted toward the implementation of specific practices. Because of the specificity of the training, training outcomes are usually relatively easy to measure.[33] Training may also occur on specialized equipment or systems and may often be provided by vendors. Such training should be carefully specified in the purchase contract.

Some common project-specific topics that may be covered for the entire team include the following:[34]

- The project plan
- Planning and scheduling for the project
- Project reviews
- Costing systems for the project
- Project engineering

- Start-up procedures
- Subcontractor management

Of course, other topics may be appropriate depending on the project and the team's training needs assessment.

Project teams may also be trained on team building, to help facilitate the coordination and effective interactions of the team. Team building may include training on effective communication between team members, team problem solving, consensus building, team conflict resolution, team decision making, and group dynamics.

Projects as training. What is trained through on-the-job experiences largely depends on the nature of the project as well as the employee's role as a project team member. In situations where the employee is in a new role on the project, experienced employees frequently train newcomers. Learning also occurs through assignment variation, where even experienced employees can gain new knowledge through the practical experiences of different assignments.

Training project managers. The skills specified in table 5–1 give a good idea of the skills needed by project managers, and thus each of them is a viable training topic. Other possible training topics include the body of knowledge needed by project managers. O*net (www.onetonline.org) lists several areas of knowledge necessary to do the job of project manager in an engineering context. These areas include engineering and technology, design, mathematics, administration and management, customer and personal service, physics, production and processing, and HRM.[35] Table 5–3 lists these areas and their descriptions. Many of the associations and standards-setting bodies discussed earlier have also specified the body of knowledge needed by project managers. The de facto standard in the oil and gas industry (and most other industries) is from PMI, in book form as *A Guide to the Project Management Body of Knowledge* (*PMBOK Guide*), now in its fifth edition. It specifies 12 areas deemed to be important knowledge bases for project managers:

- Organizational influences and project life cycles
- Project management processes
- Project integration management
- Project scope management
- Project time management
- Project cost management

- Project quality management
- Project HRM
- Project communications management
- Project risk management
- Project procurement management
- Project stakeholder management

Table 5–3. Areas of knowledge needed by project managers

Areas of Knowledge	Description
Engineering and Technology	Knowledge of the practical application of engineering science and technology, including applying principles, techniques, procedures, and equipment to design and production.
Design	Knowledge of design techniques, tools, and principles involved in the production of precise technical plans, blueprints, drawings, and models.
Mathematics	Knowledge of arithmetic, algebra, geometry, calculus, statistics, and their application.
Administration and Management	Knowledge of business and management principles involved in strategic planning, resource allocation, human resources modeling, leadership technique, production methods, and coordination of people and resources.
Computers and Electronics	Knowledge of circuit boards, processors, chips, electronic equipment, and computer hardware and software.
Customer and Personal Service	Knowledge of principles and processes for providing customer and personal services. This includes customer needs assessment, meeting quality standards for services, and evaluation of customer satisfaction.
Physics	Knowledge and prediction of physical principles, laws, their interrelationships, and applications to understanding fluid, material, and atmospheric dynamics, and mechanical, electrical, atomic, and sub-atomic structures and processes.
Personnel and Human Resources	Knowledge of principles and procedures for personnel recruitment, selection, training, compensation and benefits, labor relations and negotiation, and personnel information systems.
Production and Processing	Knowledge of raw materials, production processes, quality control, costs, and other techniques for effective manufacturing and distribution.

Source: O*net (www.onetonline.org).

Certainly, these are all viable areas of training for project managers in the oil and gas industry. However, research has shown that some are more important than others. Further, the level of importance depends on the type of project. Relative to the oil and gas industry, project integration management, project scope management, and project HRM are considered the most important knowledge areas for construction and engineering projects.[36] Nevertheless, training project managers toward certification may be particularly welcome since certification is regarded as a concrete achievement that deserves recognition.

In 1996, the title of the *PMBOK Guide* changed from *The Project Management Body of Knowledge* to *A Guide to the Project Management Body of Knowledge*. This change reflects the prevailing thought that the book covers material that is generally applicable to most projects most of the time, but that one book cannot be inclusive of all the knowledge that would be needed for every project.[37] Specialized project areas may have their own bodies of knowledge or best practices that go beyond the *PMBOK Guide*. For example, product development projects frequently use the Stage-Gate process for driving new products to market.[38]

It should be noted that the *PMBOK Guide* has received some criticism for focusing too much on hard (technical) skills and too little on soft (human) skills.[39] This criticism appears warranted, given our earlier discussion of research showing that the interpersonal skills are generally seen as most important for a project manager's success (and given the focus of this book).

How do we train in project-oriented companies?

The training methods in project-oriented companies largely follow the training methods in typical firms. As in typical companies, the method will depend largely on who is being trained and what they are being trained. However, in project-oriented organizations, the training method will also vary depending on whether the training is project-specific, is part of the assignment, and is specifically for project managers.

Project-specific training. Project-specific training can be accomplished by a wide variety of methods, including readings, lectures, seminars, roundtable discussions, e-learning, role-playing, coaching, mentoring, on-the-job training, and formal class settings. Training that is focused on specific knowledge is more likely to be in the form of formal class settings, readings, lectures, e-learning, and seminars. Training on skills is more likely to be more engaging and be on-the-job training, coaching, mentoring, role-playing, or roundtable discussions. Formal team-building programs may also be used. Team-building programs can have a very strong, positive effect on the project success, particularly when the projects are long term, bring together many higher-level employees who have not worked together before, are the first of a series of projects, have large budgets, involve numerous cultural groups, and involve numerous different technologies.[40]

Although formal team building is likely to have strong positive effect, its high cost, considerable length (it may take from 50 to 100 hours), and difficult-to-quantify results sometimes make it a hard sell.

Project-specific training may be done by vendors, contractors (especially on larger capital projects), or public training facilities (e.g., colleges, universities, technical schools, or vocational schools), or done in-house. All of them have their strengths and weaknesses.[41] Training by vendors tends to be very specific, but frequently superficial and not tailored to a broad audience. Contractors can perform training needs assessments and coordinate offsite, on-site, and vendor training, but may not have all the necessary capabilities. Public training facilities may be helpful to provide remedial training for certain groups of the project team, but such training is usually very general and frequently needs substantial supplemental training to apply to the project. In-house training enables the most control and will ensure that the training is consistent with the organization's culture, but companies are usually not able to handle the training needs of larger projects in-house. Most of the time, a firm's project management office organizes at least some of the in-house training.[42]

Projects as training. Experiential learning is generally seen as critical for most team members and even more so for project managers. Thus, depending on the project and the team member's role, an assignment to a project may be made largely for developmental reasons. Rotating team members through various functional assignments is an important way to develop project managers, but is also likely to benefit team members without project manager aspirations. Job rotations are likely to help develop employees by exposing them to the bigger picture and giving them insight into how the different jobs interact. However, the downside of job rotation is that there may be inexperienced workers in place for many jobs during their rotations.

The relationship of formal learning to experiential learning depends on the nature of the project. For example, in construction projects, formal training tends to be done after on-the-job training for project managers, with courses given to enhance their understanding of principles already experienced on the job. In information technology projects, the opposite tends to be true: Formal courses come first, followed by experiential application of the material learned.[43]

Training project managers. Although there is some disagreement, most senior company managers believe the best way to train project managers is through on-the-job experience and other experiential learning methods.[44]

These may include working with experienced project managers, working as a project team member, or job rotation through a variety of project management responsibilities. Job rotation through different practical experiences based on different assignments is also a popular way to develop project managers. Other senior managers believe that formal education and special courses are the best training methods. Still others believe that professional activities, seminars, or designated readings are the best methods.

Training of basic knowledge, such as in PMI's *PMBOK Guide*, is usually done with readings, e-learning, and classroom training. Frequently e-learning and readings are used as preparation for classroom courses, which are then followed by coaching. Online sessions can be used later to keep up with any changes in the body of knowledge.[45] Online education also has the benefit of allowing the development of innovative approaches in a flexible learning environment—a must to deal with the greater complexity in today's projects.[46] Finally, simulations may be used to help hone problem-solving skills.

Maximizing the effectiveness of training in project-oriented companies

As with training in general, in project-oriented companies the positive effects of training are maximized when learning is maximized; training transfer and performance maintenance is facilitated; and the system is monitored, evaluated, and adjusted regularly. Learning begins with participation, but because of the hectic project environment, it may be more difficult for employees to attend training. Thus, project-oriented companies must be more diligent in motivating employees to attend and actively participate in training. Sending trainers to work sites may provide the organization with more oversight and control than off-the-job training. Depending on the geographic location and nature of the workforce, increased oversight and control may be necessary. Research in project environments confirms what we know to be generally true: The more important training is perceived to be by the trainee and the more motivated trainees are to learn, the better their performance will be.[47]

Monitoring, evaluating, and adjusting the system also follow the general principles discussed in chapter 1. Training program attendance, student evaluations, participant training records, and unsuccessful course participation should all be monitored. Training course selection, course performance records, course delivery schedules, instructor performance, administrative performance, course development, and training contracts should all be regularly evaluated. Adjustments should be made based on issues shown during evaluation as well as updates that have been determined to be needed.[48] As with staffing, the number of projects and the dynamic rate of change in project-oriented companies make more frequent monitoring and adjusting more important. This includes more frequent monitoring of not only the training system but also the performance management system as well, since performance deficiencies can indicate areas of training needs.

Performance Management in Projects

Performance management in project-oriented firms differs from performance management in other firms most notably in one fundamental way: Successful completion of the project is a clear, definitive, and seemingly objective measure of performance. Project performance should tie into the pay-for-performance aspect of performance management of project employees. We will discuss this aspect in detail in the project manager section below. Project performance, however, does not necessarily have to be directly related to the performance of individual employees. Projects can have problems that in no way reflect the performance of any individual employee. Certain employees may have performed outstandingly, although the project itself was unsuccessful. Thus, although the goal-setting, goal-achievement, and the pay-for-performance aspects of performance management in projects are likely to be tied to project success, individual performance evaluation and feedback need not be. Other issues that affect performance management in project-oriented companies include the likelihood that employees will have more than one manager; the temporary nature of the project; team-based work; and greater uncertainty of roles, performance, and outcomes. We now look at how these issues affect the performance management system. We look at these issues using the framework of the key questions in performance management, consistent with past chapters.

How do we measure performance in project-oriented companies?

As with the other HR functions, performance management in project management can be relative to the project as well as the organization in general. Thus, we must determine what measures effectively assess performance at the project level. Once we have done this, we can then determine which organization-level, project-level, and individual-level performance measures we want to use. How we measure performance differs between project personnel and project managers. We begin with project personnel.

Performance for project personnel. The evaluation of individual performance of team members in projects is similar to the evaluation in other firms. However, the differences related to working in projects do have some implications for the measurement of performance. First, because teamwork is usually an important component of working in projects, some measure of an individual's contribution to the team is warranted. As with most performance measures, these may be based on behaviors or outcomes, but keep in mind that team outcomes may be relatively independent of any one team member's performance, particularly in large teams.

Second, depending on the timing, annual evaluations may be based on the employee's performance on one or more projects. If performance is based on work for more than one project, the performance scores for each project should be weighed relative to the time spent on each project. The same is true if the employee spends some time working that is not directly tied to any project. Scores of the same underlying performance dimension should be similar across projects. If not, this may indicate environmental factors that influenced the employees' performance rather than anything the employee had control over.

Third, projects can inherently involve conflicts among employees, management, and other stakeholders. Conflict may result because of different views on priorities, schedules, technical issues, administration, and costs. Measuring performance through *management by objectives* (MBO) appears to be a good way to measure performance under such conditions. In an MBO system, managers and workers agree on performance goals at the beginning of the project, and the worker's performance is ultimately evaluated by comparing actual performance relative to the agreed-upon goals. MBO in project settings may involve the project manager or functional manager setting goals in lieu of the employee's participation.[49]

Fourth, if the project assignment is largely or partly for developmental purposes (as discussed earlier in the training section) the employee's participation in developmental aspects of the role should also be included in the evaluation. In role- and extra-role-developmental activities such as extra duties, interactions with vendors, customers, or educational institutions, or involvement with professional associations should be noted. Finally, in project-based companies, additional consideration should be given to other factors such as the difficulty of the tasks (e.g., unforeseen technical challenges), out-of-role managerial responsibilities that developed on the project, and overall workload.[50]

Performance for project managers. The measure of performance for project managers is somewhat different from that of team members. The performance of the project itself should have greater influence in the evaluation of the project manager than for other team members, reflecting the project manager's greater influence on the outcome. Historically, project performance was measured on three dimensions: cost, timeliness, and quality. However, current perspectives suggest that this view should be broadened to include the stakeholder view, as discussed in chapters 1 and 7. Current Challenges Box 5-2 discusses the metrics used to measure project performance.

Current Challenge Box 5–2. Performance Metrics for Projects

Historically, project success was measured on three clear pillars (known as the iron triangle):[51]

1. Cost—are the costs of the project forecasted?
2. Time—is the project on schedule, or was the project finished on schedule?
3. Quality—is the quality of the finished project as specified, and is the performance of the completed project as desired?

However, there are some issues with these measures. First, both cost and time measures are largely best guesses that are (out of necessity) based on incomplete information. Projects that finish on time or at cost may be a better reflection of the performance of the cost/time estimator than the project team. Second, cost measures are generally estimated through financial and accounting measures, but that information is usually lagging by at least one reporting period and thus is not that helpful for timely feedback that could lead to better decisions or actions.[52] Finally, quality measures are frequently evaluated based on benchmarks from previous projects. Substantial differences among projects or new aspects of projects may diminish the appropriateness of benchmarks.

Although time, cost, and quality measures have historically been used to evaluate project performance, other measures are also possible and probably desirable. Using the stakeholder framework, as described in chapters 1 and 7, can help us determine a number of other possible project-level performance measures. Stakeholders in projects include the employees involved in the project, the government and its agencies, the parent corporation, customers, suppliers, subcontractors, end users of the project, the public (possibly represented by the media, special interest groups, and lobbyists), and the environment.[53] Thus, as we will discuss in more detail in chapter 7, the satisfaction of each of these stakeholders with respect to the project is a viable measure of performance at the project level.

One common approach that attempts to address at least some of the stakeholders is the *balanced scorecard*.[54] The balanced scorecard considers not only the typical financial measures but also customer-oriented measures, internal business measures, and innovation and learning measures. In a project environment, these would be at the project level and may include other measures related specifically to the project (such as the perspectives of subcontractors and suppliers).

A further argument to broaden the measure of project performance beyond the iron triangle is related to the complexity of many projects. Because of the dynamic nature of many projects, indices beyond the iron triangle are needed to identify potential issues. Further, because the importance of stakeholders may change during the life of the project, multiple indices are needed to address the possible issues associated with each.[55] The choice of these measures is important, because it communicates the priorities and expectations of the project to project managers and subsequently employees. Relevant measures that go beyond cost, time, and quality include safety measures, stakeholder satisfaction measures (e.g., customer satisfaction), efficient use of resources (e.g., costs of materials, tools, and equipment relative to productivity), reduced conflicts and disputes, and measures of management performance (e.g., absenteeism and turnover, employee performance measures, scheduling, team building, and reporting).

Although the stakeholder view of performance measurement has gained some support, most projects are still assessed on measures related to time, cost, or quality (such as defects and customer satisfaction).[56] These measures are frequently based on estimates using historical benchmarks. However, because of the dynamic nature of projects and the uncertainty inherent in oil and gas projects, the estimates used as benchmarks of cost and time may be inaccurate. Measures of cost may include labor-hour costs related to progress, resource costs, rework costs, total costs (labor, material, waste, equipment, etc.) related to progress, and actual costs related to budgeted costs. Time measures generally compare actual progress with scheduled progress. Progress is frequently related to achieving project milestones. Quality may be measured by rework costs, lost-time hours, tolerance levels, and so on. As shown in table 5–4, some of these measures are objective, and others are subjective.[57]

Table 5–4. Measures of project manager performance

	Objective Measures	Subjective Measures
Time Related	Progress/schedule Milestones/schedule Project cycle time Productivity	Key milestone achievements/time
Cost Related	Total costs Costs/budget Costs/progress Project profit Project net income Project return on investment Project contribution margin	Key milestone achievements/costs
Quality Related	Cost of quality/total costs Reworks Defects Error margins	Technical accomplishments Quality gates passed
Stakeholder Related	Market measures Client return rate	Customer satisfaction Responsiveness to customer Employee satisfaction Requirements performance Compliance performance Vendor progress
Safety Related	Total recordable injury rate Lost time injury frequency	Environmental management
Resource Efficiency Related	Overhead cost reduction	Resource utilization
Management Related	Turnover rate Absenteeism rate Percent overtime	Planning effectiveness Policy effectiveness Reporting effectiveness Team performance management Scheduling effectiveness Staffing effectiveness Training effectiveness Conflict resolution effectiveness

As with general project personnel, the project performance may not be entirely related to the performance of the project manager. Events beyond the control of project managers should not necessarily negatively affect their performance evaluation. Thus, consideration should be given for the difficulty of the job, the scope of the project, and the unanticipated changes that occurred on the project. Nevertheless, because project managers are responsible for overall project performance, it is reasonable that project performance measures should make up the major portion of a project manager's evaluation.

Who is involved in performance management in project-oriented firms?

The people involved in the performance evaluation differ between general project personnel and project managers. For project personnel, the functional supervisor of the employee is the one most likely to actually perform the performance appraisal. These appraisals frequently are based on considerable input from the project manager. This input may be formal or informal, and confidential or nonconfidential. Input from other team members may also help with the evaluation, particularly with respect to teamwork aspects of performance. Team member evaluations should be confidential or anonymous to get more honest feedback. Similar to 360 feedback appraisals, input should be sought from anyone who can make an accurate assessment of any relevant aspect of performance. This could include other functional managers, resource managers, customers, suppliers, and, when applicable, the employee's subordinates.

For project managers, the functional superior of the employee is the one most likely to actually perform the performance appraisal. Input may be sought from resource managers and general managers. Project team members are likely to have a considerable amount of useful information regarding the performance of the project manager. In this case, anonymity is even more important to ensure honest appraisals. Feedback from customers, clients, suppliers, and other stakeholders is likely to help obtain a more accurate assessment of the project manager's performance. However, pragmatically consideration must be given to the cost of obtaining the information, the time it will take, the reliability of the information, and the political implications of collecting the information before casting too wide a net to collect feedback from every source possible.[58]

What format should be used in project-oriented firms?

The formats used for the evaluation of project team members and project managers are generally the same as in typical firms as discussed in chapter 1. For example, essay-based appraisals are considered to be as ineffective in project-based firms as in other firms. There are, however, some differences in the usefulness of different formats in project-oriented companies. For example, ratings based on established standards tend to be less effective in project-oriented firms because of the dynamic nature of projects and the substantial differences among them.[59] Companies may develop a specific evaluation form just for project work. Although evaluations are typically done annually, in project-oriented firms, evaluations may be done after every project in addition to annually or instead of annually. A key difference in the application of any format is that in the case of multiple projects, an evaluation should be done for each project separately. In cases where the employee performed project and nonproject work during the evaluation period, there should be separate evaluations of the project work and nonproject work. Further, the overall weighing of these separate evaluations should reflect the amount of time spent on each project.

Maximizing the effectiveness of project management in project-oriented firms

As is generally the case, maximizing the effectiveness of performance management in project-oriented firms requires providing feedback; having accurate evaluations; including goal setting and rewards; and monitoring, evaluating, and adjusting the system. The nature of project-oriented companies makes some of these more difficult. Here are some examples:

- Feedback from the functional manager may be more difficult to provide in a timely manner if the functional manager is rarely at the work site of the project manager.
- Accurate evaluations may be more difficult given the moving target of project performance. Unanticipated difficulties, dynamic project environments, and increasingly large and risky projects in the oil and gas industry all make accurate forecasting more difficult, which in turn makes it hard to set accurate benchmarks. These aspects also make goal setting more difficult.
- Monitoring, evaluating, and adjusting are more important in project settings because of their dynamic nature. Flexibility in the evaluation will likely be necessary, because of external factors or new information learned during the project. Evaluations may also need to be modified to include unanticipated aspects of performance such as customer presentations or bid-proposal support.

Overall, the project environment makes some aspects of performance management easier and some more difficult. On the one hand, the definitive beginning and end of projects compartmentalize project performance and make its measurement and goal setting easier. On the other hand, the dynamic nature of projects makes these goals and metrics less stable.

Compensation in Projects

There are several differences regarding the effect on compensation and benefits between project-oriented companies and other companies.[60] First and foremost, the definitiveness of project completion and milestones create a favorable environment for rewards based on objective, concrete outcomes. Second, job descriptions, which drive base-pay levels, undervalue project-focused jobs because they tend to have less formal authority and fewer direct reports. As mentioned earlier with respect to performance measures, dual-reporting relationships may complicate the measurement of performance. Finally, project-oriented work, particularly in the oil and gas industry, frequently means extensive travel or living away from home, which generally affects benefits.

Pay level in project-oriented companies

In most companies, base pay is usually determined through a pay system that values jobs based on their job description. Jobs are valued based on their levels of responsibility, knowledge, skills, and abilities needed to do the job, education and experience needed to do the job, and other factors that the organization deems compensable. Jobs are then placed in pay grades relative to this value. These pay grades then establish a possible range of pay for anyone in that job.

Job descriptions in project management organizations may lead to issues in base pay for two reasons. First, in projects, jobs tend to have less formal authority in decision making and less formal authority over other employees, two job aspects that are usually tied to a job's value. Thus, jobs in projects may be devalued compared to purely functional jobs in the same or other organizations.[61] Second, the nature of the job can change substantially depending on the project. Thus, project-oriented companies have to either entail very vaguely written job descriptions that enable responsibilities to frequently change or frequently change the jobs of their employees, which affects salaries.[62] In addition, vaguely written job descriptions make salary determination more difficult, while changing jobs and consequently salaries can cause serious difficulties when the projects are short term or when salaries get lowered.

These issues suggest that companies must be mindful of not inadvertently having base pay be lower for project employees than functional employees. This will result in projects attracting less qualified employees or that projects may be seen as an inferior career path. Paying above the market rates will address this issue, at least when comparing pay with competitors, but of course will raise labor costs.

Pay mix in project-oriented companies

The pay mix in project-oriented companies is also affected by the differences mentioned above. In project-oriented companies, bonuses are likely to be a greater percentage of total pay for employees because a project environment provides a great opportunity to reward the achievement of objective outcomes such as project milestones and project completion. Such bonuses are likely to be highly motivational as long as the employee's individual performance has a noticeable effect on the outcome. Relatedly, the bonus-to-total-pay ratio is likely to be substantially higher for project managers than for others on the project team because the project manager will presumably have more influence on the project's outcome than other team members. Benefits are also likely to be a greater percentage of total pay because a project environment (particularly in the oil and gas industry) tends to require greater hardships such as frequent travel, foreign locations, difficult locations, and so on. Thus, the hardship premiums, travel allowances, or foreign location allowances discussed in chapter 3 are more likely to come into play.

Performance-based pay in project-oriented companies

As discussed above, because of the definitive and time-bound nature of projects, performance-based pay in the form of bonuses and recognition awards are particularly well-suited for project-oriented companies.

Bonuses. These bonuses can be tied to objective project milestones and outcomes, or more subjective aspects of performance as measured in a performance evaluation. Although objective measures at group levels are easier to come by in a project environment, subjective measures at the individual level tend to be harder to quantify because of the dynamic and interdependent nature of the work in projects.

For project managers, bonuses should be tied very closely to project performance, as described earlier with respect to the performance management of project managers. Objective measures are generally considered superior to subjective measures, but subjective measures are more likely to take into consideration extenuating circumstances that were no fault of the project manager.

There is no best way to arrive at the criteria to use to determine the type of bonus to award team members or project managers. However, as previously mentioned, projects tend to foster easier project- and team-level measures. Yet, there are still trade-offs to using project- versus team- versus individual-level measures. Various factors can help determine which type of criteria

would be most useful. Thus, the following rules of thumb should be considered when determining the criteria for bonuses in project environments:
- Project- and team-based bonuses are more effective when team members have a direct impact on the measured outcome.
- Project- and team-based bonuses tend to foster cooperation among the team members, while individual-based bonuses tend to foster competition among the team members.
- Project-based (and to a lesser extent team-based) bonuses may lead to more shirking than individual-based bonuses, particularly for large projects or teams.
- Objective project, team, or individual goals will be more accepted and seen as fairer if the team (or individual) has some input in the determination of the goal.
- Goals should be recalibrated if they are no longer feasible because of factors beyond the control of the team.

Overall, in a project environment, project- and team-based bonuses generally tend to be the most effective way to motivate and reward employees, although the caveats noted above should be kept in mind.

As with choosing the type of bonus, described above, there is no best way to determine the most appropriate size of the bonus. Again, some rules of thumb should be kept in mind when determining the size of the bonus. These include:
- Team-based bonuses are more effective when all team members get a similarly sized bonus, as opposed to some team members getting substantially greater bonuses compared to others.
- Any differences in bonus size among team members should be based on either salary level (the bonus is a similar percentage of salary) or contribution to the outcome.
- Larger bonuses will be more motivational than small bonuses, but are more likely to encourage taking shortcuts and engaging in unethical behavior to achieve the established goal.
- Larger projects should be tied to larger bonuses.

Thus, the size of bonuses should be determined with three questions in mind: Is the bonus big enough to matter? Is the bonus so big that it will lead to undesirable consequences? Is the bonus fair compared to the bonuses others are getting?

By now, it should be clear that there are many positives to using performance-based pay for project members and project managers. The goal of every project consists of outcomes that are on time and on budget. Incentives to achieve these outcomes can be powerful motivators for the various individuals involved in the management of the project. However, all the caveats mentioned above are important. Firms must recognize that there are risks associated with performance-based pay, which is why some of the largest oil and gas companies do not use variable compensation for their project managers. Enron's use of performance-based pay is an example of what can happen when there is a misalignment between project managers and operators. When Enron was created in the late 1980s, it was a gas distribution company. Enron decided to enter the international energy infrastructure business. It created multifunctional development teams that were compensated, in part, based on large incentive payments tied to the net present value of the project itself. These payments were staggered—half would be paid when the project financing was lined up and half when the project was completed. After projects were completed, operating responsibility shifted from the development team to an Enron operating group (for projects that Enron built and operated).

Enron described its international development business as having "an at-risk, entrepreneurial culture." Project managers (in this case, more accurately titled project developers) were at risk because if they did not develop new projects, their compensation could be quite low relative to that of their colleagues. The result was that Enron developers were willing to take on highly risky projects in order to earn their incentive bonuses. When these projects were turned over to Enron's operators, they were often flawed and required expensive remediation. By this point, the developers had moved on to new projects and were not accountable to the operators. By the time Enron went bankrupt, company management understood that their compensation system was deeply flawed and was creating suboptimization between the various project stakeholders.

In summary, the risks of using high levels of performance-based pay for project managers include the following:

- High pay may create a short-term "complete the project and move on" mentality with project developers.
- High pay may not build a shared interest between developers and operators.
- Projects transferred to operators may be defective because developers are too focused on completion bonuses.
- Developers may not make the effort to fully understand the broader value chain because they are compensated only for the project development.

- Developers may take shortcuts that impact process or personal safety.

Thus, although bonuses are highly motivational, they may encourage unwanted behaviors that lead to unintended negative consequences.

Recognition awards. The well-defined nature of projects also makes recognition awards particularly well-suited for project-oriented companies. Recognition awards are given in general for the achievement of milestones, and well-defined milestones are prevalent in projects. As we explained earlier in the chapter and as shown earlier in figure 5–1, every project proceeds through a common set of phases for which specific milestones can be identified.

As mentioned in chapter 1, recognition awards are an inexpensive way to reward employees for notable achievements. They are usually appreciated by employees substantially more than their cost (and effort) warrants. Thus, recognition awards should be carefully considered by every manager in a project environment. Plaques, certificates, or medals coupled with token signs of appreciation such a gift cards, meals, or merchandise may all be used effectively as a way to recognize milestone achievements. Recognition may also be used in the presence of a milestone-based bonus; however, when coupled with a bonus, tokens such as gift cards are not necessary. A good rule of thumb with recognition awards is if the achievement warrants recognition, you should recognize it!

Benefits in project-oriented companies

Benefits in project-oriented companies generally are similar to those in other companies. However, there are exceptions. When projects are in foreign countries, as is typical in the oil and gas industry, the benefits mentioned in chapter 3 come into play. Also, some projects may not be international, but still require considerable travel. To keep up morale, companies should seriously consider providing the following travel-related benefits, given that they are offered by most companies:[63]

- Reimbursement for transportation to and from airport.
- Per diem allowance or reimbursement for meals.
- Employees allowed to keep hotel points and flyer miles.
- Reimbursement of costs of Internet access.

Further, to increase the desirability of projects, and to keep employees who working on such projects happier, companies should consider providing

the following travel-related benefits, even though they are not offered by most companies:
- Reimbursements for personal phone calls.
- Travel accident insurance.
- Rental car or airfare upgrades.
- Paid dry cleaning, mini-bar snacks, or health club fees.
- Paid travel expenses for spouse.
- Paid airline club membership.

Overall, companies should consider providing any benefit that would offset costs (financial or otherwise) that employees may incur because of their participation in a project.

Maximizing the effectiveness of compensation in project-oriented companies

As is generally true, perceived fairness in compensation is important to maximize the effectiveness of compensation in project-oriented companies. Two new issues of fairness are exclusive to project-oriented companies. The first is the fairness of pay of employees assigned to projects relative to functional employees not assigned to projects. As previously discussed, salaries for project personnel may be less because of less formal authority over resources and direct reports than traditional functional jobs, as mentioned earlier. Nevertheless, because of additional other responsibilities and additional skills, knowledge, and abilities needed for them, project positions should not be seen as or paid as second-class positions. The second, as discussed earlier, is the fairness of project-based bonuses compared to the bonuses of others on and off projects. Again, every effort should be made so that the rewards are commensurate with the efforts and achievements of each team and, to a somewhat lesser extent, each individual.

Conclusion

Companies in the oil and gas industry are frequently project-oriented. In this chapter, we showed how staffing, training, performance management, and compensation differ in project-oriented companies compared to other companies. These differences are generally attributable to the application of HR functions specifically to the job of project manager and the application of HR functions within the project, in addition to within the organization. Differences in the HR function in project-oriented firms include determining assignments to projects, training for a specific project, performance management of the project, and compensation related to projects. These differences—and adapting HR to them—can have a considerable impact on projects and project-oriented firms in the oil and gas industry.

References

1. Badiru, A. B., & Osisanya, S. O. (2013). *Project management for the oil and gas industry: A world system approach*. Boca Raton, FL.: CRC Press.
2. CNN Money. (2012). *10 Most expensive energy projects in the world*.
3. Nova, R., and Rivolta, T. (2013). Large Project Management in Oil and Gas. Bain Brief. http://www.bain.com/Images/BAIN_BRIEF_Large_project_management_in_oil_and_gas.pdf.
4. Jergeas, G. (2008), Analysis of the front-end loading of Alberta mega oil sands projects. *Project Management Journal, 39*(4), 95–104. Retrieved from http://money.cnn.com/gallery/news/economy/2012/08/27/expensive-energy-projects/.
5. Lee, J. (2014), *Final investment decision: The big breakthrough*. Juneau, Alaska: Office of the Federal Coordinator.
6. Badiru, A. B., & Osisanya, S. O. (2013). *Project management for the oil and gas industry: A world system approach*. Boca Raton, FL: CRC Press.
7. Belout, A., & Gauvreau, C. (2004). Factors influencing project success: The impact of human resource management. International Journal of Project Management, 22(1), 1–11.
   Belout, A. (1998). Effects of human resource management on project effectiveness and success: Toward a new conceptual framework. *International Journal of Project Management, 16*(1), 21–26.
   Söderlund, J., & Bredin, K. (2006). HRM in project-intensive firms: Changes and challenges. *Human Resource Management, 45*(2), 249–265.
8. McKenna, M. G., Wilcynski, H., & VanderSchee, D. (2006). *Capital project execution in the oil and gas industry*. Booz, Allen, Hamilton. Retrieved from http://www.boozallen.com/media/file/Capital_Project_Execution.pdf.
9. Huemann, M., Keegan, A., & Turner, J. R. (2007). Human resource management in the project-oriented company: A review. *International Journal of Project Management, 25*(3), 315–323.
10. Fabi, B., & Pettersen, N. (1992). Human resource management practices in project management. *International Journal of Project Management, 10*(2), 81–88.
    Kerzner, H. R. (2013). *Project management: A systems approach to planning, scheduling, and controlling*. New York, NY: Wiley.
11. Turner, R., Huemann, M., & Keegan, A. (2008). Human resource management in the project-oriented organization: Employee well-being and ethical treatment. *International Journal of Project Management, 26*(5), 577–585.
12. McComb, S. A., Green, S. G., & Dale Compton, W. (2007). Team flexibility's relationship to staffing and performance in complex projects: An empirical analysis. *Journal of Engineering and Technology Management, 24*(4), 293–313.

13. Keegan, A., & Turner, J. R. (2003). Managing human resources in the project-based organization. In J. R. Turner (ed.), *People in project management* (pp. 1-12). Aldershot, UK: Gower.
14. Keegan, A., & Turner, J. R. (2003). Managing human resources in the project-based organization. In J. R. Turner (ed.), *People in project management* (pp. 1-12). Aldershot, UK: Gower.
15. ShaleNET. (2013). *A guide to careers in the oil and natural gas industry*. Retrieved from http://careerguide.shalenet.org/Content/guides/CareerGuide2013.pdf.
16. Fabi, B., & Pettersen, N. (1992). Human resource management practices in project management. *International Journal of Project Management, 10*(2), 81-88.
17. Fisher, E. (2011). What practitioners consider to be the skills and behaviours of an effective people project manager. *International Journal of Project Management, 29*(8), 994-1002.
    El-Sabaa, S. (2001). The skills and career path of an effective project manager. *International Journal of Project Management, 19*(1), 1-7.
    Pettersen, N. (1991). What do we know about the effective project manager? *International Journal of Project Management, 9*(2), 99-104.
    Kerzner, H. R. (2013). *Project management: A systems approach to planning, scheduling, and controlling*. New York, NY: Wiley.
18. El-Sabaa, S. (2001). The skills and career path of an effective project manager. *International Journal of Project Management, 19*(1), 1-7.
    Pettersen, N. (1991). What do we know about the effective project manager? *International Journal of Project Management, 9*(2), 99-104.
19. Pinkerton, W. (2003). *Project management: Achieving project bottom-line succe$$*. New York, NY: McGraw Hill Professional.
20. Keegan, A., & Turner, J. R. (2003). Managing human resources in the project-based organization. In J. R. Turner (ed.), *People in project management* (pp. 1-12). Aldershot, UK: Gower.
21. Maurer, I. (2010). How to build trust in inter-organizational projects: The impact of project staffing and project rewards on the formation of trust, knowledge acquisition and product innovation. *International Journal of Project Management, 28*(7), 629-637.
22. Dinsmore, P. C. (1990). *Human factors in project management*. New York, NY: American Management Association.
23. Hill, G. M. (2007). *The complete project management office handbook*. Boca Raton, FL: CRC Press.
24. Heimerl, C., & Kolisch, R. (2010). Scheduling and staffing multiple projects with a multi-skilled workforce. *OR spectrum, 32*(2), 343-368.
25. Engwall, M., & Jerbrant, A. (2003). The resource allocation syndrome: The prime challenge of multi-project management? *International Journal of Project Management, 21*(6), 403-409.
26. Crawford, L. H. (2003). Assessing and developing the project management competence of individuals. In J. R. Turner (ed.), *People in project management*, p. 9-34. Aldershot, UK: Gower.
27. Crawford, L. (2005). Senior management perceptions of project management competence. *International Journal of Project Management, 23*(1), 7-16.
28. Kerzner, H. R. (2014). *Project management-best practices: Achieving global excellence*. New York, NY: Wiley.

29. Fabi, B., & Pettersen, N. (1992). Human resource management practices in project management. *International Journal of Project Management, 10*(2), 81–88.
30. Pinkerton, W. (2003). *Project management: Achieving project bottom-line succe$$*. New York, NY: McGraw Hill Professional.
31. Hill, G. M. (2007). *The complete project management office handbook*. Boca Raton, FL: CRC Press.
32. Fabi, B., & Pettersen, N. (1992). Human resource management practices in project management. *International Journal of Project Management, 10*(2), 81–88.
33. Kerzner, H. R. (2013). *Project management: A systems approach to planning, scheduling, and controlling*. New York, NY: Wiley.
34. Dinsmore, P. C. (1990). *Human factors in project management*. New York, NY: American Management Association.
35. O*net online. (n.d.). *Summary report for architectural and engineering managers*. Retrieved from http://www.onetonline.org/link/summary/11-9041.00.
36. Zwikael, O. (2009). The relative importance of the PMBOK® Guide's nine knowledge areas during project planning. *Project Management Journal, 40*(4), 94–103.
37. Hodgson, D., & Cicmil, S. (2006). Are projects real? The PMBOK and the legitimation of project management knowledge. In D. Hodgson & S. Cicmil (Eds.), *Making projects critical*, 29-50. New York, NY: Palgrave MacMillan.
38. Cooper, R. G. (2008). Perspective: The Stage-Gate® idea-to-launch process—Update, what's new, and NexGen systems*. *Journal of Product Innovation Management, 25*(3), 213–232.
    Cooper, R. G., Edgett, S. J., & Kleinschmidt, E. J. (2002). Optimizing the stage-gate process: What best-practice companies do—I. *Research-Technology Management, 45*(5), 21–27.
39. Pant, I., & Baroudi, B. (2008). Project management education: The human skills imperative. International Journal of Project Management, 26(2), 124–128.
40. Dinsmore, P. C. (1990). *Human factors in project management*. New York, NY: American Management Association.
41. Pinkerton, W. (2003). *Project management: Achieving project bottom-line succe$$*. New York, NY: McGraw Hill Professional.
42. Dai, C. X., & Wells, W. G. (2004). An exploration of project management office features and their relationship to project performance. *International Journal of Project Management, 22*(7), 523–532.
43. Turner, R., Keegan, A., & Crawford, L. (2003). Delivering improved project management maturity through experiential learning. In J. R. Turner (ed.), *People in project management* (pp. 45–63). Aldershot, UK: Gower.
44. Kerzner, H. R. (2013). *Project management: A systems approach to planning, scheduling, and controlling*. New York, NY: Wiley.
45. Tjahjana, L., Dwyer, P., & Habib, M. (2009). *The program management office advantage*. New York, NY: AMACOM.
46. Thomas, J., & Mengel, T. (2008). Preparing project managers to deal with complexity: Advanced project management education. *International Journal of Project Management, 26*(3), 304–315.

47. Tabassi, A. A., Ramli, M., & Bakar, A. H. A. (2012). Effects of training and motivation practices on teamwork improvement and task efficiency: The case of construction firms. *International Journal of Project Management, 30*(2), 213–224.
48. Hill, G. M. (2007). *The complete project management office handbook*. Boca Raton, FL: CRC Press.
    Loo, R. (1996). Training in project management: A powerful tool for improving individual and team performance. *Team Performance Management, 2*(3), 6–14.
49. Kerzner, H. R. (2013). *Project management: A systems approach to planning, scheduling, and controlling*. New York, NY:Wiley.
50. Kerzner, H. R. (2013). *Project management: A systems approach to planning, scheduling, and controlling*. New York, NY: Wiley.
51. Atkinson, R. (1999). Project management: Cost, time and quality, two best guesses and a phenomenon—It's time to accept other success criteria. *International Journal of Project Management, 17*(6), 337–342.
52. Bassioni, H. A., Price, A. D. F., & Hassan, T. M. (2004). Performance measurement in construction. *Journal of Management in Engineering, 20*(2), 42–50.
53. Atkinson, R. (1999). Project management: Cost, time and quality, two best guesses and a phenomenon—It's time to accept other success criteria. *International Journal of Project Management, 17*(6), 337–342.
54. Kaplan, R. S., & Norton, D. P. (1996). Using the balanced scorecard as a strategic management system. *Harvard Business Review, 74*(1), 75–85.
55. Marques, G., Gourc, D., & Lauras, M. (2011). Multi-criteria performance analysis for decision making in project management. *International Journal of Project Management, 29*(8), 1057–1069.
56. Bassioni, H. A., Price, A. D. F., & Hassan, T. M. (2004). Performance measurement in construction. *Journal of Management in Engineering, 20*(2), 42–50.
    Cox, R. F., Issa, R. R., & Ahrens, D. (2003). Management's perception of key performance indicators for construction. *Journal of Construction Engineering and Management, 129*(2), 142–151.
57. Toor, S. U. R., & Ogunlana, S. O. (2010). Beyond the "iron triangle": stakeholder perception of key performance indicators (KPIs) for large-scale public sector development projects. *International Journal of Project Management, 28*(3), 228–236.
    Kerzner, H. R. (2013). *Project management: A systems approach to planning, scheduling, and controlling*. New York, NY: Wiley.
    Center for Business Practices. (2005). *Measures of project management performance and value*. Retrieved from http://www.pmsolutions.com/audio/PM_Performance_and_Value_List_of_Measures.pdf.
58. Martin, V. (2006). *Managing projects in human resources, training and development*. London, UK: Kogan Page.
59. Kerzner, H. R. (2013). *Project management: A systems approach to planning, scheduling, and controlling*. New York, NY: Wiley.
60. Turner, R., Huemann, M., & Keegan, A. (2007). *Human resource management in the project-oriented organization*. Newtown Square, PA: Project Management Institute.
    Kerzner, H. R. (2013). *Project management: A systems approach to planning, scheduling, and controlling*. New York, NY: Wiley.

61. Kerzner, H. R. (2013). *Project management: A systems approach to planning, scheduling, and controlling*. New York, NY: Wiley.
62. Fabi, B., & Pettersen, N. (1992). Human resource management practices in project management. *International Journal of Project Management, 10*(2), 81–88.
63. Society for Human Resource Management. (2013). *2013 Employee benefits*. Retrieved from www.shrmstore.shrm.org.

6

The Unconventional Workforce of the Oil and Gas Industry

The Workforce of the Oil and Gas Industry

The workforce of the oil and gas industry has a number of features that are unconventional compared with the workforce of most other industries. Although some of these features also apply to other international high-capital-intensive industries such as mining, taken together, they paint a picture of a workforce that is very different from that of most other industries and at least somewhat different from any other. These include the prevalence of contractors, the use of rotators, an aging workforce, a male-dominated workforce, difficult work conditions, and an impending skills gap. Although these factors vary somewhat by country, globally they are relatively stable. These features have implications for the staffing, training, performance management, and compensation of employees. Before we discuss the implications of these features for the management of human resources (HR), we provide some detail of each of these features that make the industry so different from most others.

The Prevalence of Contractors

In the oil and gas industry the term *contractor* means a firm that is contracted to provide a product or service to another oil and gas firm. A subcontractor is a contractor to a contractor. Contract workers can cover a broad range of employment relationships, such as:

- Short-term workers with a direct or commercial contract
- Part-time workers with a direct or commercial contract

- Agency workers
- Day laborers
- Informal workers
- Workers in any employment relationship with a contractor or subcontractor to an oil and gas company[1]

Thus, contractors can be independent contractors; part of a service team; or part of supplier-provided maintenance, engineering, or technical teams. Although we lump all these types together for ease of discussion, the type of employment relationship even within the general umbrella term of *contractors* is likely to have implications for the management of workers.

Contractor firms as well as individual contract employees work in all aspects of the oil and gas value chain. The use of contractors in the oil and gas industry has been a characteristic in the industry since the earliest days, but has increased a great deal over the last 15 years. Figure 6–1 shows the hours worked by contractors and company employees across 107 countries from 1985 to 2012. The figure shows that about two-thirds of the hours worked in 1985 were by company employees, while this number dropped to less than one quarter of the hours worked in 2012. The large presence of contract workers in oil and gas is a global phenomenon and is as high as 100% in some countries (e.g., Russian labor law requires all oil and gas industry workers to be on contracts). Contractors are prevalent in exploration, drilling, production, construction, transport, and catering upstream. They are prevalent downstream in all aspects of refinery and production work, including planning, building, equipping, and maintaining facilities.[2] The largest contract firms design and build facilities or provide oil field services or technologies. Table 6–1 lists 10 of the largest oil field service or technology providers and 10 of the largest construction companies in the oil and gas industry.[3] Interestingly, almost all the service or technology providers are headquartered in the United States, specifically in Houston, Texas.

There are several reasons why contractors are so prevalent in the oil and gas industry:

- They provide flexibility in talent.
- They provide cost flexibility.
- They allow fast adjustments to short-term needs.
- They have a specialized skill set that is not required long term.
- They are highly skilled and may prefer to work as contractors or consultants.
- They often cost less.

- They can be evaluated in the short term and hired full-time later.
- They can bring insights from outside the organization.
- They may reduce the company's liability.
- They provide a valuable service, given that no single company can perform all activities across the oil and gas value industry chain.

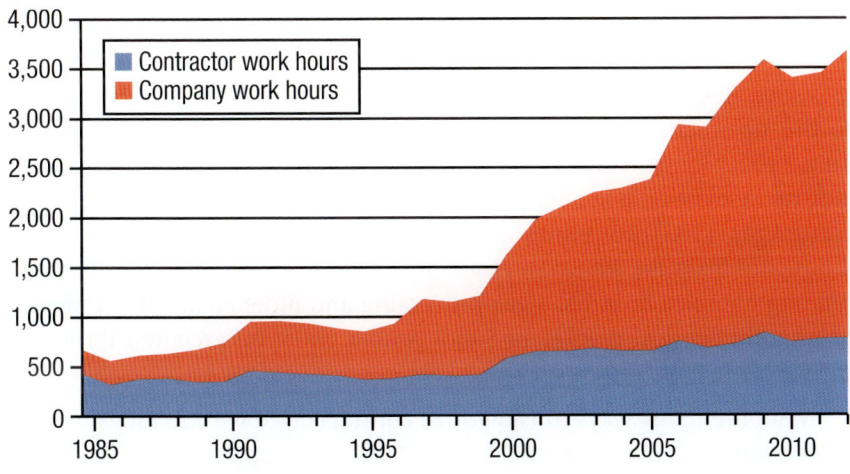

Fig. 6–1. Trends in global oil and gas industry work hours (in millions of hours worked)

Source: Safety Performance Indicators—2012 Data, Oil & Gas Producers Association, Report No. 2012s, June 2013, p. 1–1. 2012 data represents 107 countries and a total of 3,691 million hours of work data.

Table 6–1. Large oil and gas contractor firms

Construction Firms	Oilfield Services and Technologies
Bechtel (USA)	Schlumberger Ltd (USA)
Technip (France)	Halliburton (USA)
Aker Solutions (Norway)	Saipem (Italy)
Chiyoda Corporation (Japan)	Transocean Ltd (USA)
SNC-Lavalin Group (Canada)	Baker Hughes (USA)
J. Ray McDermott (USA)	Fluor (USA)
JGC Corporation (Japan)	BJ Services Company (USA)
Hyundai Heavy Industries (S. Korea)	Smith International (USA)
Foster Wheeler (USA)	National Oilwell Varco (USA)
Deailim Industrial Co. (S. Korea)	Weatherford International (USA)

Source: Arabianoilandgas.com.

The popularity of contractors is a result of the value of the above benefits and their costs, which may depend on the type of work performed, the location, and the availability of labor. Contracts are particularly common in projects, as discussed in chapter 5.

Construction Contracts for Oil and Gas Projects

As we discussed in the previous chapter, oil and gas projects, whether in the upstream or downstream, are typically complex, expensive, and long-lived. Construction of the project, whether it is a floating, offshore jack-up rig to be built in Singapore and operated in the North Sea, a liquefied natural gas (LNG) processing facility (a train) off the north coast of Australia, a pipeline from the Peruvian cloud forest over the Andes to the Pacific, a refinery complex in India, a gas-processing and pipeline facility in Algeria, or a petrochemical cracker in Saudi Arabia, must be executed through a construction contract.

There are many ways to categorize contracts, but we will focus on two: ownership and management. Each of the different contract structures dictates organizational arrangements, and that often means very different professional and contractual staffing and HR competencies.

Contracts based on ownership

This type of contract is generally based on a private company (i.e., a nongovernmental body) owning the project from construction—*build*—through operation. In many cases, the firm may design, construct, and then operate the facility under a long-term contract with a host government. In other cases, once construction is completed, ownership is transferred to the end-user owner, often a government. And with all large projects around the world, who is responsible for which part of financing is always of interest and the object of negotiations.

Some examples of these ownership-based construction contracts are build, own, operate (BOO); build, own, operate, transfer (BOOT); build, operate, transfer (BOT); and design, build, finance, operate (DBFO). Regardless of the specific form, all of these require a level of capital investment for ownership that is beyond what most of the oil and gas industry today can support.

Contracts based on management

Contracts based on engineering and construction management, by far the most commonly used in the global oil and gas industry today, are based on a contractual construction arrangement, in which the contractor will not own the asset constructed. Many of these projects today are *turnkey*, meaning that the contractor delivers to the owner a facility, project, or plant that is fully functional and ready for operation. Currently, the two most common forms of this type of contract are engineering, procurement, and construction (EPC) and engineering, procurement, construction, and management (EPCM).[4]

EPC. Under an EPC contract, the contractor provides the engineering design of the project, purchases (procures) all of the necessary equipment and material, and constructs and delivers a functioning asset.[5] The EPC contractor is responsible for the schedule, coordination of all interested parties and subcontractors, and on-time completion.[6] Under an EPC, the contractor performs many construction activities itself. This is the single most common contract used in the global oil and gas industry today.

EPC contracts can be either *cost-reimbursable* or *fixed-price, lump-sum* (or in some cases, *lump-sum, turnkey*). The financial risks to the contractor are obviously significantly higher with the fixed-price, lump-sum contract, as potential cost overruns or design or technology flaws reduce potential profitability. Cost-reimbursable contracts shift much of the financial burden and risks associated with project completion to the owner.

The EPC world of firms, like the upstream of the oil and gas industry itself, is essentially two-tiered, with the top tier made up of a small number of global majors—Fluor, Bechtel, Jacobs—and the second tier composed of more specialized service, industry, geographic companies.

EPCM. Under an EPCM contract, the contractor performs all of the EPC duties with the important exception that the contractor does not perform any of the construction activities itself. Instead, the contractor acts on behalf of the owner, managing the subcontracting and construction activities.

The EPCM contract is in essence a services agreement; the contractor manages others, is responsible for coordinating the schedule and for safety during construction, but does not perform construction activities itself. As a result, EPCM contracts are fee-based, such that the EPCM contractor earns a series of fees for services rendered.

Contractors and Safety

As discussed in chapter 4, the use of contractors entails particular safety and health concerns. There is evidence that contractors in the oil and gas industry tend to have a higher likelihood of being involved in accidents both upstream and downstream. Some possible reasons for this difference are that companies are more reluctant to invest in the training of short-term employees (of the contractor), contractors are less familiar with the company's safety procedures, contractors may be less committed to the company's safety procedures, contractors may be less familiar with the specific workplace, and contract workers are more likely to be in the most dangerous jobs. Companies generally use one of three following standards regarding health, safety, and environment (HSE) policies when working with contractors:

- The contractor is under the direct supervision of the company and follows its HSE policies.
- The contractor follows its own HSE policies, which are verified by the company.
- The contractor follows its own HSE policies, which are not verified by the company.

Which one of these standards to use depends on company policies, legal requirements, the type of contractor, and the location of the contractor. The standards to be used are typically specified in the contract. Before they can be hired, contractors will usually have to prequalify based on defined safety factors that include the following:[7]

- The contractor's commitment to safety, as demonstrated by an ongoing safety program that is supported by its top management.
- The completeness of the contractor's safety programs and their appropriateness for the work site and the safety standards of the owner.
- The contractor's safety staffing plan. The plan describes the on-site person or persons appointed by the contractor and subcontractor who will be responsible for safety. It also describes their expertise and authority.
- A description of the safety orientation program to be provided by the contractor to all contractor employees on-site.
- The contractor's enforcement and disciplinary action program regarding safety violations.
- The contractor's policy and programs regarding alcohol, controlled substances, and firearms.

- A description of the contractor's employee training program.
- A description of the contractor's employee selection process and whether it includes security screening.

Note that although companies have instituted a number of policies to improve the safety record of contractors, the accident rate of contractors continues to be higher than that of company employees.[8]

Rotators in the Oil and Gas Industry

The oil and gas industry upstream, that part of the industry that is about the exploration, development, and production of the base hydrocarbon, often finds itself at the ends of the Earth. As upstream operators say, "You go where the oil is." And that is often in Papua New Guinea, the arctic, the frozen shallow waters of the north part of the Caspian Sea, far offshore in the South Pacific or Brazil, or in deepest Sub-Saharan Africa. Finding the skills and knowledge locally in these remote locations is difficult. The result has been one of the more unique HR solutions—the use of the rotational employee, or rotator.

Many industries have been confronted with a similar staffing dilemma for centuries. The solutions, as described in chapter 3, are typically the use of expatriates and the continued long-term development of locals to eventually move into these key skilled roles. But the global oil and gas industry is specifically challenged in even these solutions due to three key factors:

1. The extreme remoteness and complexity of many operational sites on- and offshore
2. The technical complexity and experience needs of the skills required
3. The magnitude of at-risk capital, sometimes for decades

The rotator has continued to be the solution.

Rotators defined

Rotational employees are workers who regularly commute to a work site on a fixed schedule, such as 28 days on and 28 days off, or 28/28 (which is the most common; others include 35/35 and 21/21 days). Although globally the focus of much attention in the industry is on the cross-border rotational employee (e.g., a petroleum engineer who lives in the United States but rotates regularly to Equatorial Guinea in West Africa), rotators often work in the same country in which they reside (e.g., a systems analyst who lives in Arizona who commutes regularly to the North Slope of Alaska). When working, rotators often work seven days a week and live in compounds or company-sponsored structures where transportation, housing, and meals are provided and subsidized or paid for via a per diem compensation program. They then return to their home country for an equal amount of time (e.g., 28 days), during which they are in principle completely off (no work requirements).[9]

The rotator is therefore quite different from the traditional expatriate or parent-country national (PCN). Expatriates usually become long-term residents of the foreign country, living and working full-time in the host country. Rotators are more similar to full-time, long-distance commuters.[10] They do not take their families and do not establish permanent housing other than that provided on-site by the employer. But they do spend roughly half their lives living in host countries, entailing large salary requirements on employers (with sizable country premiums in many cases), as well as travel and accommodation expenses (frequent round-trip airfares for their monthly commutes).

Skills and roles also differ between PCNs and rotators. The use of PCNs has declined in recent years as more and more companies have increased their localization efforts and recruited, developed, and retained more local talent. Most oil-producing countries have targets for the percentage of local employees. In many cases, PCN use continues only at the top of the organization, at the senior management and leadership levels. But, in the upstream oil and gas industry, there is a large-scale need for highly skilled and experienced employees, which may not be met through local hiring for decades—if ever. Many of today's new oil and gas (specifically, LNG facilities) developments are located in remote areas where there is no existing population —and there will never be, other than those needed for the oil and gas development. As a result, it is not unusual to find professionals who have been rotators for oil and gas developments in Sub-Saharan Africa, for example, for 20 years or more.

The use of rotators also creates an additional employer burden not incurred through traditional expatriate or commuter employment: rotator substitution. Because the rotator's role is a technical and critical one, when one worker rotates out, a parallel replacement or substitute must rotate in. This means that most rotator employment positions have two people of equal skill and competence, as well as a workflow that ensures an effective handoff between rotations.

Current rotator employment challenges

Unlike many other forms of expatriate employees, the use of the rotator is on the rise, not the decline.[11] All indicators and surveys indicate a growing need for rotators as the global oil and gas upstream goes further afield in search of hydrocarbons. As rotator employment continues to rise, many companies have aggressively moved to control these activities more from corporate headquarters in an attempt to manage the risks and challenges of rotator use more formally. In addition to simply tracking rotator use more closely, companies are facing rising complexity over work permits, immigration restrictions, and security clearances and concerns—in short, every dimension of compliance, control, and governance.

Although there is a general movement toward centralized policy and program management across the industry, there is a clear distinction between subsegments. Roughly 80% of all upstream firms use their centralized global mobility function, whereas most of the oil field services firms, drilling firms, and EPC firms split their global management of rotators between HR and the individual business units. Regardless, as the rules surrounding foreign national employment in host countries continue to grow and become more onerous, centralized management and governance are clearly the trend.

In addition to the organizational employment challenges of increased rotator use, there is also growing concern over a talent and willingness shortfall. The rotator is a unique combination of skill, experience, and willingness to spend roughly half of their life in remote work environments far from home—and, some would also say, a sense of adventure. Although often enjoying significant country and employment premiums, money is often not enough to expand the available talent pool for recruitment. In addition, as industry demands have grown for experienced technical rotators, retention is a growing concern. See Current Challenge Box 6-1 for a more detailed discussion of this issue.

Current Challenge Box 6–1. Retaining Talent

In a labor shortage situation, it is more important than ever to keep the talent you have. A recent survey found that almost 9 out of 10 upstream workers are likely to remain within the oil and gas industry, but only 6 out of 10 believe that they are likely to remain with their current organization.[12] The same survey found that only one-third of upstream workers thought that their organization used programs or incentives to retain productive workers. Both technical and managerial employees believed that compensation and career growth were by far the two most important reasons that attracted them to their current employer. Relatedly, a recent global oil and gas survey found that more than half of the respondents believed that a lack of training and development opportunities would lead them to consider leaving their employers.[13] Thus, compensation and development should be the pillars of any retention strategy.

However, other factors matter too. Benefits, flexibility, a great culture, small perks, career planning, promotion from within, opportunities to work abroad, and support and recognition from managers have all been considered important by at least some employees. Be an employer of choice by focusing on the little things as well as the big ones. Recognition is relatively cheap but seen as important. Address employee concerns when possible. Tailor rewards and retention strategies to individual employee needs. For example, cultural differences are likely to change the factors that are important for retention. Thus, what works in the United States may not work in the United Arab Emirates. Use different retention strategies in different places.

Another way to increase retention is to have plans in place that will motivate valued employees to delay retirement. Flexible work arrangements, additional bonuses, changing job status (part-time, consultant, contractor, etc.), and being given the opportunity to mentor others have all been found to persuade employees to delay retirement in the oil and gas industry.[14]

Finally, the importance of rotators means that they deserve special attention with respect to retention. Yet, many industry employers report that little current effort is being made to retain rotators, including no real effort at transitioning rotators into nonrotational positions. As illustrated by figure 6–10, the primary concern going forward seems to be the loss of rotators to competitors while on assignment. The predominance of this concern emphasizes the unique skill set of the rotator and may also signify the added value an individual rotator gains by operational experience on the ground (or above the water) in some of the world's most challenging oil and gas developments at the far ends of the earth. Thus, rotators, like all valued employees, should be the focus of a retention strategy that is proactive in retaining those employees who would be the most difficult, and expensive, to replace.

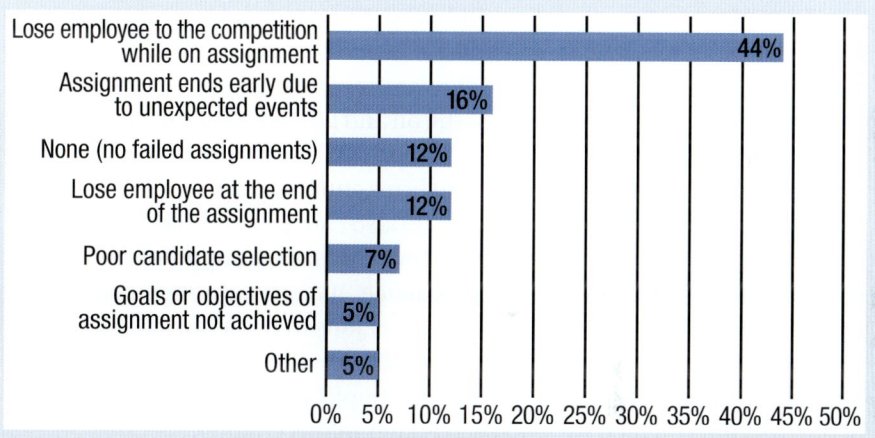

Fig. 6–10. Causes of failed rotational assignments
Source: Data drawn from *Oil & Gas Industry rotator assignment survey report*, Deloitte, 2013.

Characteristics of the Current Workforce

An aging workforce

The workforce in the United States is generally getting older. For example, US employment of those 65 and older increased by 101% from 1977 to 2007, while total employment of all ages increased only 59%. In the United States, the percentage of workers 55 and older is expected to grow from 18% in 2008 to 24% in 2018.[15] Further, this is a global phenomenon. A recent UN report details the global nature of this phenomenon and estimates that the median age of the world population will increase by seven years between 2010 and 2050.[16] The oil and gas industry appears to be particularly affected by this phenomenon. For example, the median age in petroleum refining is 44, two years older than the median age of the US overall workforce. Some surveys indicate that the average age of workers in the oil and gas industry may be as high as 49 or 50. Workers hired in the 1980s are now poised for retirement. During the 1990s oil prices were low, and hiring of petroleum engineers and geoscientists slowed down, which means that the pool of replacements is not as deep as it should be. This has created an age gap, which we will discuss later in this chapter. The aging workforce in the oil and gas industry also applies globally. In Bahrain, the average age of the oil and gas workforce is 50 ; in India, it is over 50.[17]

There are several reasons for this phenomenon in general and why it is more prevalent in the oil and gas industry. The two most important reasons for the aging of the world's population in general are declining birthrates and increasing life expectancy rates. There are several reasons why the aging of the workforce is particularly high in the oil and gas industry. First is the talent gap, which will be discussed shortly. The lack of enough new entrants into the industry with the needed skills leads to firms enticing older employees to stay longer. The global recession of 2008–2009 also caused many older employees to postpone their retirement. Other factors are a lack of supply of young, new entrants because of less interest in science, technology, engineering, and mathematics (STEM) careers, the growth of jobs in the attractive high-technology sector, and oil's negative reputation.[18] The presence of an aging workforce in the oil and gas industry has implications for the management of the workforce, which we will discuss later in this chapter.

Male-dominated

Historically, the oil and gas industry has been male-dominated. Although the percentage of women in the industry is difficult to estimate—and quite variable depending on the job, sector, country, and employer—various sources suggest that the industry is about 80% male. Female employment in the US oil and gas industry is detailed in figure 6–2. Broken down into upstream, downstream, and petrochemical sectors, the percentage of female employees rarely rises above 25%. The upstream demonstrates the lowest female share across the various regions of the United States, averaging 15%. Although the downstream and petrochemical sectors are higher at 25%, these are obviously industrial sectors with continued male-dominated employment.[19] The UK's oil and gas sector, one of the oldest and most established oil and gas countries, demonstrates much of the same employment rates. Female employment in the oil extraction and development sectors, including refining, is roughly 20% or less. The manufacturing sector of the UK economy is, however, not much better, at 25.7%, although both values pale in comparison with the average across all British industry of 46.2% female.[20]

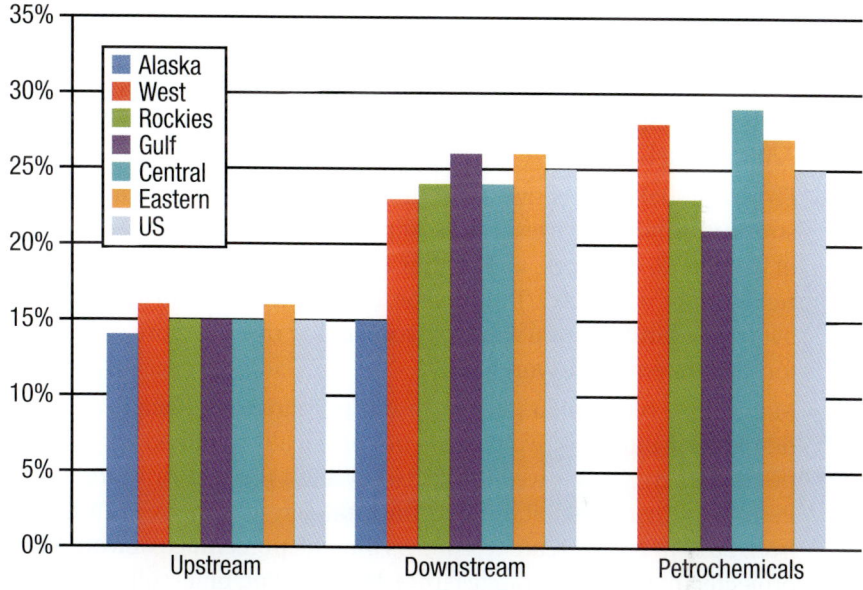

Fig. 6–2. Female employment by sector and region (percent of female workers in US oil and gas industry, 2010)
Source: Minority and Female Employment in the Oil & Gas and Petrochemical Industries, IHS Report, API, March 2014, p. 63.

These figures clearly show that globally, the oil and gas industry is male-dominated. However, the degree of male dominance is slowly lessening. Figure 6–3 provides a small window into how female employment rates have been changing in recent years, reporting the percentage of female workers by the nature of the organization and occupational segment in the industry in 2006 and in 2011.[21] Curiously, international oil companies (IOCs) seem to be the lowest in female employment. National oil companies have clearly made more progress in female employment, as shown here, but still fall far below what would be considered anything close to female employment rates in manufacturing or service sectors as a whole. However, the most recent data are more encouraging. In early 2013, almost half of the new oil and gas jobs went to women, compared to less than one-third in the previous quarter.[22] Nevertheless, women are still greatly underrepresented in the oil and gas workforce. The reasons for this are many:

- Although women's labor participation rate has been growing globally, women make up less than 30% of the STEM workforce in most countries. This is partly because women have lower university participation rates than men in STEM subjects (compared to higher rates in non-STEM subjects) and because women are far less likely to pursue a career in a STEM field once receiving a first degree in science (27% versus 54%).[23] Note that this can vary dramatically by country. For example, Asian universities tend to have STEM participation rates of women that are twice as high as North American schools.

- From a global perspective, the cultural barriers for women are still strong in some countries. For example, in Saudi Arabia the majority of women do not work, and in Kuwait, women did not have the right to vote until 2005. Thus, in some countries the industry has to overcome fundamental views of women that make it more difficult for them to enter the workforce.
- Many of the blue-collar jobs in the industry are so physically demanding that most women are not physically qualified to perform them. This is well illustrated by a 2014 report for American Petroleum Institute that projects the total new jobs in the industry from 2010 to 2030. Figure 6-4 shows that the percentage of women projected in these jobs ranges from 2% for skilled blue-collar jobs to 66% for office and administrative services jobs.[24] However, advances in technology and automation have transformed some of the manual jobs in the industry into jobs of controlling machinery electronically—and this is likely to continue and further reduce barriers to entry for female recruits.
- Low participation rates of women may be due to discrimination based on historical precedence. Male-dominated industries tend to foster cultures and attitudes that support the status quo, making overt and covert discrimination more likely. See Current Challenge Box 6-2 for a more detailed discussion of this topic.

Further, the dominance of men in the industry increases as one moves up the organizational hierarchy. A 2011 study found that the oil and gas industry had the lowest percentage of women directors (9.6%) of any industry.[25] Although the percentage of female leadership is improving, it still has a long way to go. In 2000, the percentage of female group leaders at BP was 9%. That number was 17% in 2012, and BP set a goal of 25% by 2020.[26] The percentage of women in leadership positions also varies by country and geographic region. For example, a 2013 survey found that less than half of Europeans believed that women had equal opportunities as men for advancement to management in the oil and gas industry, compared with almost two-thirds of Asians.[27] This underrepresentation of women in the industry in general, as well as in leadership positions, has many implications for staffing, training, performance management, and compensation, as we will discuss shortly.

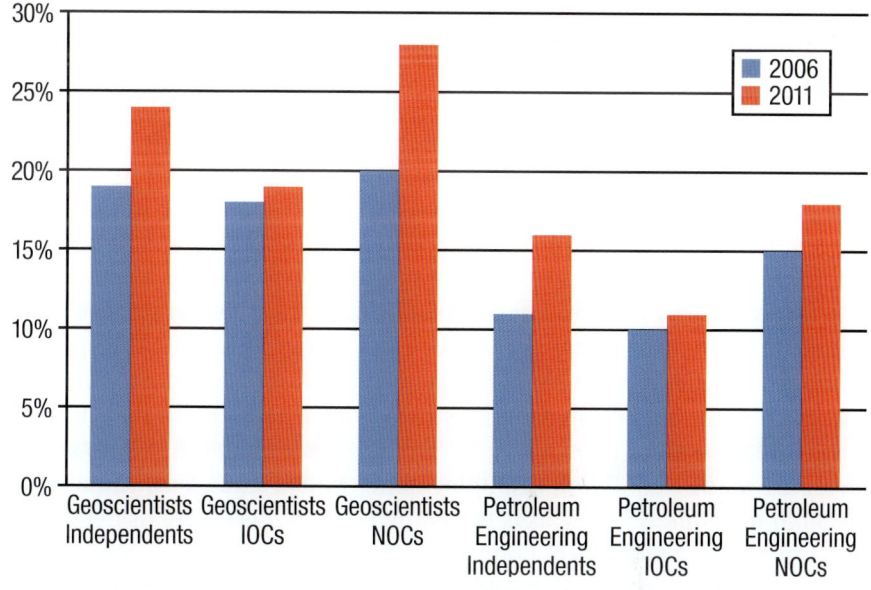

Fig. 6–3. Percentage of female workers by occupation and company type
Source: Current and future skills, human resource development and safety training for contractors in the oil and gas industry, ILO, Sectoral Activities Department, Geneva, 2012, p. 15.

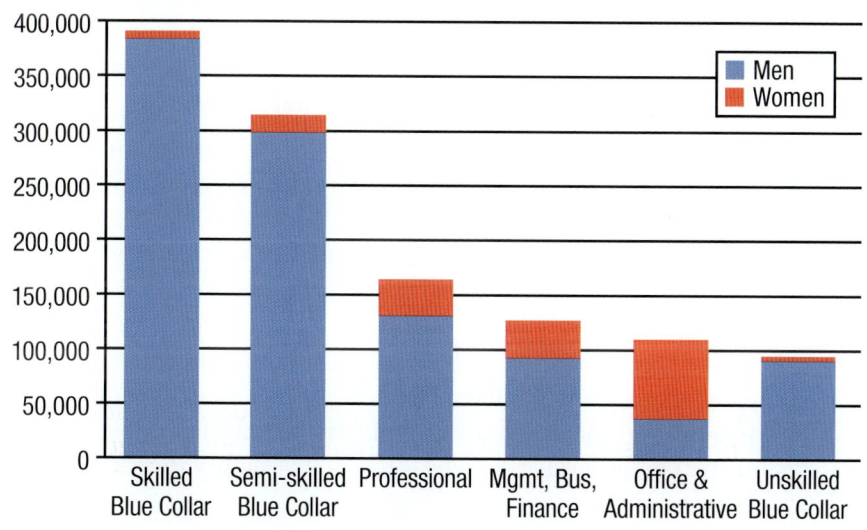

Fig. 6–4. Projected job openings in oil and gas, 2010–2030
Source: IHS 2014 Report: Minority and Female Employment in the Oil & Gas and Petrochemical Industries.

Current Challenge Box 6–2. Gender Harassment and Sexual Harassment

The ratio of male to female workers appears to be a factor in the prevalence of gender harassment and sexual harassment. So are the culture, policies and procedures, and everyday practices that tend to emerge in organizations in male-dominated industries.[28] Gender or sexual harassment occur when employees experience conduct that creates a hostile, intimidating, or offensive work environment that is related to their gender or is sexually oriented. Gender or sexual harassment may occur by or to either gender (although incidents of males harassing females are the most common, particularly in male-dominated industries). Sexual harassment also includes a type of harassment known as quid pro quo discrimination, where sexual favors are made a condition of employment. Quid pro quo discrimination is illegal in the United States, as is harassment when it is severe or pervasive, and unwanted. Further, US companies are liable for the actions of their employees if they knew or should have known about the harassment and took no action. So a clear, definitive harassment policy can not only protect the organization against expensive lawsuits, but also help to provide a more pleasant work atmosphere for all employees.

Employers should create a set of clear policies, procedures, and practices that inform employees about the rules, help detect harassment, handle complaints, protect those involved, and fairly discipline employees when necessary. Specifically, employers should do the following:

- Establish clear procedures and policies regarding harassment that build in checks to detect harassment (e.g., exit interviews), specify to whom complaints should be made (and alternates), set requirements to cooperate with investigations, and establish investigation procedures. These policies should be consistent with the company's general grievance procedures.

- Inform employees of the rules by proactively communicating the company's harassment policies, clarifying what actions constitute harassment, and specifying the disciplinary actions that will result. This can be easily achieved through mandatory awareness training sessions.

- Protect those involved by giving the accused due process and the opportunity to respond to complaints, and by preventing retaliation to accusers and whistleblowers.

- Fairly discipline harassment perpetrators through a clear framework that may apply to other policy violations.[29]

Harassment policies work better when they reflect the beliefs and actions of managers and organizational leaders. Thus, it is important for every manager to not only support the policies but also model following them without exception.

Difficult working conditions

Another unconventional aspect of the oil and gas industry is that many of the workers are located in remote locations. These locations are usually remote because the living (and working) conditions are so difficult that under ordinary conditions no one would want to live there. At the extremes, this includes the desert in the summer, the arctic in the winter, the deepest jungles, and offshore platforms. Needless to say, nature's elements in these locations can be challenging, to put it mildly. For example, facilities on platforms out on the open sea are generally unshielded from extreme weather such as high winds, water, ice, cold, and heat. Living conditions are compact and functional, similar to ship cabins. Although there are some conveniences, such as satellite TV, gyms, and cinemas, the remoteness, lack of privacy, monotony, and loss of time and space markers make life on a platform much more difficult than normal.[30] As we discussed earlier, this remoteness is largely responsible for the prevalence of rotators in the industry, as already described. However, the difficult working conditions per se have implications for HR beyond the fact that the employees are on a rotator's schedule.

The skills gap

There is much evidence that a severe talent shortage is imminent in the oil and gas sector. In fact, many believe that this shortage is already evident: 9 out of 10 senior HR managers in large, international gas companies see this issue as one of the top concerns facing the industry.[31] A recent survey by Hays found the skills shortage was the main concern for oil and gas employers worldwide by a wide margin over other historically troubling issues such as economic instability and security and safety.[32] This shortage is believed to be particularly severe for jobs that require considerable training and experience such as highly skilled engineering jobs, project manager jobs, and higher-level management jobs. Figure 6–5 shows the results of a recent survey by Mercer asking whether companies anticipate a talent gap in experienced and/or new employees in certain jobs in the next five years.[33] The results showed that more than half of the respondents believed there was an impending talent gap for petroleum engineers, plant/operations engineers, plant/operations technicians, plant/operations managers, and geoscientists.

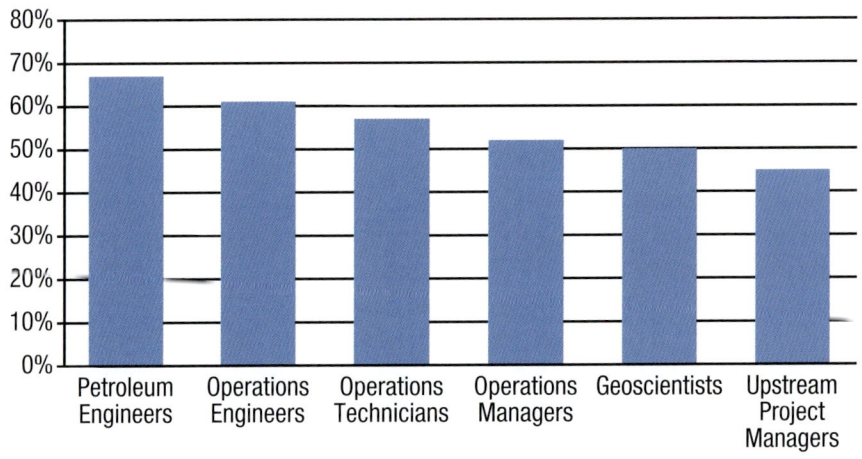

Fig. 6–5. Percent of companies anticipating a talent gap by job
Source: Mercer's 2013 Oil & Gas Talent Outlook and Workforce Practices Survey.

There are many reasons for these shortages, many of which have been already discussed in this book. The demand factors include the aging workforce (and the imminent retirements because of it) and the current and forecasted rapid industry growth. The talent shortage is also due to a number of supply factors, including the age gap, the highly specialized nature of the jobs, fewer STEM graduates, and the industry's unpopularity. See table 6–2 for a detailed discussion of the factors associated with the skills gap. For example, the age gap refers to the dearth of employees who are now between 40–50 years old, the prime age when employees are usually ready to be moved into positions that require a lot of experience. This is a result of the downturn in the industry in the mid-1980s, which caused a reduction in hiring for over a decade. Now, the older generation of managers, geoscientists, and petroleum engineers who were hired before the 1980s are approaching retirement age, with too few potential replacements from within because the age gap also reflects an experience gap.[34]

Table 6–2. Supply and demand factors of the talent shortage

Demand Factors	Supply Factors
The Aging Workforce: The average age of the oil and gas workforce (particularly in high skill jobs) is relatively high (estimates range from 42 to 50). Considering that many in the industry retire as young as 55, it is clear that a large percentage of the work force will be retiring in the next decade or two. Replacing these retirees is frequently referred to as "the great crew change".	**The Age Gap:** The downturn in the industry in the mid-1980s caused a reduction in hiring for over a decade. This has resulted in a large gap in employees who are now between 40-50 years old, the prime age to be moved into positions that require a lot of experience.
Rapid Industry Growth in Demand: The increase in demand from emerging economies is increasing global oil demand. Further the growing middle classes in the BRICs (Brazil, Russia, India, and China) suggest that demand will continue to increase in the long-term.	**Highly Specialized Jobs:** Many of the jobs in the oil and gas industry require specialized training and high levels of experience. The large range of skills needed in today's highly technical environment require highly trained and experienced workers. Much of this training is industry specific, so quick fixes are not possible
Rapid Growth in Supply in Some Sectors: The increase in supply from new oil sands, oil shale, and deeper water drilling suggest that supply will be able to keep up with demand without major price pressures. Thus, this greater supply of oil and gas will contribute to a greater demand of labor.	**Fewer STEM graduates:** Many developed countries are seeing declines in science, technology, engineering, and mathematics graduates, particularly those focused on petroleum careers. High tech sectors and other more positively viewed sectors are attracting many of the STEM graduates.
	Industry attractiveness: The industry's historic reputation as environmentally unfriendly, old, dangerous, and cyclical makes it more difficult to attract labor to the industry.

The imminent replacement of this large sector of retiring experienced workers is widely known in the industry as the great crew change.[35] GE Oil & Gas estimated in 2013 that half of its workforce, about five million oil and gas workers, would be eligible to retire in 2015.[36] Nearly four out of five industry professionals believe that the great crew change will have a noticeable effect on how they do business. Although all sectors of the industry are likely to be hit (logistics may be an exception), the geoscience, reservoir engineering, and petroleum engineering sectors will probably be the most affected.[37] This potential and imminent reduction in experienced labor, coupled with expected growth, means that companies must be prepared to hire new employees in the short and long term. Although the shale boom in the United States has made that domestic market more attractive for young professionals in that country, that is not happening globally, because of forces such as low oil prices and the industry's reputation.[38] Surviving the great crew change will depend on more than just finding replacements. Companies will need to make sure that the knowledge and expertise of those leaving is handed down to their replacements. This continuity of knowledge can be improved by planning for it, identifying key areas where it needs to happen, and incorporating it into training and career development plans.[39]

This skills shortage leading to the great crew change is likely to result in increased pressure on wages as companies try to recruit the employees of other companies. For example, senior project managers now earn 52% more in the oil industry than in other industries. If not addressed, skills shortages are likely to delay projects, increase risk taking and increase accident rates because it will constrain growth, delay expansion, constrain innovation, and blow out budgets because of more costly alternative staffing methods.[40]

In the next section we discuss how staffing, training, performance management, and compensation can all be instrumental in reducing the negative effects of the skills shortage, the great crew change, and other issues due to the unconventional aspects of the workforce.

Managing the Unconventional Workforce of the Oil and Gas Industry

We have shown that the oil and gas industry has an unconventional workforce in that it has a prevalence of contractors, often uses rotators, is aging, has an impending talent gap, is male-dominated, and frequently works in difficult conditions. All of these factors have implications for how to manage HR in the industry. In the rest of this chapter, we use the framework described in chapter 1 to describe how staffing, training, performance management, and compensation relate to this unique workforce.

Staffing the Unconventional Workforce of the Oil and Gas Industry

Many of these different features of the unconventional workforce of the oil and gas industry have important implications for staffing. The talent gap and remote locations make staffing more difficult. The aging workforce and male domination make targeting different groups more important. The use of contractors and rotators also changes some aspects of staffing, including key staffing decisions.

Labor markets

We address two major topics in this section. The first is how the great crew change and talent shortage is leading to expanded labor markets. The second is the labor markets for rotators. We begin with the impact of the talent shortage and of other aspects of the oil and gas workforce on the expansion of the labor market.

Expanding the labor market. The talent shortage and the remote locations of many jobs make attracting applicants more difficult, which frequently means that the labor markets must be expanded. This difficulty in attracting qualified applicants has turned the labor markets of many jobs in the oil and gas industry into a worldwide market, particularly for North American and Western European firms. They must now seriously consider Eastern Europe, India, and China as important locations because of the availability of highly trained engineers in those regions. Further, hiring engineers in those areas is frequently more cost-effective than hiring their Western European or North American counterparts.[41] Another reason for a global perspective is that the geographic location of the supply of engineers seems to be shifting to Asia and the countries of the former Soviet Union. Nearly three-quarters of geoscience graduates and four-fifths of petroleum engineering graduates come from those regions.[42] There are, of course, hotspots, where oil and gas companies are prevalent and the demand for labor is largely oil driven. Some of these locations include Houston and Denver in the United States, Edmonton, and Calgary in Canada, Singapore, Cairo, Kuala Lumpur, Mumbai, Rio De Janeiro, Perth, Jakarta, and Dubai. However, although the labor supply will tend to gravitate toward these hotspots, the increased demand may necessitate that firms move beyond them to other potential sources of labor.

The talent shortage requires companies to expand their labor market not just geographically, but also by considering nontraditional recruits. This includes women, minorities, immigrants, and in some areas native citizens and nonnative locals.[43] It may also include targeting younger workers to achieve a balance relative to the older workforce, and older workers because of their experience and availability. Overall, the talent shortage, the underutilization of women, and the aging workforce all require firms to tap into labor markets that they may not have previously considered and to target the recruiting to make it more effective.

There are several other ways to expand the labor markets. First, oil and gas firms should consider looking at recruits from other industries. Second, many firms just focus on a few universities that they have been recruiting from for many years. Widening the choices of the schools they recruit from would increase the potential labor pool. Third, companies have tended to focus on narrowly defined majors and degrees. Expanding the number and types of majors or degrees considered could also provide firms with more recruiting options. Finally, firms may consider targeting foreign students. Some firms have shied away from international students for local work because of the inherent immigration and work visa issues. However, the additional effort needed to employ these students may be worth the benefit of having an increased labor pool. These aspects will be discussed in greater detail in the recruiting section later in this chapter.

Rotator labor markets. Rotator use is growing. According to surveys, 65% of employers expect their rotator populations to grow in the next five years.[44] This demand will be intensified by the aging populations among rotators originating from the United States and Canada, home to a large proportion of the traditional rotator talent. As illustrated in figure 6–6, rotator use today is representative of the size, scale, and current growth of projects in global oil and gas development. The recent shale gas and oil development boom in the United States has now made the that country a major base of deployment for rotator use. Australia's demand for rotators reflects a multitude of LNG project developments, while offshore Brazil and Angola continue to demand more and more technical talent.

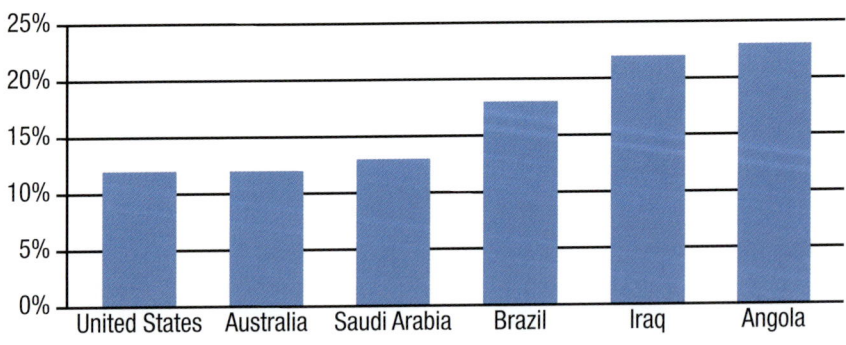

Fig. 6–6. Top five countries of current rotator use (in percent of workforce)
Source: Data drawn from *Oil & Gas Industry rotator assignment survey report*, Deloitte, 2013.

The sourcing of rotators, not surprisingly, is dominated by the United Kingdom (29%), United States (27%), and Canada (17%). Other significant sources of highly demanded technical talent include India, Norway, and Australia. The concentration of rotator employment, however, is somewhat surprising, as seen in figure 6–7: 19% of employers surveyed reported having more than 2,500 rotators in their employment, an enormous concentration of rotator use by many of the world's largest IOCs, national oil companies (NOCs), and oil and gas field services firms.

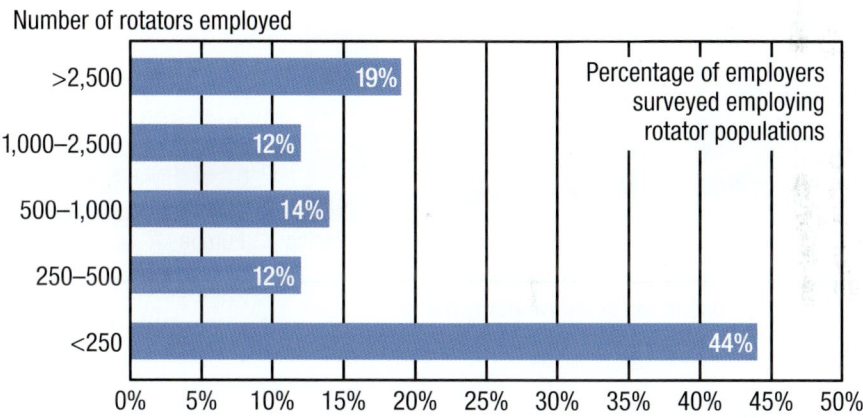

Fig. 6–7. Number of rotators worldwide
Source: Data drawn from *Oil & Gas Industry rotator assignment survey report*, Deloitte, 2013.

Note also the sourcing-deployment challenge. Australia and the United States are both countries that are sources of rotators and areas of deployment. Understandably, this is the result of both countries having long histories of developing oil and gas properties both at home and abroad, and the recent growth of new oil and gas developments in both countries (LNG projects in Australia, shale oil and gas development in the United States). This is contributing to a growing talent gap in these geographic markets, which we have already discussed.

The expected growth of the global oil and gas industry is also the primary driver of future rotator demand. Figure 6–8 reflects this employment driver, as 70% of rotator employers surveyed indicated that Africa will be the destination of rotator use growth in the coming three-year period alone, followed by the Middle East, Latin American, and Asian Pacific in roughly equal likelihoods. This, of course, will have further implications for the rotator labor market.

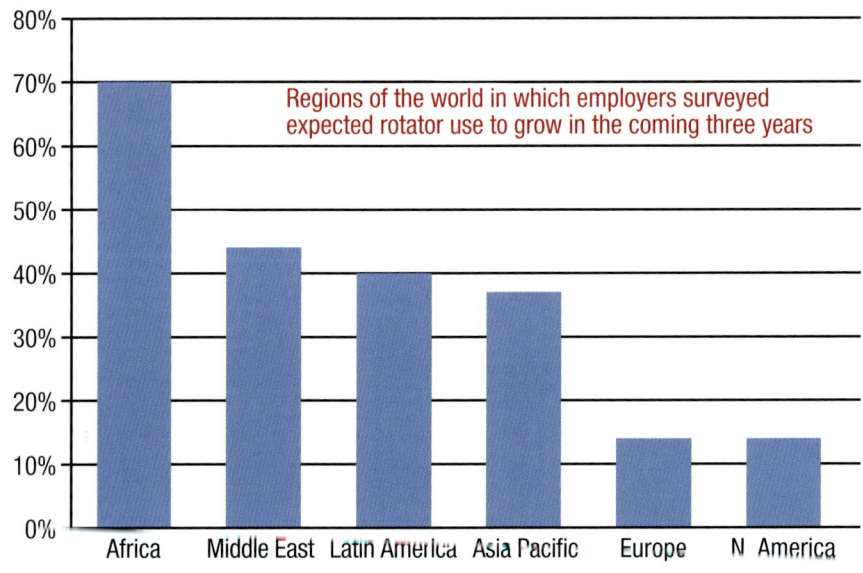

Fig. 6–8. Regions of expected rotator growth
Source: Data drawn from *Oil & Gas Industry rotator assignment survey report*, Deloitte, 2013.

Recruiting

Expanding the labor market to help address the talent shortage and the great crew change requires new ways to think about recruiting—not only who we recruit, but also how we recruit. Currently, almost half of the outside hires in the oil and gas industry come from other oil and gas companies. About 20% of the hires of come from outside the oil and gas industry, and slightly more than 10% of new hires come from colleges and universities. Both the percentage of new hires from colleges and universities and those from outside the industry are expected to increase in the future. We now discuss a number of ways to improve recruiting by targeting specific groups and using recruiting methods relevant for each group.

Targeting employees from other industries. Although there is considerable specialized knowledge in the oil and gas industry, depending on the job, other industries may be able to provide workers with an adequate knowledge base. Industries that have been considered as viable sources for employees include chemical manufacturing, transportation manufacturing, forestry, automotive, defense, and aerospace. Those who served in the military are also frequently considered because they are familiar with extended travel, working in remote locations, and rotator schedules. However, there is no need to limit recruiting to specific industries. The targeting can be done on a job-by-job basis. For example, school bus drivers are currently being recruited in hot spots of the oil and gas industry as truck drivers, a high-demand job in some areas.

Targeting employees from other oil and gas companies. Because half of the outside hires in the oil and gas industry come from within the industry, this target audience cannot be ignored. Although poaching was considered taboo many years ago, it is now far more common. Still many oil and gas firms use third-party recruiters to make poaching more palatable, since many companies work on joint projects with their competitors.[45] Many viable recruits can be found in passive candidates at other firms, that is, those who are not actively looking for another job. Passive candidates can be reached through marketing, referrals, and headhunters. Looking at the employees of vendors is also becoming more popular. Although hiring the employees from other firms is a viable recruiting method, it does not do much good if other firms hire away your employees as well. Thus, the emergence of poaching makes retention more important than ever in the industry. See Current Challenge Box 6-1 for a detailed discussion of this topic.

Targeting young people. Targeting young people can help replenish the losses from the aging workforce. It can also help firms keep a good stock of replacements for retirees in the short and long term. Millennials (also known as Generation Y, born 1982–1993) have a number of specific needs that are applicable to the oil and gas industry:

- Long-term career development
- Sense of purpose and meaning in work
- Availability and access to mentors
- Work/life flexibility
- Tech-savvy work environment
- Open social networks[46]

Thus, organizations need to have programs and policies in place that meet these needs and tout them to young job candidates. Money is always important, but offering a competitive salary is not enough. To attract young workers, firms must show that they offer career options (such as travel opportunities or different work schedules), mentoring, flexibility, and meaningful work. Further, the recruiting method should be targeted toward a young audience. Social media, such as Facebook and Twitter, and Internet job sites are the dominant way young people look for jobs today. Internships are another good way to attract younger workers. Internships provide a great, realistic job preview and provide both employees and employers with ample information to help them make a good decision regarding the fit of the recruit.

Targeting women. Targeting women can help address the talent shortage by tapping into an underused resource. Targeting women can also help reduce the male dominance of the workforce. Women in the oil and gas industry value career customization, work/life balance scheduling (such as flextime or telecommuting) and benefits including child care facilities, maternal leaves, career breaks, and part-time work or job sharing.[47] Professional associations, university groups, and energy conferences focusing on women are a great place to target female recruits. Some of the most popular associations include Society of Women Engineers (www.societyofwomenengineers.swe.org) with 27,000 members, the Society of Petroleum Engineers women's network (www.spe.org) with 13,000 members, and the Women's Energy Network (www.womensenergynetwork.org) with 3,000 members.

Other options. If the talent shortage becomes as prevalent as many believe, no options for attracting needed talent should be overlooked. Recruiting internally may help fill critical needs, although it may also create new needs. Promoting qualified top performers faster than usual may also be a way to address critical shortages. Outsourcing the function to specialized recruiting and staffing companies is always an option, but is likely to lead to higher costs. Finally, technology may be used to replace the need for some workers. For example, Shell is exploring the use of artificial intelligence software to determine training needs, while Baker Hughes is using artificial intelligence software to monitor rigs and write some safety reports.[48]

Selecting

The selection tools used are generally the same for this unconventional workforce as in other industries, as we mentioned in chapter 1. Nevertheless, there are some differences. The talent shortage may mean that selection criteria need to be somewhat relaxed to meet desired hiring levels. Of course, relaxing selection criteria (particularly if the criteria have been doing a good job in detecting qualified candidates) increases the risk of hiring unqualified candidates. Also, for rotator jobs, some criteria need to be introduced to make sure that the candidates are a good fit for rotator schedules. This may just be simple interview questions that assess the candidates' past experience or willingness to take on such assignments. Finally, the selection of contractors can play an important role in the make-up of the oil and gas workforce.

The selection of contractors or contractor firms follows the principles of general selection: Who do you think will perform the best? Of course, the definition of good performance depends on the job. Surveys of oil and gas firms have found a number of factors that affect the selection of contractor firms in oil and gas EPC:[49]

- Business integrity
- Safety and environmental performance
- Labor conditions and human rights
- Quality of key personnel
- Project management capability
- Price
- Quality of project control systems
- Detailed engineering capability
- Construction capability
- Experience with similar work
- Experience with same geographic area
- Responsiveness and flexibility

Although these apply to contractor firms, many of the same principles—capabilities, knowledge, experience, and quality of work—apply to individuals as well.

Combining tools

The talent shortage should lead firms to generally prefer compensatory versus multiple-hurdle approaches to combining tools. This is because multiple-hurdle approaches are likely to reduce the applicant pool much more than compensatory approaches. In a shortage situation, reducing the applicant pool is less desirable than under normal conditions. Nevertheless, multiple hurdles should not be abandoned if it would result in the hiring of candidates who do not have acceptable levels of the critical skills needed for the job.

Maximizing the effectiveness of staffing

The unconventional workforce introduces two new issues to consider in maximizing the effectiveness of staffing. First, since fairness is a perception, firms must consider how employees will view the fairness of the staffing function. For example, women may perceive some questions as unfair or older workers may perceive that a certain test is unfair. Thus, the system not only should be objectively fair, but should also consider how different employee groups may interpret the fairness of it. Second, staffing may require creative thinking to maximize its effectiveness in the context of a talent shortage. Because of the difficulty in attracting talent, outside-the-box alternatives should be considered. These may include outsourcing, investing in new technologies, using virtual teams, and reorganizing the division of work to help ease the need for jobs that are critically short-staffed.

Training the Unconventional Workforce of the Oil and Gas Industry

Many of the different features of the unconventional workforce of the oil and gas industry have important implications for training. The great crew change makes training more important, because in the long term it can be used to provide employees with the skills needed to meet the demands of difficult-to-fill positions. The aging and male-dominated aspect of the workforce make targeting different groups through different training methods even more important. The use of contractors and rotators also change some aspects of training. We now discuss these differences in the framework of the key training decisions.

Who gets trained?

The age gap, skills shortage, and great crew change all suggest that massive investment must be made into training younger workers. In the long term, to have a stable supply of trained workers for the skilled engineering jobs needed in the oil and gas industry, universities must have the capacity to educate enough students to meet demand. Although it appears that the number of petroleum-related programs and degrees in the United States has slowly increased in the last few years, it is still not nearly the number seen before the downturn in the mid-1980s. To help foster a steady supply of entry-level skilled labor and have access to them, companies should work on developing relationships with universities around the world, by awarding scholarships, sponsoring grants, upgrading laboratories, creating internships, making donations, and attending career fairs.[50]

Although some of the necessary training can be done through universities, most oil and gas employees expect the employer to assume a large portion of the responsibility of preparing them for a new job.[51] Thus, new hires should also be the focus of training. Organizations must have a process in place that accesses the gap between the knowledge new hires have and the knowledge and skills they need to perform the job. Some of these gaps may be organizational-level knowledge (e.g., the policies, procedures, and rules of the company). In such cases, centralized training of all new hires in these areas should be considered.[52]

The male-dominated aspect of the industry makes it particularly susceptible to gender and sexual harassment incidents. Gender harassment is the harassment of employees because of their gender. Sexual harassment is harassment with a sexual connotation. Both are illegal in the United States and the EU. Thus, clear policies and procedures related to these issues must be established, and all employees should be trained on these policies. Current Challenge Box 6-2 discusses these issues in greater detail. Further, gender harassment has an effect on the ability to recruit women into the oil and gas workforce. A recent survey found that 58% of the respondents believed that for women to take on more roles in the field in the oil and gas industry, companies had to implement strict rules against harassment and discrimination in the workplace, as well as create an environment that accommodates female requirements.[53]

What gets trained?

What gets trained frequently depends on who gets trained, which we discussed above. It also depends on which aspect of the workforce's unconventionality we are addressing. Related to the skills gap, technical training of the skills that are in short demand is clearly a viable way to fill open technical jobs, although this is likely a longer-term solution. However, much of the concern about the skills gap, related to both the age gap and aging workforce, is regarding the higher-level management and leadership jobs that require more than technical skills. For these jobs, soft skills are essential: critical reasoning, teamwork skills, ethics, communication, negotiation, and problem solving. Further, these skills are rarely taught in technical degree programs, and thus must be obtained either through advanced broader degrees (such as an MBA) or taught by employers.[54] The importance of soft-skill development to the careers of highly technical workers was acknowledged by the Society of Petroleum Engineers, as shown by the formation of the Soft Skills Council in 2011.

There are a few other content areas that may need extra focus because of the unconventionality of the workforce. First, one content area that needs extra focus because of the rising demand for new hires is safety. The inexperience of new workers has been shown to be the cause of many serious accidents.[55] Thus, specialized safety training that addresses this issue is also clearly warranted. Second, as mentioned earlier, the male-dominated aspect of the industry calls for a greater focus on gender and sexual harassment training. Finally, the prevalence of contractors may require additional training regarding various aspects of the organization, including policies, practices, and procedures, as well as the organization's culture. This will help create a better working relationship among firm employees and various contractors.

How do we train?

The unconventional nature of the workforce of the oil and gas industry has minor implications for its training methods. As is generally true, training can be classroom based or on the job. It can be delivered face-to-face or online. Online training provides an option for cost-effective training for rotators and employees in remote locations. Further, online also allows specialized courses that would not be possible in person because of the limited demand for jobs that are more specialized. Other methods that can be used to help bridge the skills gap include formalized coaching, knowledge management programs, and job rotation, which have all been shown to reduce the time it takes for employees to be trained.

One training method that appears to be particularly well-suited to help reduce the talent shortage in the long term is mentoring. Mentoring is becoming more prevalent in the oil and gas industry. Total created its Professor Academy, where retired geologists, engineers, and managers mentor young talent. Petrobras created Petrobras Mentor, which pairs older workers with young talent. Mentoring helps with the crucially important issue of transferring the knowledge gained by the experienced employees to the younger employees.[56] Without this transfer, knowledge that had been acquired over decades of experience could be lost to the organization forever upon the departure of the older worker. The positive effects of mentoring seem to be recognized around the world. A survey of managers in the oil and gas industry in Russia found that almost one-third thought that mentoring was the best way to improve the skills of employees.[57] Only specialized training was seen as more effective. Related to the male dominance of the industry, the vast majority of young female employees believe that having a senior female mentor is important to career success. A recent survey found that 63% of the respondents believed that mentorship programs led by women in the energy industry would have the greatest impact on attracting female engineers into the industry after graduating.[58] Thus, mentoring can be a valuable tool that can be adapted to specialized segments of the workforce.

Maximizing the effectiveness of training

The unconventionality of the workforce has several implications for maximizing the effectiveness of training. First, as discussed in chapter 1, effective training requires effective learning. Different targeted groups are likely to learn in different ways. For example, younger workers may be far more comfortable than older workers with training software based on mobile devices. Organizations should be mindful that training should be tailored to the individuals being trained. Second, remote locations may make monitoring and evaluation of training and its transfer to the workplace more difficult. Thus, organizations must make sure that feedback loops are in place in these contexts. Finally, the importance of mentoring warrants further description on ways to maximize its effectiveness. A study in Australia's upstream oil and gas sector found that the following aspects maximize the effectiveness of mentoring[59]:

- The public commitment and support of upper management
- A program plan, with a statement of purpose, closure policies, and a program coordinator
- A selection process that takes mentoring participation into account
- Formal mentor and mentee preparation
- A well-thought-out matching process
- Coordinated support mechanisms for mentors and mentees
- Evaluation and assessment processes

Thus, each of these points should be carefully addressed when preparing a mentoring program. Although this may require more effort, the increased effectiveness of the program will be well worth it.

The Performance Management of the Unconventional Workforce of the Oil and Gas Industry

The different features of the unconventional workforce of the oil and gas industry have less of an impact on performance management than they do on staffing and training. Nevertheless, there are still some implications for performance management. For example, the attitude of younger workers toward performance management and performance appraisals appears to be different from older workers. Compared to baby boomers, younger workers expect constant, immediate, and fair feedback. These expectations of feedback include regular formal evaluations as well as informal feedback, including recognition when deserved.[60] We now discuss these differences in the framework of the key performance management decisions.

How do we measure performance?

The workforce issues discussed earlier in this chapter can have an impact on the fundamental question: "What is performance?" For example, addressing staffing shortages in the long term should be part of the performance evaluation done on every manager. Do they have a succession plan in place? Are they sending their subordinates to the appropriate training? The male-dominated aspect of the industry and the need for better gender balance suggest that managers should also be evaluated on their progress toward increasing the numbers of women in their ranks and the prevention of gender and sexual harassment. Objective measures such as the percentage of female new hires, the percentage increase in female employees, the percentage of women in the applicant pool, and the percentage of workers who have taken gender harassment and sexual harassment training courses may be used to measure performance in this area. Subjective measures that assess the manager's efforts in these areas may also be used to accommodate situations where the outcomes are not as good as hoped through no fault of the manager.

Depending on the type of contractor, the performance of contractors is frequently specified in the contract and should be measured as specified. Subsequent contracts should take into account anything learned in the previous contract regarding the measurement of performance. For example, were there any aspects of performance that were not specified and, therefore, not properly completed? Were there problems with contractor safety performance? Were there problems with the handover to the owner's operations team?

Who is involved and in what format?

The prevalence of contractors changes the format and procedures of performance evaluation as compared with employees in general. This is largely because of the legally defined nature of the relationship through a contract. Thus, performance as specified through the terms of the contract should be evaluated in a formal, clearly specified framework. The use of rotators may also have some effect on who is involved in the performance evaluation. Although supervisors are generally believed to be the best source for assessing the performance of employees, for rotators in remote locations, supervisors may be a continent away. Thus, greater reliance may need to be given on feedback from co-workers or nonsupervisory managers who are working closely with the rotator.

Maximizing the effectiveness of performance management

The effectiveness of performance management of employees generally follows that of contractors. The terms of performance should be clearly defined early. In the case of contractors, the terms are specified in the contract. In the case of employees, the terms should be part of their job description and the employees should be given goals related to them. They should then be evaluated on those goals and receive rewards when achieving them. Although the unconventionality of the workforce may slightly change what those goals should be or how they can be achieved, the fundamental principles remain the same.

Compensating the Unconventional Workforce of the Oil and Gas Industry

Many of the features of the unconventional workforce of the oil and gas industry have important implications for compensation. The talent gap makes compensation more important, because it can be used to attract and retain employees with critical skills. The aging and male-dominated aspects of the workforce make targeting different groups through different benefits more important. The use of contractors and rotators also affects some aspects of compensation. We now discuss these differences in the framework of the key compensation decisions.

Pay level

Pay level has important implications for the talent shortage. Although it is frequently denied by survey respondents, pay level is important to people and affects many of their behaviors. Thus, it is not surprising that companies use pay levels to attract and retain workers in key jobs. It is also not surprising, given the talent gap in jobs like petroleum engineer, that companies are paying dearly to attract new graduates. Petroleum engineering was the highest paid college major in 2013, at $96,200, a lot higher than the second place job of computer engineering, at $70,300.[61] Economic theory would suggest that these astonishingly high wages are a temporary phenomenon of the talent shortage. As more university students clamor for petroleum engineering degrees and the high wages that come with them, supply will increase, eventually stabilizing wages to levels equal with majors of similar needed skills. But in the short term, such high wages for new hires can create havoc for organizations with respect to perceptions of fairness. This will be discussed in detail later. Also, this presumes that universities can quickly gear up to meet an increasing demand for petroleum engineering degrees, which may not be true.

Pay level is also related to the prevalence of contractors. The pay level of contractors relative to permanent employees depends on at least a couple of factors. First is the type of contractor. For example, pay for temporary employees tends to be less than for permanent employees, possibly because they lack seniority pay in their pay scales, they are more likely to have less skilled jobs, and are generally not unionized. The second factor is the country and its corresponding pay laws and unionization levels. For example, pay level seems to be the same for contractors and company employees in Ecuador and Australia, but higher for company employees in Canada and Norway.

Pay mix

The pay mix for contractors and rotators is somewhat different from other employees. With respect to contractors, one aspect of the pay mix that may differ is the availability of various bonuses. Again depending on the country's laws and collective bargaining agreements, contractors may get bonuses (sometimes called a "loading" of the rates) as compensation for not receiving the benefits permanent employees get. The bonuses may be compensation for not receiving health insurance, annual leave, vacation pay, and so forth. Special bonuses may also be awarded for seasonal activities such as refinery turnarounds or as offshore supplements.[62] Thus, the pay mix of contractors may be more bonus-heavy than the pay mix of permanent employees. The same is true for rotators, but in their case the bonuses come from large premiums in many cases and travel and accommodation expenses (frequent round-trip airfares for their monthly commute).

Pay mix may also be related to the composition of the workforce. For example, different employee groups have different preferences for bonuses versus salary and different preferences for benefits. Thus, when trying to attract and retain more females, making the pay mix benefit-heavy with many work/life balance and family-friendly benefits makes sense. This may also attract younger workers, who are more open to foregoing some salary for greater upside potential in bonuses.

Performance-based pay

As we discussed relative to performance management, because the workforce composition has implications for performance measures, it should have implications for the measures of performance-based pay. For example, providing bonuses tied to efforts to overcome the talent gap or for achieving a greater gender balance will motivate managers to work toward those outcomes. These bonuses need not be large monetary bonuses; even small recognition awards such as plaques with token rewards (such as gift cards) can have good motivational effects.

Bonuses may also be a factor in attracting workers relative to the talent shortage. Signing bonuses (with a minimum stay requirement) and returner bonuses for retired employees or women who left the workforce will increase the attractiveness of joining or rejoining the company. Being proactive with returner bonuses for women who have left the workforce will also help address gender imbalances.

Benefits

Benefits can have important implications for recruiting workers. Some benefits help attract workers and then encourage them to further develop themselves to benefit both themselves and the organization. Examples of these include tuition reimbursement for postgraduate degrees and reimbursement of fees to professional organizations. Other benefits help attract workers in general or targeted groups. For example, younger workers prefer wellness facilities, social activities, and fair and nonprejudicial maternity and paternity leaves.[63] Women are particularly attracted to flexible work hours, flexible career ladders, and family-friendly benefits. Exxon's new 385-acre campus in North Houston is designed to accommodate 10,000 employees. Its perks include an on-site wellness center (including fitness, medical, and occupational health facilities), a diverse array of on-site dining facilities, a child-development center for children six weeks to prekindergarten, and Wi-Fi throughout the campus.

Such youth- and family-friendly benefits may help improve the image of the oil and gas industry and allow it to compete against firms in other industries, such as Google, which is generally seen as the US firm providing the most desirable benefits.

Benefits can also have important implications for retaining workers. Benefits such as vacation days, which increase as employees' seniority increases, can help to retain employees. See Current Challenge Box 6–1 for a more detailed discussion of employee retention.

The benefits provided to contractors and rotators tend to differ from those of permanent employees. Contractors tend to get fewer benefits, while the opposite is true for rotators. Rotator benefits packages continue to expand and, in fact, appear to be rising as the demand grows rapidly and the talent pool does not. Figure 6–9 provides an overview of the multitude of package components for rotators. It is also clear that benefits packages also reflect the changing nature of rotator deployments, as greater percentages are now in onshore activities as opposed to the relative dominance of offshore deployment just a few years back. The benefits of rotators also raise some tax issues, which are important enough that they deserve some discussion.

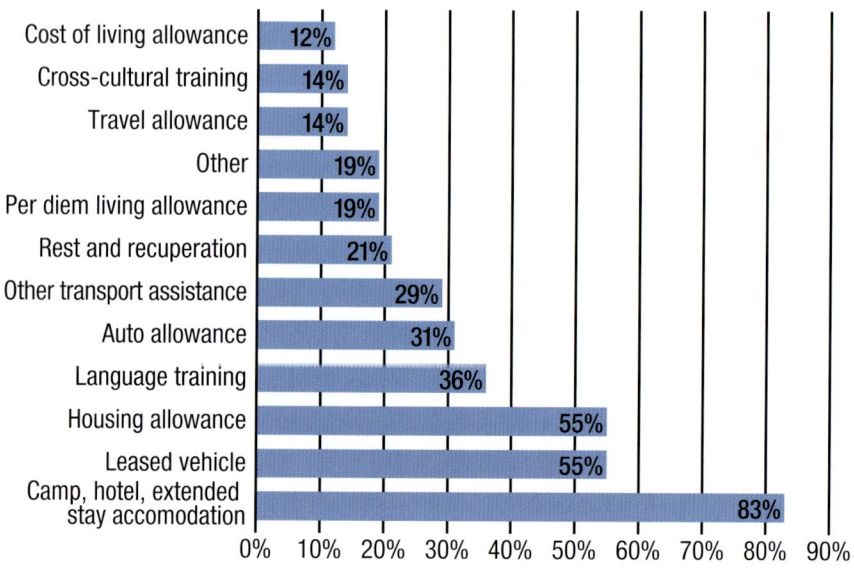

Fig. 6–9. Percent of firms providing rotator benefits and allowances
Source: Data drawn from *Oil & Gas Industry rotator assignment survey report*, Deloitte, 2013.

US tax law covers short-term business travelers and traditional long-term expatriate employees in great detail, but US outbound rotators do not fall into either category, which presents a number of potential tax challenges.[64] Two of the largest challenges for US citizens are the foreign earned income exclusion and travel expenses. The typical long-term expatriate who moves from the United States to a foreign country for more than one year of employment qualifies for foreign earned income and housing exclusions. That is, under the Internal Revenue Service (IRS) Code §911, $97,600 of an expat's foreign earned income can be excluded from gross income for tax purposes. To qualify, the employee must be deemed to have the foreign country as their "tax home," with the "183-day rule" a common provision in tax treaties between the United States and other countries.[65] But this does not apply to most rotators, who are considered to retain the United States as their "abode" and therefore their tax home.

The second tax challenge is whether benefits such as travel and housing expenses covered by the employer are considered compensation and therefore taxable by the US IRS as taxable wages. Travel expenses associated with commuting to and from rotational shifts, which can number 10 or 12 per year and exceedingly costly to the individual, could easily approach $60,000 or $70,000 per year. Other employee benefits such as transfers, housing, meals, and other per diem expenses can be quite high as well. In many cases, if the per diem expenses do not exceed the government's specified per diem rates, these compensating payments may not be taxable. If they do exceed government rate schedules, the difference between paid and scheduled amounts will constitute taxable income.

These are two of the most prevalent tax and income challenges for rotators, but there may be many more depending on the specific countries and projects involved. If the rotators are on their own in the preparation of their US tax documents, they may make mistakes regarding these and other tax and income issues. If many errors occur, the employer of the rotators may suffer reputational damage for compliance failure.

Maximizing the effectiveness of compensation

As we have discussed throughout this book, perceptions of fairness are a critical aspect of maximizing the effectiveness of compensation. Many of the implications of the unconventional workforce that we have discussed have serious implications for fairness. Targeting benefits or bonuses toward certain groups, providing sign-on bonuses, and setting high pay levels for new hires can lead to strong perceptions of unfair treatment by the current workforce. For example, high pay levels for new hires can lead to inexperienced new employees earning almost the same (or even more) than more experienced and more productive employees. This is known as pay compression, and it is quite demoralizing to those who experience it. There is an expectation that greater performance and greater seniority will be related to more pay for those in the same job and in the same company. Of course, raises could be given to senior employees to keep a reasonable pay difference between them and new employees, but this could greatly raise labor costs. Although there is no easy answer to address compression, being aware of it is the first step to dealing with it. Giving high-performing senior employees other rewards to help differentiate them such as recognition awards, time off, flexible work schedules, and so forth, can help differentiate them from the new hires for minimal costs.

Conclusion

We have shown how the oil and gas industry has an unconventional workforce in that it has a prevalence of contractors, often uses rotators, is aging, has an impending talent gap, is male-dominated, and frequently works in difficult conditions. We have also shown how all of these factors have implications for how to manage HR in the industry. Specifically, staffing can be used to help overcome the talent shortage and increase the presence of women in the industry. Training can also be used to overcome the talent shortage in the long term, as can compensation. We have shown how the use of contractors and rotators affects many of the different HR functions. Overall, the unconventional workforce of the oil and gas industry creates many challenges; however, we have shown how many of these challenges can be overcome through sound HR management practices.

References

1. Graham, I. (2010). *Working conditions of contract workers in the oil and gas industries.* (Working Paper No. 276). Geneva, Switzerland: International Labour Office, Sectoral Activities Department.
2. Graham, I. (2010). *Working conditions of contract workers in the oil and gas industries.* (Working Paper No. 276). Geneva, Switzerland: International Labour Office, Sectoral Activities Department.
3. *World's 10 largest oilfield services companies.* Retrieved from http://www.arabianoilandgas.com/article-5728-worlds_10_largest_oilfield_services_companies/.
World's 10 largest oil and gas contractors. Retrieved from http://www.arabianoilandgas.com/article-5773-worlds_10_largest_oil_and_gas_contractors/.
4. Loots, P., & Henchie, N. (2007). *Worlds apart: EPC and EPCM contracts: Risk issues and allocation.* London, UK: Mayer Brown.
5. Schramm, C., Meibner, A., & Widinger, G. (2010). Contracting strategies in the oil and gas industry. *Pipeline Technology,* 1 (Special Edition), 33–36.
6. Turner & Townsend. (2001). *EPC-v-EPCM.* http://pceuae.com/d/82_Va7mhI.pdf.
7. API. (2011). Contractor safety performance process, API standard (draft 3rd ed.). http://gost-snip.su/download/api_std_2220_2011_draft_contractor_safety_performance_proces.
8. International Labor Organization. (2012). Current and future skills, human resources development and safety training for contractors in the oil and gas industry (Issues paper). Geneva, Switzerland. Retrieved from http://www.ilo.org/wcmsp5/groups/public/---ed_dialogue/---sector/documents/meetingdocument/wcms_190707.pdf.

9. Flexpatriate assignments: Trends and policy implications (2012, December 4). *Re:locate Magazine*. http://www.relocatemagazine.com/articles/2266flexpatriate-assignments-trends-and-policy-implications.
10. Sandlin, M. (2013, December 13). On the move: International rotator assignments are in demand and on the rise. *Chron.com*. http://www.chron.com/jobs/article/On-the-Move-International-rotator-assignments-5062263.php.
11. Sandlin, M. (2013, December 13). On the move: International rotator assignments are in demand and on the rise. *Chron.com*. http://www.chron.com/jobs/article/On-the-Move-International-rotator-assignments-5062263.php.
12. Tennant, J. (2012). Making informed human resources decisions based on workforce outlook. *World Oil, 233*(9). http://www.worldoil.com/magazine/2012/september-2012/supplement/making-informed-human-resources-decisions-based-on-workforce-outlook.
13. SPE Research. (2012). *Training and Development Survey*. Retrieved from http://www.spe.org/career/docs/12Training-and-Development-Study.pdf.
14. Deloitte. (2012). *Talent 2020: Surveying the talent paradox from the employee perspective—The view from the oil and gas sector*. Retrieved from http://www.deloitte.com/assets/Dcom-SouthAfrica/Local%20Assets/Documents/dtt_Talent_2020-Surveying_talent_paradox.pdf.
15. Tennant, J. (2012). Making informed human resources decisions based on workforce outlook. *World Oil, 233*(9). http://www.worldoil.com/magazine/2012/september-2012/supplement/making-informed-human-resources-decisions-based-on-workforce-outlook.
16. *World population ageing* 2013. (2013). United Nations Department of Economic and Social Affairs Report. Retrieved from http://www.un.org/en/development/desa/population/publications/pdf/ageing/WorldPopulationAgeing2013.pdf.
17. Torr, R. (2010, May 5). Aging workforce a threat to oil sector. *McClatchy–Tribune Business News*, B1.
18. Mah, B. (2011, June 15). Labour shortage bigger challenge than ever; Aging workforce, global competition stretch worker supply. *Edmonton Journal*, p. C6.
19. API. (2014). *Minority and female employment in the oil & gas and petrochemical industries* (IHS Report). Retrieved from http://www.api.org/~/media/Files/Policy/Jobs/IHS-Minority-and-Female-Employment-Report.pdf.
20. Forde, C., MacKenzie, R., Stuart, M., & Perrett, R. (2005). *Good industrial relations in the oil industry in the United Kingdom* (Working Paper). Geneva, Switzerland: Sectoral Activities Programme, International Labour Office.
21. International Labor Organization. (2012). *Current and future skills, human resources development and safety training for contractors in the oil and gas industry* (Issues paper). Geneva, Switzerland: Author. Retrieved from http://www.ilo.org/wcmsp5/groups/public/---ed_dialogue/---sector/documents/meetingdocument/wcms_190707.pdf.
22. Unger, D. J. (2013, May 9). Unconventional energy: Rise of women in oil and gas industry. *Christian Science Monitor*. Retrieved from http://www.csmonitor.com/Environment/Energy-Voices/2013/0509/Unconventional-energy-rise-of-women-in-oil-and-gas-industry.
23. Lee, S. M. (2012). *Nurturing the future generations for the oil and gas industry* (Task Force 2 Report). Vevey, Switzerland, International Gas Union.
24. IHS. (2014). *Minority and female employment in the oil & gas and petrochemical industries*. Retrieved from http://www.api.org/news-and-media/

news/newsitems/2014/mar-2014/~/media/Files/Policy/Jobs/IHS-Minority-and-Female-Employment-Report.pdf.
25. Tennant, J. (2012). Making informed human resources decisions based on workforce outlook. *World Oil*, *233*(9). http://www.worldoil.com/magazine/2012/september-2012/supplement/making-informed-human-resources-decisions-based-on-workforce-outlook.
26. Dupre, R. (2013). *Women fill 40% of vacancies in oil, gas*. Retrieved from Rigzone website: https://www.rigzone.com/news/oil_gas/a/127452/Women_Fill_40_of_Vacancies_in_Oil_Gas/?all=HG2.
27. BP & Rigzone. (2013). *Global diversity and inclusion report*. Retrieved from http://www.bp.com/content/dam/bp/pdf/Press/Full-report-Diversity-and-Inclusion-BP-Rigzone.pdf.
28. International Labor Organization. (2012). *Current and future skills, human resources development and safety training for contractors in the oil and gas industry* (Issues paper). Geneva, Switzerland: Author. Retrieved from http://www.ilo.org/wcmsp5/groups/public/---ed_dialogue/---sector/documents/meetingdocument/wcms_190707.pdf.
29. Jackson, S. E., Schuler, R. S., & Werner, S. (2011). *Managing human resources*. CengageBrain.com.
30. BP. (2014). *Life on remote platforms: Hard work and close quarters*. Retrieved from http://www.bp.com/en/global/corporate/about-bp/what-we-do/extracting-oil-and-gas/life-on-remote-platforms.html.
31. Pyron, D. (2008). Solutions to the recruitment and retention challenges. *Talent and Technology*, *2*(2): 4–6.
32. Chazen, G. (2014, July 17). Terrifying oil skills shortage delays projects and raises risks. *Financial Times*, 1.
33. Mercer's. (2013). 2013 *Oil & gas talent outlook and workforce practices survey*. New York, NY: Mercer.
34. Chazen, G. (2014, July 17). Terrifying oil skills shortage delays projects and raises risks. *Financial Times*, 1.
35. Lopez, J., & Moughon, J. (2014). *The great crew change: How to minimize the emerging workforce shortage in the oil and gas industry*. Houston, TX: Blueprints for Business.
36. Mainwaring, J. (2013). *GE Oil & Gas tackles the "great crew change."* Retrieved from Rigzone website: http://www.rigzone.com/news/oil_gas/a/128806/GE_Oil_Gas_Tackles_the_Great_Crew_Change/?pgNum=0.
37. Haidar, T. (2013) *The great crew change: An extinction level event in the making for oil and gas? Oil and gas IQ*. Retrieved from www.oilandgasiq.com.
38. Arnsdorf, I. (2014, May 22). Crew Change: Millennials Hit the Oil Patch, *Bloomberg Business*. Retrieved from http://www.bloomberg.com/bw/articles/2014-05-22/millennials-look-to-cash-in-on-shale-energy-boom.
39. Senterfit, S. (2014, April 28). A practical approach to knowledge continuity during "the great crew change." *Oil & Gas Monitor*. Retrieved from http://www.oilgasmonitor.com/practical-approach-knowledge-continuity-great-crew-change/7082/.
40. Mills, R. (2014, June 1). The great crew change will come at a cost. *The National*. Retrieved from http://www.thenational.ae/business/industry-insights/energy/the-great-crew-change-will-come-at-a-cost.
41. Parmesh, M. (2008). The need for new recruitment strategies. *Talent & Technology*, 2(2): 21–23.

42. Tennant, J. (2012). Making informed human resources decisions based on workforce outlook. *World Oil, 233*(9). http://www.worldoil.com/magazine/2012/september-2012/supplement/making-informed-human-resources-decisions-based-on-workforce-outlook.
43. Williams, B. (2007). HR Council tackling Canadian petroleum staffing challenge. *Offshore, 2*, 8–10.
44. Deloitte. (2013). *Oil & gas industry rotator assignment survey report.* https://www2.deloitte.com/content/dam/Deloitte/uk/Documents/tax/deloitte-uk-global-mobility-transformation-oil-gas-rotator.pdf.
45. Clock winds down for experienced oil and gas hands. (2006). *Oil and Gas Investor*, 6(10): 6–16, 18–23.
46. Deloitte Research. (2005). *The talent crisis in upstream oil and gas: Strategies to attract and engage generation Y*. Retrieved from http://www.deloitte.com/assets/Dcom-Canada/Local%20Assets/Documents/CA_ER_TalentCrisisUpstreamOilGas_18_11_2005.pdf
47. Sprunt, E. S. (2008). Retaining women in the oil and gas industry. *Talent & Technology*, 2(2):12–16.
48. Davies, S. (2014, December 04). Shell trials virtual assistant Amelia for back-office role. *Financial Times*, 27.
49. Study: Project management capability key to selecting engineering contractors. *Oil and Gas Journal*. Retrieved from http://www.ogj.com/articles/print/volume-101/issue-27/processing/study-project-management-capability-key-to-selecting-engineering-contractors.html.
50. Pyron, D. (2008). Solutions to the recruitment and retention challenges. *Talent & Technology*, 2(2): 4–6.
51. SPE Research. (2012). *Training and development survey*. Retrieved from http://www.spe.org/career/docs/12Training-and-Development-Study.pdf.
52. Ryder, J. A. (2007). Complex human resource challenges: Call for new approaches. *Talent & Technology*, 1, 14–16.
53. Gulf Intelligence. (2014, February). *Women in energy—On the right track*. White paper presented at the International Petroleum Technology Conference, Kuala Lumpur.
54. Ogbonna, F. C. (2010). *The role of petroleum engineering education in the enhancement of oil and gas production* (SPE Report 140631). Houston, TX: Society of Petroleum Engineers.
55. Eaton, L, Casselman, B., & Power, S. (2010). Deepwater drillers face labor dilemma. *Wall Street Journal*. Retrieved from http://www.wsj.com/articles/SB10001424052748704268004575417383606531658.
56. Bairi, J., Manohar, B. M., & Kundu, G. K. (2013). Knowledge acquisition by outsourced service providers from aging workforce of oil and gas industry. *Journal of Information and Knowledge Management Systems, 43*(1), 39–56.
57. Gaysina, L. M. (2011). Deficiency in an abundance country: Shortage of highly-qualified personnel in Russian petroleum industry. *Oil and Gas Business*, 6. Retrieved from http://www.ogbus.ru/eng/authors/Gaysina/Gaysina_1e.pdf.
58. Gulf Intelligence. (2014, February). *Women in energy—On the right track*. White paper presented at the International Petroleum Technology Conference, Kuala Lumpur.

59. APPEA. (2008). *A mentoring model for the Australian upstream oil and gas industry*. Retrieved from http://www.businessgroupaustralia.com.au/wp-content/uploads/2009/08/bga-appea-mentoring-model.pdf.
60. Oil and Gas UK Next Generation Task Group. (2009). *Report to industry*. Retrieved from http://www.oilandgasuk.co.uk/cmsfiles/modules/publications/pdfs/EM004.pdf
61. Adams, S. (2013). The college degrees with the highest starting salaries. *Forbes*. Retrieved from http://www.forbes.com/sites/susanadams/2013/09/20/the-college-degrees-with-the-highest-starting-salaries/.
62. Graham, I. (2010). *Working conditions of contract workers in the oil and gas industries* (Working Paper No. 276). Geneva, Switzerland: International Labour Office, Sectoral Activities Department.
63. Oil and Gas UK Next Generation Task Group. (2009). *Report to industry*. Retrieved from http://www.oilandgasuk.co.uk/cmsfiles/modules/publications/pdfs/EM004.pdf.
64. Spotlight on the United States: Under the radar? The challenges of managing US rotational cross-border employees. (2013, January). *Global Oil & Gas Tax Newsletter: Views from Around the World*, pp. 2–8.
65. Frase, M. J. (2007, March). International commuters. *HR Magazine, 52*: 24–25.

Case 2

Skills Shortages in the Oil and Gas Industry: The Case of Australian LNG Development

> *The federal government has rejected claims from Chevron that Australia's high-cost economy is threatening the nation's biggest energy project, Gorgon, even as the Maritime Union of Australia demands a 26 per cent pay rise and more than 100 other benefits for its members, including Qantas Club memberships and iTunes store credits. As Chevron's $52 billion Gorgon project became embroiled in the -election campaign, trade union officials accused Chevron of seeking to dodge responsibility for poor labour productivity and high costs.*[1]

To illustrate many of the issues described throughout this book, we now present a case showing the employment challenges raised by the skill shortages in the development of Australian liquefied natural gas (LNG). Australia is in the midst of a boom in LNG and coal seam gas (CSG) projects. Between 2009 and 2014, seven separate projects reached final investment decisions (FIDs), moving from planning to construction. Seven of the 13 major LNG projects under construction globally were in Australia. This massive development in the oil and gas sector in Australia, totaling more than $200 billion in devoted capital, posed a series of major challenges for sourcing of labor. By early 2014, several projects were experiencing cost overruns and project delays as a result of labor costs and shortages.[2] As many projects suffered cost blowouts of 20%, 25%, or more, future projects were being delayed or canceled. The future of Australia's LNG industry was threatened.

The Labor Challenge

The labor challenge was a combination of four major forces: (1) rapid expansion, pitting one project's needs against another; (2) skill and experience demands, such that even with slower growth and less project competition, many labor shortages would require imported labor; (3) productivity challenges associated with construction, timing, and acquiring the right skills at the right times; and (4) remote locations far from population centers.[3] As a result, labor costs and shortages had skyrocketed. Some examples are the annual salaries of welders at A$400,000 (US$363,636), cooks at A$350,000 (US$318,182), and even laundry workers earning A$325,000 (US$295,455).[4]

The staffing challenges faced by these projects had expanded to include many different parties, such as the project company sponsors, the Australian and provincial governments and their various labor market entities, major Australian labor unions, and educational service providers in different parts of the country.[5] But regardless of the stakeholder, Australia was proving to be living case study of the rapid expansion of the oil and gas industry in a highly industrialized country and labor market.

Australia and project locations

The first staffing challenge for the growing LNG industry is that the majority of the Australian projects are located far from Australia's population centers. As illustrated by figure C2–1, the projects are spread across three major states—Western Australia, Northern Territory, and Queensland. With the exception of Queensland, the two other states are sparsely populated and therefore have little in the way of infrastructure. Even in the case of Queensland developments, most of Australia's population and workforce live far to the south along the coast of New South Wales. Geographic labor mobility was a critical problem.[6]

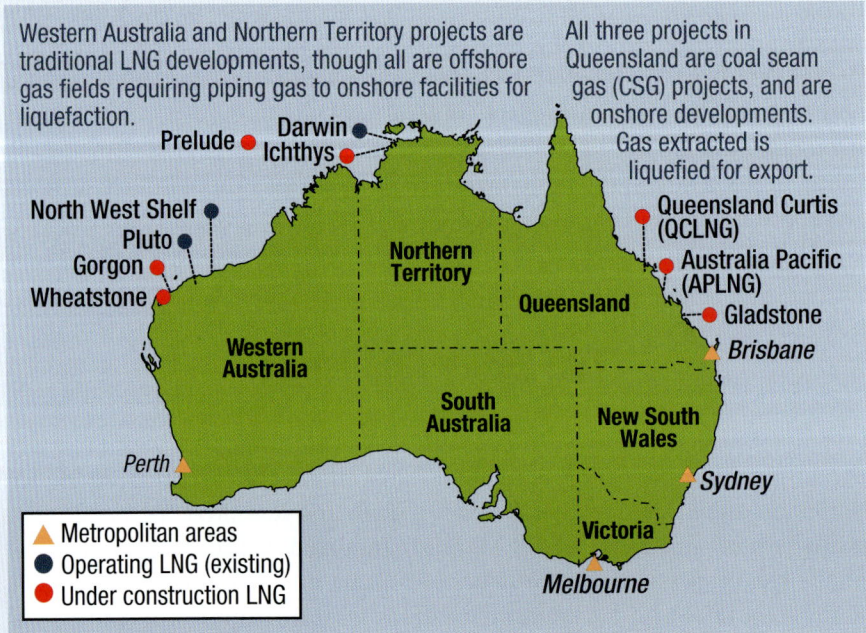

Fig. C2–1. Australian LNG development: existing and under construction

All three projects under construction in Queensland were CSG/LNG projects, as seen in table C2-1. CSG is produced from coal beds typically 200–1,000 m below the surface. This is considerably shallower than the shale gas developments typical of the United States. CSG operations also use hydraulic fracturing to a much lesser degree (roughly 40% of the time) compared to shale gas in North America. These projects are obviously also onshore developments. For example, the Queensland Curtis LNG project (QCLNG) taps CSG onshore, then pipes the gas to liquefaction facilities on nearby Curtis Island. Each of the projects employed roughly 5,000 workers in the construction phase, and would employ anywhere from 700 to 1,000 throughout their production lives.

The LNG projects in the Northern Territory and Western Australia were conventional LNG developments, all offshore. Each offshore development would have production facilities sitting on the ocean floor anywhere from 100 to 450 km off the Australian coast. The raw gas produced would then be piped to processing and liquefaction facilities onshore via pipelines laid across the ocean floor. Then it would undergo liquefaction by LNG trains. An LNG train is a liquefaction unit in which natural gas is converted to a liquid state by chilling the gas to below 160°C. Once liquefied, the LNG could then be transported by special LNG tankers (keeping it chilled to liquid) to any destination in the world. All current Australian projects had signed long-term sales and purchase agreements (typically 15 to 20 years) with customers in Japan, Korea, and Taiwan.

There was one exceptional project on the list. Prelude was to be a floating LNG project. The world's largest floating vessel is currently under construction in Korea, which when finished, would be permanently moored approximately 475 km offshore and would be capable of all gas processing, liquefaction, and offloading functions without any onshore facilities. This would be the world's first floating LNG project, and many believed it might be the partial solution to the ever-increasing costs of LNG development.

Table C2–1. Australian LNG project developments

Project	Capacity (mt/day)	FID Cost (billion A$)	Current Cost (billion A$)	Cost Increase (%)	Type	Trains	Project Startup	Location
North West Shelf	16.3	—	25.0	—	Conventional	5	1989	Carnarvon Basin, Western Australia
Darwin LNG	3.7	1.7	1.8	2.9%	Conventional	1	2006	Darwin, Northern Territory
Pluto Train 1	4.3	11.2	14.9	33.0%	Conventional	1	2007	Carnarvon Basin, Western Australia
Gorgon	15.0	34.0	54.0	58.8%	Conventional	3	2015	Barrow Island, Western Australia
Queensland Curtis (QCLNG)	8.5	15.0	20.4	36.0%	CSG/LNG	2	2014	Gladstone, Queensland
Gladstone (GLNG)	7.8	16.0	18.5	15.6%	CSG/LNG	2	2015	Gladstone, Queensland
Wheatstone	8.9	29.0	29.0	0.0%	Conventional	2	2016	Carnarvon Basin, Western Australia
Ichthys	8.4	32.0	44.2	38.1%	Conventional	2	2016	Darwin, Northern Territory
Australia Pacific (APLNG)	4.5	18.0	24.7	37.2%	CSG/LNG	2	2016	Gladstone, Queensland
Prelude	3.6	10.0	13.0	30.0%	Floating LNG	1	2017	Browse Basin, Western Australia
Total or average	81.0	166.9	245.5	28.0%				

Source: Constructed by authors. Notes: CSG is coal seam gas. LNG is liquefied natural gas. All values in Australian dollars. Current cst estimates are unofficial, as projects are not yet complete. Although not official, Gorgon's final cost is expected to be revised upwards to A$60 billion.

Staffing and skill requirements

Although Australia has seen similar isolated project challenges with its mining industries over the years, the size, rapid pace of development, and high skills and experience needs had proven exceptional. In addition to the projects already running and under construction depicted in figure C2-1, there were eight other major Australian LNG projects under some stage of planning and evaluation. And oil and gas labor demands were not limited to Australia itself. At the same time, huge LNG projects were also underway just to the north in New Guinea and Papua New Guinea, as well to the far north in Singapore.

Australia may be a country of considerable geographic size, but its population of 23 million (in 2013) and a total workforce of 11.6 million are not large by global standards. The rapid build-out of the gas sector would have both national and state impacts of substance.

The employment demands differed between projects in Queensland and those in Western Australia. The Queensland projects involved CSG—the extraction of natural gas from coal—and therefore had additional technical requirements as well as operational needs (e.g., the drilling and development of more wells). In Queensland, most of the LNG projects now under construction involved at least two trains, where each train would require between 550 and 650 professional and technical staff during operations, and an additional 3,900 workers per train throughout the 20 to 30 phases of the projects.

The Queensland Curtis LNG project (QCLNG) was representative of the complexity of CSG/LNG employment needs. QCLNG was scheduled for start-up in late 2014. In its Social Impact Plan filed with the Queensland authorities, it estimated that 800 workers would be required for initial well development activities alone, with an additional 400 workers for well development works continuing into the far future. For its gas operations and maintenance, QCLNG estimated that it would continue to employ about 750 workers indefinitely.

Energy Skills Queensland estimated that the CSG component of these LNG projects in Queensland would alone employ an additional 6,000 workers. The Australian Petroleum Production & Exploration Association forecasted that, including development, production, operations, and maintenance staff, the CSG/LNG sector would generate as many as 18,000 new jobs by 2017.

The LNG projects in Western Australia, although simpler in principle as a result of using more traditional gas reservoirs as their basic feedstock, all had additional complexities that added to cost and staffing requirements. Projects on the Western Australia coast would require approximately 300 additional production workers if a *greenfield* project (a completely new investment project), or about 150 additional workers if a *brownfield* development (an expansion of a preexisting facility). With at least seven additional LNG trains already under construction and scheduled for start-up by 2017, an additional 2,400 workers will be needed as the projects move into their production phase. As illustrated by figure C2–2, the estimated employment impacts in Western Australia were large. An additional concern was the impact these major developments have on small remote communities. Public services quickly became inundated by the influx of workers, and the costs of living in the areas have seen rapid and large increases.

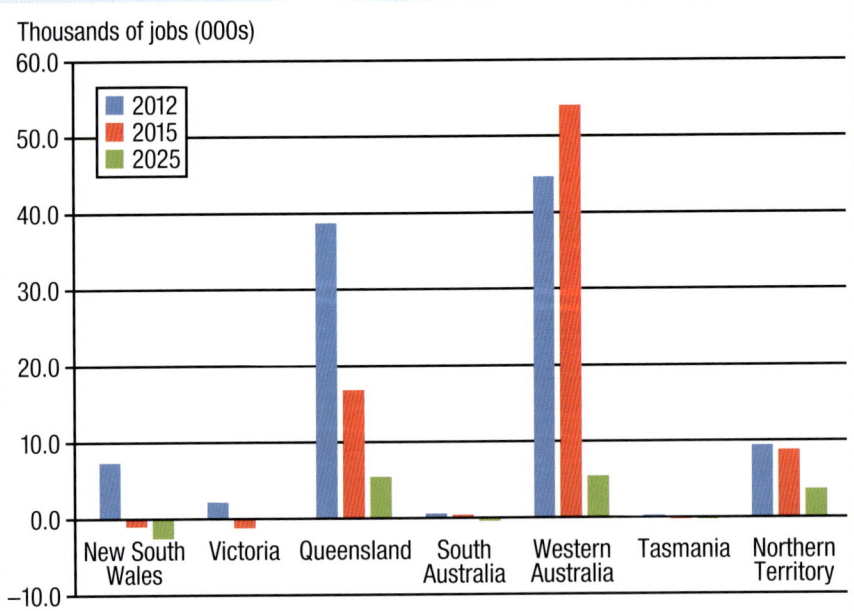

Fig. C2–2. Estimated employment impact of gas development by state
Source: Constructed by authors based on data estimates by Deloitte Access Economics.

The industry's impact in a sparsely populated portion of the country was already dramatic. Australia's Department of Education, Employment and Workplace Relations (DEEWR) reported in 2013 that it expected total sector employment Australia-wide to increase from 16,700 workers in 2011 to more than 28,000 in 2017. The industry's continuing staffing costs were, however, not improving according to DEEWR[7]:

> Surveyed employers indicated that employees leaving the company were often taking up employment elsewhere in the offshore oil and gas sector. This suggests a high level of movement within the sector rather than recruitment from other sectors.

Gorgon's cost blowout

The Gorgon project had become a symbol of the risks associated with leading-edge technology and the increasing burden of staffing costs. A joint venture of Chevron (50%), ExxonMobil (25%), and Shell (25%), the project's cost was originally estimated at US$34 billion. By 2014 this cost had risen to over $54 billion, a $20 billion cost blowout, and further cost increases and project delays were anticipated.

The Gorgon project was based on an expansive series of gas fields between 130 and 200 km off the Northwestern Australia coast. The gas would be produced from subsea wellheads at a depth of roughly 250 km, and then transported via trunk lines to gas processing facilities on Barrow Island, 50 km off the Australian coast. On Barrow Island (largely an ecological preserve), the gas would be processed, removing carbon dioxide, water, mercury, and other nonmethane components. The methane would then be liquefied by three LNG trains and exported. The captured carbon dioxide was to be geosequestered, that is, reinjected for permanent storage in a series of caverns more than 2,000 m below Barrow Island.

The causes of Gorgon's rather high-profile cost blowout were numerous and heavily debated. Chevron, the project operator, has cited a combination of causes:

- High Australian dollar
- High wages
- Low productivity
- Weather delays
- Logistical challenges

Most of these causes were relatively well documented, although there was considerable debate over the high-wage and low-productivity arguments. Representatives of several Australian government agencies had gone on record arguing that high wages were indeed harming LNG development in Australia, arguing that Australian wages were 30% higher than comparable jobs in the closest LNG supplier sector, Canada.

Several Australian unions refuted Chevron's claims. The Maritime Union argued that union workers were being unfairly blamed for low productivity and resulting cost blowouts. The union argued that multiple levels of bureaucracy and mismanagement by Chevron were largely to blame.[8]

> For example, a KJV [Kellogg Joint Venture Gorgon] report showed that it takes 132 days to load and unload a barge on the non-unionised Barrow Island wharf, but it takes 95 days to load and unload the same barge at the fully-unionised wharf at the Australian Marine Complex. Instead of blaming workers for the cost blowouts on Gorgon, Chevron management should be taking a good look at themselves and their management practices.[9]

Industry sources, however, agreed that the combination of higher wages, lower Australian productivity, and the extreme locational and logistical challenges had combined to drive costs upwards at Gorgon and all other LNG projects currently under development. As noted in Table C2-1, these cost blowouts were exceptionally large and had been the subject of much debate across the global oil and gas industry as major oil companies reconsidered their capital commitments to similar new projects.

Strategic promise and challenge

The development of the gas industry for Australia is a clear priority for Australia.[10] As noted by a McKinsey analysis in 2013, the development of LNG for export from Australia would soon surpass traditional export earnings sources such as coal (steam and thermal), iron ore, gold, meat, wool, alumina, wheat, machinery, and transport equipment. Deloitte Australia estimated that in the coming years Australia would average one export tanker's departure daily, and assuming that an average tanker contains 140,000 cubic meters of LNG, averaging A$33 million per tanker in export earnings, the resulting tax contributions to Australia would be A$8.7 million.

Australia was also confronted with a more fundamental issue—competitiveness.[11] A recent McKinsey study has analyzed the competitiveness of Australia's LNG sector against other major competitors like Canada and the prospective exports of the United States. Its conclusion is sobering: that a multitude of issues are contributing to lower productivity and higher cost. Specifically regarding labor rates and productivity, McKinsey identified the following four factors for improvement:

1. **Constructing residential communities closer to LNG sites.** This would reduce travel time by workers, increasing average work time by 14%, increase productivity through reduced work pressures on long shifts, and increase stability in team composition.

2. **Optimizing shift patterns.** McKinsey's studies indicate that Australian workers spend more time traveling than workers in other countries due to shift patterns. Optimizing shift patterns for travel time could result in significant cost reductions and productivity improvements.
3. **Improving site productivity including lean construction.** Improved training, supply chain structures, and workplace logistics could increase productivity and reduce waste. Much of this efficiency gain would come from reduced construction times.
4. **Utilizing skilled labor.** Ensuring a sufficient supply of skilled workers would reduce above average wage increases in the gas sector and increase productivity through utilization of a more experienced, skilled workforce. The study specifically noted that Brazil had seen major productivity improvements through increased work permit approvals for foreign skilled and experienced workers.

Employment Structures and Strategies

Some parts of the business community and union groups have been critical of the use of overseas workers and firms to undertake significant elements of new project development. In considering local content issues, it is important to recognize the complexities and commercial realities of developing world-scale oil and gas projects.[12]

Australia, like any country experiencing an oil or gas boom, was confronted with a multitude of labor interest and initiatives. As a result of both the lessons of history and the new lessons experienced at home, Australia was constantly exploring a mix of labor force development, employment, training, and recruitment options.[13]

Indigenous skills development

Without exception, all of the LNG projects had identified the promise of the indigenous labor force. Although requiring early and substantial investment in training, the investment in indigenous workers was considered a win-win solution. The indigenous populations, many traditionally isolated in the very same locations as the LNG projects themselves, could gain marketable and sustainable skills that would support meaningful and economically sustainable employment for many years to come. The LNG projects themselves would benefit from a local and sustainable workforce that was both competitively priced and increasingly purposeful in productivity.

Capex versus sustainable skills

The early construction and development skills required by the LNG build-out required a highly skilled set of employees not readily available within the regions or the country. But this capital expense (capex) phase was relatively short-lived, and many of these highly skilled and experienced workers would not be needed past the completion of the phase. The greater and longer-term employment needs could be met with local employees if projects, industries, states, and the country as a whole invested in developing these workers early and often enough to feed the growing LNG industry.

The FIFO workforce

With few exceptions, the LNG projects of Australia would require a multitude of highly skilled workers, in both the capex and ongoing production phases of operations, who would ultimately be either FIFO (flown-in and flown-out) or DIDO (drive-in and drive-out). Labor mobility, both inside a country and cross-countries, is critical for successful staffing strategies. FIFO workers were seen as a continuing and productive solution to staffing challenges, allowing highly skilled and sought-after skilled workers to operate in regional areas, remote areas, without the need to relocate or spend extended periods away from their families.

FIFO workers are often located in regional or metropolitan areas, including their families, and commute to work by plane, living on-site in villages or compounds when on-site. The workers are typically compensated some allowance for commute time over and above their base salary, and have meals and accommodations provided when on-site. As opposed to the oil and gas industry's typical rotator schedules, FIFO workers would often work two weeks on (seven-day weeks) and one week off. Although obviously more costly than local labor (if it was truly available), it has been a cost-effective solution for filling hard-to-find skills and experienced workers who may ordinarily be unwilling to work in such remote locations. Ironically, in some cases more traditional workers in major metropolitan areas have been seen to occasionally move aggressively to obtain FIFO work opportunities, presumably for the combined additional income and skill development opportunities.

Project sponsor employment investments

Given the duration of LNG projects, with extended business lives of more than 50 years in some cases, project sponsors with a multitude of project investments have begun developing a skilled workforce from a grassroots level. Chevron, one of the largest players in the Australian LNG sector development, has invested $12 million in training apprentices and employing university graduates as part of its project programs.[14] The company has initiated a number of apprenticeships, some targeted specifically at Australian Aboriginals.

Another example can be found in the baseline equipment and service supplier industry that will be supporting the LNG industry for many years to come. General Electric's subsidiary GE Oil & Gas has ongoing personnel initiatives that include working with schools to promote and chart career paths in the oil and gas industry, offering summer internship programs and providing flexible work packages for its women employees to increase retention rates.

Nontraditional employment sources

One of the more obvious but unused alternatives until recently was the attempt to either employ or retain older workers. Older workers represent a large but largely untapped resource, and one that often is not as challenged by compensation or family-related concerns.

Other alternatives

The seriousness of the employment shortage in the Australian LNG industry has motivated a number of major strategic and corporate organizational solutions not often seen in other sectors or even in the oil and gas industry.

Company-sponsored "upskilling." Many of the LNG companies were funding and supporting their own long-term skill development programs. One example was the Global FLNG Training Consortium in Western Australia, a partnership between Shell, the Challenger Institute, and Curtin University. This was a multiyear training program to develop LNG technicians in Western Australia, one of many similar programs in the three major LNG states.

Integrated LNG project development. With so many large-scale and remotely located projects, the industry is reconsidering the typical stand-alone development. Although Australian LNG project locations are located far from metropolitan areas, they are clumped in two or three major regions of the country, sometimes less than 100 km from each other. This has caused a substantial reconsideration of the structure of the projects, with many expansions and new developments on the way incorporating the shared use of capital intensive segments such as gas processing, liquefaction, and export terminals. Shared use of common facilities would not only reduce capital investment commitments drastically, but they would also reduce skilled labor needs in these same remote locations. Given the rapid—and in the opinion of some market analysts, overbuilt—scaling up of the LNG capacity in Australia and the South Pacific in general.

Enterprise migration agreements. The government of Australia has responded with a multitude of programs and immigration law changes to support the industry—and its ability to train, hire, and retain Australian workers.[15] The Enterprise Migration Agreement (EMA) initiative was one such program. For projects with capital expenditures exceeding A$2 billion and expected employment of 1,500 people (which included nearly every existing or proposed LNG project in Australia), the government would provide a series of benefits to the projects to prevent them from suffering increasing cost and staffing shortfalls.

The basic elements as described by the Australian Department of Immigration and Border Protection were as follows:[16]

- An EMA is negotiated with the project owner and sets the terms by which workers outside Australia will be engaged on the project, as well outlining training commitments that must be met by the project.
- The terms set out in the EMA will include the occupations, qualifications, and English language skills of the workers from outside Australia.
- Subcontracting employers, with the endorsement of the EMA holder, will sign onto labor agreements under the terms of the EMA, ensuring that responsibility for sponsorship obligations rest with the direct employer of the worker from outside Australia.
- EMAs would take a project-wide approach to meeting skill needs, no longer requiring each individual subcontractor to negotiate separate labor agreements. The "bulk" of negotiations would occur under a single umbrella with the project owner/sponsor. Workforce needs and planning would be expedited.
- Under an EMA, occupations ordinarily not eligible for standard migration programs under Australian law would be sponsored, provided the project could justify a genuine need—skill set including experience—that could be met from the Australian labor market.

Although the EMA programs have run into selective union opposition within Australia, the LNG project sponsors, largely multinational or foreign corporations working within Australia, have been strong supporters of the program given their global experience. Union opposition has been focused to date on the possibility of foreign sourcing of workers who are compensated below domestic rates. Although this might be the case for guest worker populations in a number of oil and gas developments in some Mideast countries, this has largely not been a problem in Australia. The problem in Australia has generally been the opposite: Foreign workers recruited for rotations in Australia have tended to be at extremely high and rising costs.

Summary

The massive growth in the Australian LNG industry is now facing significant cost and structural challenges. The need for skilled and experienced labor in remote areas of Australia has resulted in strong cost and management pressures to the point that many prospective investments are now being delayed or indefinitely postponed. If the true potential of the industry for Australia is to be seized, skilled labor costs and availability challenges will have to be resolved, and soon.

References

1. Macdonald-Smith, A., & Skulley, M. (2013, August 19). Labor: LNG costs not our fault. *Financial Review*. Retrieved from http://origin-www.afr.com/p/markets/market_wrap/labor_lng_costs_not_our_fault_780Fq5CWRvMMRsnDwnzdkN.
2. Koh, Q. (2012, February 4). Australian LNG industry faces severe labor crunch. *Rigzone*. Retrieved from www.rigzone.com/news/oil_gas/a/123976/Australian_LNG_Industry_Faces_Severe_Labor_Crunch.
3. Macdonald-Smith, A. (2014, February 12). How Australia can win the LNG war. *Financial Review*. Retrieved from www.afr.com/business/construction/how-australia-can-win-the-lng-war-20140212-ixt5j.
4. Chamber of Shipping of British Columbia, (2014, April 10). Labour costs hit Australian LNG developments. Retrieved from http://www.cosbc.ca/index.php/international/item/327-labour-costs-hit-australian-lng-developments.
5. Skills shortages are the primary challenge for oil and gas employers in Australia. (2014, February 26). Hays.com. Retrieved from www.hays.com.au/press-releases/skills-shortages-are-the-primary-challenge-for-oil-and-gas-employers-in-australia-145255.
6. *Geographic Labour Mobility Study*. (2013, August 13). SubmissionProductivity Commission, Canberra, Australian Petroleum Production & Exploration Association, Melbourne, Australia.

7. *Resources Sector Skill Needs, Report 2012*, Australian Workforce and Productivity Agency, Commonwealth of Australia 2012, p. 20.
8. Diss, K., & Lannin, S. (2013, December 12). Chevron reveals second Gorgon gas project cost blowout. *ABC News*. Retrieved from www.abc.net.au/news/2013-12-12/gorgon-gas-project-cost-blowout/5151982.
9. MUA says workers not to blame for cost blowouts in Chevron's Gorgon LNG project. (2013, December 12). Retrieved from www.itfglobal.org/presidents_page/?p=79
10. "Australia's skills shortage: Where to next? (2013, May). *Gas Today*. Retrieved from http://gastoday.com.au/news/australias_skills_shortage_where_to_next/81434.
11. Ellis, M., Heyning, C. & Legrand, O. (2013, May). Extending the LNG boom: Improving Australian LNG productivity and competitiveness. *McKinsey*, p.x.
12. Ellis, M., Heyning, C., & Legrand, O. (2013, May). Extending the LNG boom: Improving Australian LNG productivity and competitiveness. *McKinsey*, p. viii.
13. *Harnessing our comparative energy advantage.* (2012, June). Australian Petroleum Production and Exploration Association Limited, Deloitte Access Economics Retrieved from https://www.deloitteaccesseconomics.com.au/uploads/File/APPEA%20-%20Main%20report%20-%20Final%20-%2022%20June%202012.pdf.
14. Chan, S. (2012, January). Tackling the skills shortage head on. *The Australian Pipeliner*. Retrieved from http://pipeliner.com.au/news/tackling_the_skills_shortage_head_on/65634.
15. Lannin, S. (2013, December 16). Foreign workers may fill oil and gas skills shortage, report finds. *Resources*. Retrieved from http://news.yahoo.com/foreign-workers-may-fill-oil-131657474--spt.html.
16. *Enterprise migration agreements* (Fact Sheet 48A). Retrieved from http://www.immi.gov.au/media/fact-sheets/48a-enterprise.htm.

7

The Proactive Stakeholders of the Oil and Gas Industry

The Influence of Stakeholders in the Oil and Gas Industry

There are few industries that possess the complexity of stakeholder interests as that of the global oil and gas industry. As noted in chapter 1, in addition to the traditional stakeholders of firms—owners (private or public), creditors, employees (managerial and nonmanagerial), suppliers, and customers—the oil and gas industry is subject to much more intense interest and scrutiny from external stakeholders such as the community and government.

There are several reasons why these stakeholders are particularly vocal, visible, and influential. First, as chapter 8 will explore in greater depth, governments are typically the owner of the hydrocarbons, controlling access to the oil and gas within the country, in addition to serving as the regulator for the industry. In the end, little happens within a nation's boundaries in its oil and gas sector without some major degree of government involvement. This involvement includes the state's interests in safety and security of energy supplies, the economic implications of energy development, and the impacts on civil society, including the environment.

Second, the historically poor reputation of the industry has increased the visibility and scrutiny focused on firms operating in the oil and gas sector, evoking considerable criticism for any perceived negative outcomes. For example, any nontrivial rise in gas prices is sure to attract a lot of attention from the news media and, subsequently, customers in general. Third, the broad reach of the industry means it touches nearly every person on the planet as a customer or community member, thus everyone on earth is a stakeholder.

Stakeholders in the Oil and Gas Industry

Stakeholders are any organization or group of people that have a vested interest in the firm. Stakeholders in the oil and gas industry include:

1. Communities and society—all the communities that have a vested interest in the behaviors of the firm as well as the media and society at large
2. Owners and investors—owners, shareholders, and suppliers of capital to the firm
3. Employees—executives, managers, nonmanagerial employees, and contractors
4. Customers—firms or individuals who purchase any of the company's products
5. nongovernment organizations (NGOs)—any group or organization that is not part of a government and is nonprofit
6. Unions—unions that represent employees of the firm as well as other unions in the firm's business environment
7. Governments—any governments that have a vested interest in the firm's behaviors

The role of government is so influential in the oil and gas industry that we address it in its own chapter (chapter 8) rather than here. However, in addition to governments, firms in the oil and gas sector have other vocal, highly visible, motivated, and passionate stakeholders. We now discuss each of these stakeholders in detail.

Communities and society

Oil and gas companies can impact the lives of virtually everyone in the country where they are located, directly or indirectly, positively or negatively. Clearly oil and gas discoveries and the subsequent entry of oil and gas companies can positively impact a country through its powerful role in economic development—including the facilitation of transportation, the creation of new infrastructure, the creation of new jobs, a tremendous income source, and the improved availability of energy, among other benefits. However, an abundance of oil and gas does not guarantee positive benefits for a country's citizens. Many of the biggest oil and gas–producing countries, particularly in the middle east, have relatively low life expectancies, low levels of adult literacy, low enrollment rates in formal education, low gross domestic products (GDPs), high income inequality, low political freedom, and high levels of corruption.[1] Although it would be overstating to say that oil and gas resources cause all these negative outcomes, "the curse of oil" does not necessarily widely benefit a country's inhabitants. One step that countries and all stakeholders can take to avoid the curse of oil is to create a revenue management plan (RMP). This is discussed in greater detail in Current Challenge Box 7-1.

Current Challenge Box 7–1. Long-Term Revenue Management

Many nations are well aware of the value or wealth that may result from the development of their oil and gas, but they are also aware of the possibility of the benefits of its development being spread unevenly across its citizenry, at the present time and even over into the future. How can revenue be managed such that all stakeholders benefit, even in the event of changes in the government? Although this issue has been around a long time, it is still of concern given the instability of many governments around the world. One solution is a *revenue management plan* (RMP). The RMP adopted by the government of Chad prior to the development of its oil and gas in 1999 is one example of how all stakeholders, as a group, can try to protect their interests for the long term (fig. 7-1). The Chadian RMP was designed to create a system of assurances, both inside and outside of Chad, that the funds generated by the development of its oil would go to the Chadian people, present and future.

The funds generated by the consortiums of companies producing the oil and selling it on world markets would first be deposited in a London bank account. This was done to ensure that 10% of the funds would be deposited in the Future Generations Fund, which was created to save these earnings for the citizens of Chad in the future after the oil was produced and depleted, beyond the immediate reach of the Chadian government. The balance of the funds would then be passed to the Chadian government treasury, which had agreed to distribute the funds to specific social sectors (80% of the funds), the general government budget (15%), and the specific local area directly impacted from the oil development and production—the Doba region (5%). To ensure that the RMP was followed, a separate monitoring group was organized: the Collége de Contrôle et de Surveillance des Resources Pétrolieres. Staffed by appointed and independently elected representatives, the college would be a nongovernmental body whose primary purpose was to ensure that the all parties complied with the RMP.

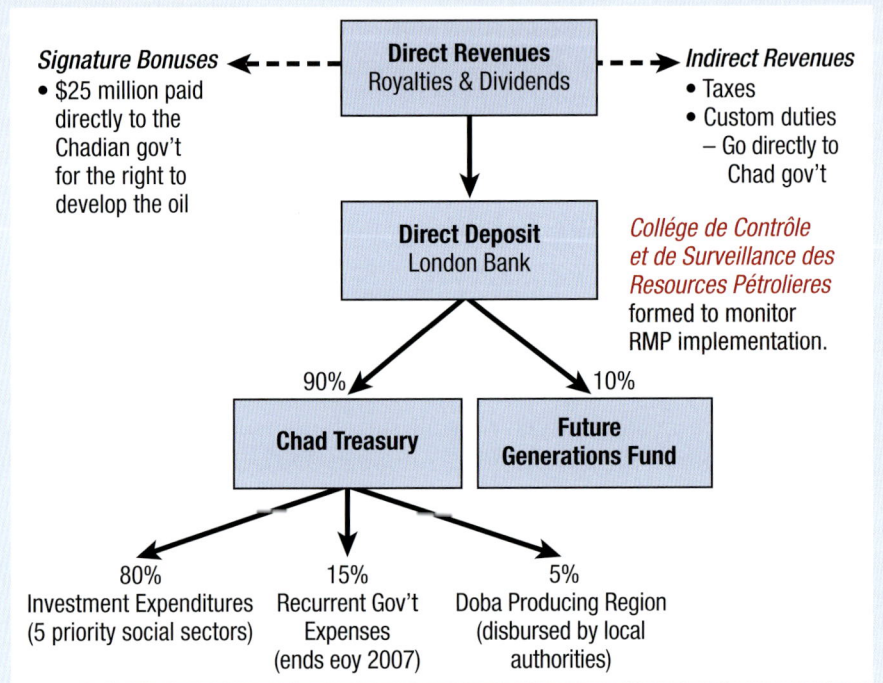

Fig. 7–1. Chad revenue management plan (RMP)

RMPs have been recently introduced in Ghana and Uganda.[2] Because RMPs frequently involve a new governing body of oversight, this adds yet another stakeholder to the mix, whose satisfaction can be at least partially addressed through the human resource functions.

Moving away from the country level, benefits and costs of oil also occur at the community level. As at the country level, positive economic outcomes from oil discoveries are likely to result at the community level, with relatively high levels of income, low unemployment, and economic revivals occurring in oil rich communities. However, negatives may also occur at the community level, including environmental issues and economic restructuring, which leads to economic winners and losers. For example, oil and gas discoveries in Ghana led to the perception of negative outcomes by many in the surrounding communities, including decreases in fish catch, loss of jobs, and reduced income.[3] In Midland, Texas, the boom caused by fracking has greatly improved local economic conditions, but also created a strain on local living conditions because of housing shortages, overcrowded schools, more traffic accidents, and more noise.[4] The general negative reputation of the oil and gas industry is also likely to spread to the community level and affect the community's perceptions and concerns.

Community concerns. Community concerns can be grouped into three general categories that may be captured under the larger umbrella of *corporate social responsibility* (CSR): (1) health and safety, (2) environmental concerns, and (3) social and economic concerns.

Health and safety concerns include:

- Safety and security: the increased risk to the community and workers resulting from catastrophic events, labor unrest, the presence of security forces, and an increase in crime.
- Health: the possibility of the increased health risks to the community, employees and their dependents, including risks due to damage to the environment and risks attached to product safety.

Environmental concerns include:

- Climate impact: the organization's effect on climate in the long term, including concerns about the organization's greenhouse gas emissions, use of energy, type of energy used, and flaring and venting gases.
- Ecosystem impact: the organization's effect on the community's ecosystem in the long term, including concerns about the organization's direct and indirect effect on biodiversity and the ecosystem related to the terrestrial, freshwater, and marine environment of the local community and beyond. Table 7–1 lists examples of the activities oil and gas companies perform that can directly and indirectly lead to negative effects on biodiversity and ecosystems.[5]
- Local environmental impact: the organization's effect on the local environment in the short term. Local environmental concerns encompass the organization's air emissions, spills to the environment, discharge to water, and waste disposal. Figure 7–2 shows the environmental expenditures by the US oil and natural gas industry by sector. Refining, exploration, and production account for more than three-quarters of the expenditures. Figure 7–3 shows the environmental expenditures by emissions medium. The cost of reducing air and water emissions accounts for more than half of the total expenditures.[6]

Table 7–1. Examples of possible direct and indirect effects on biodiversity and ecosystems

Activities	Direct Effects	Indirect Effects
The movement of equipment, materials, supplies, and people Land clearing Revegetation	Introduction of non-native species	Consumption or displacement of native species
Land clearing New infrastructure	Deforestation Loss of habitat Wildlife disturbance Soil compaction	Soil erosion Waterway contamination Loss of species Loss of productive capacity
New Infrastructure Increased population Creating access to undeveloped areas	Air, water, and soil pollution Increased demands on resources Contamination	Habitat conversion, degradation, or fragmentation Degradation of ecosystem functions
Material spills or leakage	Air, water, and soil pollution	

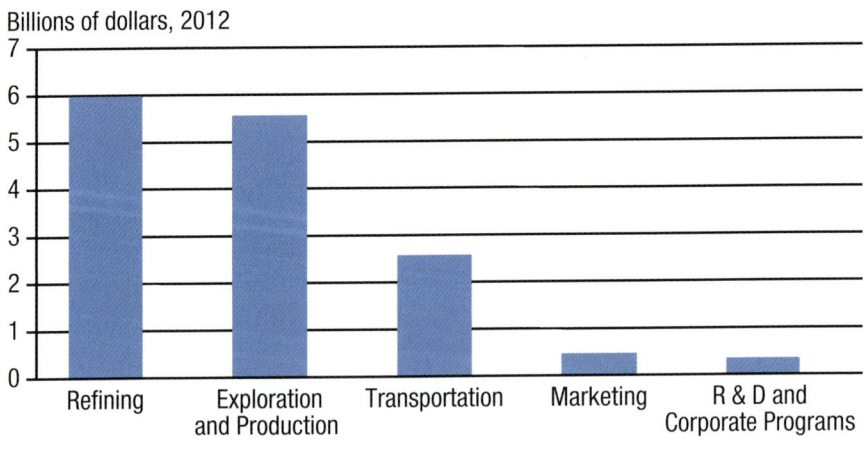

Fig. 7–2. Environmental expenditures by US oil and gas industry by sector
Source: American Petroleum Institute (API).[7]

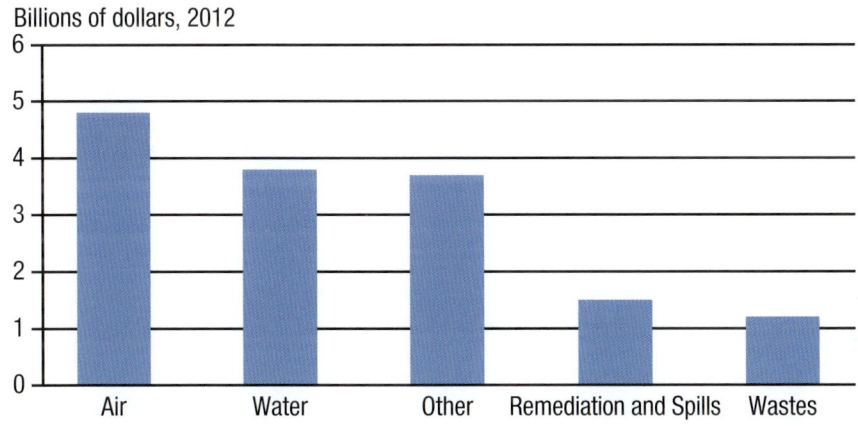

Fig. 7–3. Environmental expenditures by US oil and gas industry by emissions medium
Source: API.[8]

Social and economic concerns include:

- Lack of voice: the community's desire for transparency, consultation, information disclosure, and participatory actions with the company regarding matters of the community's interest.
- Fair compensation: compensation of land acquisition and property rights including the basis, rates, and type of compensation.
- Resettlement: the possible need for resettlement and the terms and process of the resettlement.
- Impact on cultural resources: the possible impact on land having archaeological, paleontological, historical, religious, or natural aesthetic value.
- Employment issues: the fair treatment of community members for consideration of employment as well as their treatment once they are employed.
- Other community impacts: the effects of economic development on the community including issues related to immigration; inflation; social equity; supplies of food and water; and overburdening infrastructure, schools, and other service and recreational facilities.[9]

Many of the social and economic concerns are related to the more general notion of human rights. Human rights generally include the rights to life, freedom of expression, privacy, and education in general. In a work context, human rights include favorable work conditions, freedom of association and collective bargaining, unforced labor, and nondiscrimination. The concern for human rights generally goes beyond just the community level and is shared by many stakeholders including employees, customers, unions, NGOs, and governments. Although following laws and regulations will address many of the human rights concerns, the widely diverse contexts and business partners of oil and gas firms (including governments with poor human rights records) requires a broader approach than just following regulations. Publications such as the European Commission's *Oil and Gas Sector Guide on Implementing the UN Guiding Principles on Business and Human Rights* can help firms achieve this.[10]

The perceptions by the community of the benefits and the costs of oil and gas projects depend on the project. For example, the community in Northern Louisiana around the Haynesville Shale Formation saw many positives regarding the unconventional shale gas development including increased jobs, more tax revenues, better services, and greater economic opportunities for local business and landowners. The perceived negatives of the community included increased road damage; more noise; more traffic accidents; perceived threats to public health, animals, and the rural landscape; and degradation of water resources, a cost more likely to be associated with hydraulic fracturing.[11]

The perceptions by the community of the benefits and the costs of oil and gas projects may also greatly vary within the same community, because some people will receive more benefits, while others may bear more of the costs. In the Haynesville example above, most people in the community believed that the benefits of development surpassed the costs; however, there were some who strongly disagreed that the benefits were worth the risks. Another example of this is the *NIMBY* (not in my backyard) attitude; that is, community members generally believe that there are many positives about a project and support it, but do not want it located near them because this would increase the negatives for them. For example, most community members are pro-wind power, but most would not like a large windmill obstructing the view from their home. Also, depending on the location, the community may also include native peoples or indigenous groups, who may require special consideration because of the different impact projects may have on them.

Addressing concerns. There are many different mechanisms that can be used when addressing community and societal concerns. Depending on the country, regulations may provide guidance on how a company should handle the concerns of local communities. Interacting with the community early in the process of the project and throughout the life of the project is important to project success. Early interactions should identify and help create plans of actions to address the community's concerns and needs. These interactions can also help establish realistic expectations and commitments.[12] A recent new method of interacting with a community is the use of social media. For example, Chesapeake Energy has regional Facebook pages targeted at each of its shale plays around the country. This allows the company to communicate directly with the community in each region, giving the community voice and allowing it to address concerns.[13]

Addressing community concerns has become an important consideration of oil and gas companies when conducting their core business. For example, environmental impacts are routinely minimized or mitigated by integrating biodiversity conservation objectives and impact mitigation practices into the early phases of projects. Dealing with indirect, longer-term environmental impacts usually requires early and continuous engagement with all applicable stakeholders. However in addition to working with and reacting to stakeholders, many oil and gas companies choose to undertake proactive measures to support communities, such as funding training, creating education initiatives, supporting infrastructure development, funding health centers, and supporting local charities.[14] Social and community investment programs can be implemented through international NGOs, national or local NGOs, institutions, or consultants, or through their own community development departments or foundations.[15]

Finally, concerns regarding transparency can be addressed directly by voluntarily reporting CSR or sustainability performance. Doing so will also indirectly help address all the other community concerns because employees will know that their performance relative to each concern mentioned above will be widely publicized. Annual sustainability reports also help instill a culture in the organization that recognizes the importance of stakeholders. A publication by the International Petroleum Industry Environmental Conservation Association (IPIECA), API, and the International Association of Oil and Gas Producers (IOGP), *Oil and Gas Industry Guidance on Voluntary Sustainability Reporting*, provides guidance on the desirable content of such reports.[16]

Owners and investors

Ownership in the oil and gas industry differs dramatically across the globe. At one end of the spectrum is the United States, where a private individual may own oil and gas assets and potentially profit from their development. At the other end of the spectrum is a multitude of countries in which the state owns all oil and gas assets, decides who and when and how they may be developed, and takes nearly all profits or returns from their development. And in between are a number of different cultures and countries and legal systems in which ownership has varying meanings and benefits. Yet, even in a country like the United States, there are many restrictions over oil and gas development (including, at the time of this writing, the prohibition against exporting oil from the United States).

A second financial investor often overlooked is the creditor, the financial services provider who provides capital in the form of debt to governments, national oil companies, private oil companies, and all others involved in the industry in order to undertake the massive investment required today in the global oil and gas industry. These creditors traditionally require repayment in a timely and predictable manner, and they are paid before private owners are. In return for the use of their capital, they are paid interest—which has been extremely low in major markets across the globe in the past decade.

The capital provided by creditors may or may not be secured. If secured, their rights to take ownership of assets because of nonperforming loans is often severely restricted in the global marketplace where energy is often treated more like a national good than a product for commercial sale. As stakeholders, they may have environmental and social requirements in place (affecting other stakeholders) that are applicable to any project they fund. For example, both the World Bank and the International Finance Corporation have many policy directives, advisory documents, reports, and publications related to social and environmental assessments applicable during the project development phase.[17]

As with the community, social media is emerging as a new way of interacting with investors and potential investors. For example, Payson Petroleum, a small, independent operator from Dallas, uses social media to educate potential investors. The company has a YouTube channel with over 400,000 views, where many of the videos are of the CEO and director of Client Relations discussing questions related to oil and gas investments.[18]

Employees

Another important stakeholder group (particularly in the context of this book) comprises all employees, including top management, supervisors, full-time and part-time employees, and individual contractors. Employees tend to have three overarching concerns:[19]

- Pay and benefits. Employees are concerned with receiving fair pay, the perception of which is largely driven by how their pay compares to others in their organization and workers in the same job in other companies. Most employees are also concerned with benefits such as health insurance, sick leave, retirement plans, vacation days, holidays, and the ability to balance work and family. Annual employee surveys can help companies monitor the current satisfaction level of their employees.

- Quality of work life. Employees want to enjoy a good quality of life while on the job. For rotational oil and gas workers, this could include time off the job while on the company compound. Quality of work life is largely affected by the nature of the job, but can be enhanced by designing work that empowers workers, by training and development, by increasing job variety and responsibilities, by creating physically and psychologically safe and healthy working environments, and by promoting team work.
- Career security: Employees are concerned with stability in the job and, relatedly, careers. This can be achieved even in organizations that have made layoffs, by providing workers with skills that are in demand and providing laid-off workers with job placement services.

Clearly, dissatisfaction among employees can lead to many different detrimental outcomes for companies, many which would negatively impact other stakeholders. For example, employees who are unsatisfied or those who believe that they have been treated unfairly are more likely to participate in withdrawal behaviors or counterproductive work behaviors, including voluntary absenteeism (e.g., calling in sick when they are not really sick), quitting, sabotage, theft, and shirking (goofing off), as well as exhibiting lower performance and giving unsatisfactory customer service. Workers, through negative or positive actions, can have a direct effect on all other stakeholders.

Customers

Customers in the oil and gas industry vary by segment, and in an industry that has a highly developed value chain (as described in chapter 2), the sequence of supplier–customer–supplier is often fairly rigid. Upstream customers may include other firms that purchase crude oil, joint venture partners, governments, or the firm's own subsidiaries, and so forth. The same is true for midstream. Customers for downstream firms may include other firms and governments, but are generally individual consumers. Naturally, customers are an important stakeholder because without demand, prices will drop (although energy's demand is not as strongly tied to price as most other products), negatively affecting many other stakeholders. Further, other stakeholders may affect customers and vice versa. Negative publicity from the media, NGOs, communities, unions, or employees can tarnish a firm's reputation, negatively affecting customer perceptions of the company or product. The firm's reputation can also be damaged by lawsuits, regulation violations, court settlements, and other governmental actions. As with communities, social media are emerging as the next step in proactive customer engagement. Facebook, Twitter, YouTube, and LinkedIn are all viable new channels to use to communicate with customers, address concerns before they escalate, and proactively publicize positive news.

NGOs

NGOs are organizations that are not part of government. Although their goals can vary widely, they tend to have longer-term developmental, environmental, or human rights objectives.[20] Because they tend to focus on improvements in quality of life they usually have a good reputation with the general public. Although many NGOs are seen to be combative with oil and gas companies, many have experience in dealing with issues important to other stakeholders and may provide useful information in identifying, evaluating, and addressing such issues. Thus, there may be opportunities for partnerships with NGOs or at least positive interactions between NGOs and oil and gas firms. This is consistent with the fact that some NGOs are funded by foundations or charitable trusts related to oil companies. For example, the Pew Charitable Trusts (www.pewtrusts.org) is an independent nonprofit NGO formed from trusts established by the children of the founder of Sun Oil Company, Joseph N. Pew, and his wife, Mary. Other oil-related NGOs include the BP Foundation, the Baker Hughes Foundation, the Fluor Foundation, the Dominion Foundation, the Suncor Energy Foundation, the Ryder Charitable Foundation, the Koch family foundations, CBT Charity, and Spindletop Charities. The goal of the Pew Charitable Trusts is to improve public policy, inform the public, and stimulate civic life through a rigorous, analytical approach. Reports issued by the trust frequently have implications for energy companies. For example, a recent report, *Arctic Standards: Recommendations on Oil Spill Prevention, Response, and Safety in the US Arctic Ocean*, specifies principles to guide arctic policy and management decisions in the oil and gas exploration and production in the highly challenging arctic. One of the principles is to "ensure that local communities have a meaningful voice in decision making" reflecting the usual strong ties between NGOs and community stakeholders.[21] We discuss the challenge of the arctic in more detail in Current Challenge Box 7–2.

Current Challenge Box 7–2. Drilling in the Arctic

The arctic has only recently been accessible for off-shore drilling, and thus is a current focus of a number of environmentally oriented NGOs. From the environmental groups' perspective, the arctic is a pristine, rare, and fragile ecosystem that would suffer tremendously with an oil spill because of the tremendous cleanup challenges there. These challenges include extreme remoteness, inaccessibility for nine months of the year, complete darkness for three months, high seas, subfreezing temperatures, dense fog, unpredictable ice hazards, and lack of existing communications, logistics, and information infrastructures. Further, the vulnerability of arctic species, ecosystems, and cultures would increase the damage caused by a spill. In 2012, the US Bureau of Safety and Environmental Enforcement granted Shell Oil permission to drill off the Alaskan coast. However, subsequent legal challenges, operational issues, and capital concerns led Shell to cancel its drilling plans in 2013 and 2014.[22]

Another challenge with the arctic is that some of the arctic is in Russian territory, which adds complexities because of the current strained relationships between Russia and the West (the EU and the United States). In late 2014, ExxonMobil halted its first arctic exploration well (University-1) in Russia's Kara Sea. The shutdown was in response to new US and EU sanctions on Russia regarding technology transfer. Ironically, a recent report by the National Research Council effectively responding to spills in the arctic would require the United States and Canada to expand bilateral agreements with Russia to include arctic spill scenarios and to conduct joint response exercises. The report also suggests that effective responses will require investments in better response tools, including improved real-time forecasting systems, research on oil spill response countermeasures in the arctic environment, and enhanced logistics and infrastructure. Related to stakeholders, the report also suggested that arctic mapping priorities should be undertaken with a stakeholder view and that local officials and village response teams should be included in planning and decision making.[23]

Unions

As representatives of nonmanagement workers, unions are another important stakeholder. Unions are outside organizations that represent (nonmanagerial) employees and deal with management over issues relating to their work. Terminology in this field is sometimes confusing. On the most general level, the term *industrial relations* is used to described the multidisciplinary field encompassing all dimensions of the employee relationship. Below this, the term *employee relations* is often used to refer to all nonunion relationships between employer and employee, while *labor relations* refers to union relationships. We will, however, use the term *labor relations* more generally, because in many countries, what legally constitutes a union is often in debate.

The International Labour Organization (ILO), part of the United Nations, adopted the Declaration on Fundamental Principles and Rights at Work in 1998, committing its member states to respect and promote four principles and rights of work:

1. The freedom of association and the effective recognition of the right to collective bargaining
2. The elimination of forced or compulsory labor
3. The abolition of child labor
4. The elimination of discrimination in respect of employment and occupation

The first declaration relates specifically to unions. The ILO's Declaration is "universal" in that member states believe that these principles should apply to all people in all countries regardless of the level of a country's economic development. The ILO also states that these rights apply to all workers, employed or unemployed, permanent citizens or migrant workers. Although not universally accepted at this time, these four principles are common across most of the global business marketplace today, and in some way they serve as a baseline or benchmark when viewing any specific country or industry. Labor unions seek to help employees improve the terms and conditions of their employment. In addition to sharing the interests of the workers, unions are also concerned with their own success and survival. Unions typically effect change through one or more of the following three methods:

1. Collective bargaining. Collective bargaining allows employee groups to negotiate with management to create agreements and provisions that regulate the employment relationship.
2. Legislation. Unions undertake lobbying and political action initiatives in pursuit of policies that regulate labor relations and employee working conditions and benefits.
3. Mutual assistance. Unions provide or negotiate with the employer for insurance (for example, supplemental disability insurance) and other benefits (for example, training sessions) that are desired by employees and may provide opportunities for individual employee growth and career advancement.

Oil and gas developments, upstream and downstream, are often subject to a combination of the strong country, industry, and socioeconomic forces that drive labor relations and industry practice. These include foreign firms dominating the in-country oil and gas industry; the high numbers of foreign workers and employees; the involvement of national oil companies in the domestic industry, which may constitute a major source of government employment; and as mentioned in chapter 5, a strong project orientation of the industry, which increasingly utilizes contractual employment.

Union prevalence. In the United States, the unionization rate of private workers is now below 7%, while that of government workers is above 35%. Oil and gas companies generally have low unionization rates, particularly upstream, but other related industries such as construction and transportation have higher rates. US union membership rates by industry are reported in figure 7-4. The membership rate for oil and gas extraction (including mining and quarrying) is 5.4%, while it is 14.1% for construction and 19.6% for transportation and warehousing.[24] These are US numbers, and of course rates vary by country. Nevertheless, although membership rates may be generally low, when unions are present, they may have considerable power over the company. This is because in the process of getting the crude from the ground to the consumer, some union members are likely to be involved. For example, in the United States, the United Steelworkers union claims that 77% of the oil in the United States is processed by employees represented by unions.[25] A union's power generally stems from its ability to call a strike, and strikes can inflict a great deal of economic hardship on the company (as well as on the workers), particularly when the workers are difficult to replace.

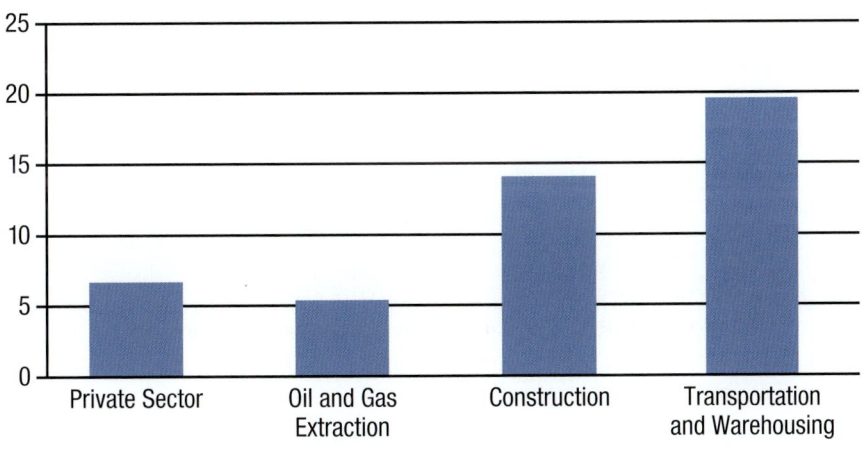

Fig. 7–4. US union membership rates by industry
Source: US Bureau of Labor Statistics.

Collective bargaining. Once a union represents workers, the union strives for a collective bargaining agreement. Collective bargaining in the oil and gas industry worldwide is increasingly becoming more decentralized, generally at the company division level or even the plant level. However, many exceptions exist, including most Nordic countries, where centralized negotiations are typical. For example, recent agreements in Norway covered more than 30 oil and gas companies.[26]

Typical collective bargaining agreements cover wages and total compensation (including benefits and services), hours of employment, institutional issues such as union security clauses, strike provisions and managerial rights, and administrative issues such as grievance procedures, job security, discipline procedures, and work rules (production standards, health and safety issues, training issues, etc.). Collective bargaining agreements in the oil and gas industry tend to cover the standard issues, although there are some differences, as noted in table 7–2.[27] Most of these differences relate to the variations in working conditions, working locations, or the work itself. For example, the greater importance of health and safety in the industry is recognized by unions, and thus, safety- and health-related provisions are more likely to be extensive in this industry. Collective bargaining may vary a great deal across countries, partly because of the differences in national labor laws. For example, the right to organize (or the right to representation) at work, and more specifically the right to strike against private employers, is usually recognized as a right in most countries. However, because of the importance of oil and gas on a country's commerce and welfare of the population, strikes (and in some cases organizing workers) in the oil and gas industry have met with considerable government resistance in a number of countries including Brazil, Norway, Sudan, and Nigeria. To illustrate differences, we briefly describe labor relations in the United States and in Nigeria.

Table 7–2. Collective bargaining issues in the oil and gas industry

Issue	Industry Perspective
Wages	Wages tend to be higher than comparable jobs in other industries, most likely because of the unique working conditions. This is particularly true in the upstream sector.
Working time	Working hours tend to be less on average and more flexible than in other industries. Shiftwork and rotational schedules are more common.
Occupational safety and health	More important than in other industries. Unique issues due to unique locations, such as offshore.
Use of contractors	Prevalence of contractors makes this a more important issue than in other industries. Also relates to sub-contractors. May be related to safety issues.
Medical services	More important than in other industries because of nature of work, working conditions, and working locations

Unions in the United States. Although the first unions in the United States were formed in the 1790s, it was not until pro-union legislation and the creation of the National Labor Relations Board in 1935 that unions and the right to strike were clearly determined to be legal. Since then, a number of other federal laws have clarified union, management, and union member rights including the process of determining representation and rules regarding collective bargaining. Collective bargaining is required on mandatory issues such as wages, hours, and other conditions of employment. Unions and management are not allowed to bargain about any illegal issues. Unions are allowed to strike, but only if they are not currently covered by a collective bargaining agreement and if they previously have notified the *Federal Mediation and Conciliation Service*, which provides conflict resolution services.

Major unions now representing oil and gas workers in the United States tend to be *trade unions* that cover groups of workers with specific skilled trades such as electric workers, metal workers, and pipefitters. There is no prevalent *industrial union* that covers a broad class of workers in the oil and gas industry (as, for example, the UAW does in the auto industry) although the United Steelworkers absorbed what was once the Oil, Chemical, and Atomic Workers Union (the United Steelworkers merged with the Paper, Allied-Industrial, Chemical and Energy Workers International Union in 2005, which was formed from a previous merger including the Oil, Chemical and Atomic Workers Union in 1999).

An illustrative case: Unions in Nigeria. The Nigerian labor movement traces its origins to the 1890s, when the first unions were formed in Nigeria for civil service, education, and railway workers under British rule. In 1931 the colonial government codified union recognition with the passage of the Trade Union Ordinance. Although British rule ended with Nigerian independence in 1960, it remained on the books as the primary labor law in practice for many years. Eventually, with the passage of the Trade Union Act of 1978, labor union practices were formalized. Since that time, the government's relationship with labor has been characterized by frequent strikes, military interventions, imprisonment of labor leaders, and constant fights to protect labor's rights.

There is no more critical sector in the Nigerian economy than the oil and gas sector. Two major unions operate in the Nigerian oil and gas industry: the National Union of Petroleum and Natural Gas Workers, one of the 29 industrial unions currently affiliated to the Nigeria Labour Congress, which represents the interests of the junior or blue-collar workers in the industry; and the Petroleum and Natural Gas Senior Staff Association of Nigeria, which represents the interests of senior white-collar workers, affiliated with the Trade Union Congress.

Nigeria has officially ratified all ILO core fundamental principles. Current Nigerian law recognizes the right of workers to organize and to collectively bargain. Employers are required to recognize unions upon their legal establishment. Collective agreements are still negotiated on the company level, as no industry-wide negotiations are possible because employers (the oil and gas companies) cannot formally collaborate in business practices including labor relations. But, contrary to ILO principles, Nigeria does not allow strikes in the oil and gas industry. Under current labor law, collective bargaining agreements require a no-strike clause prior to their implementation. Nigeria also prohibits any union activities that may be construed as politically motivated. Many international observers believe that Nigeria's approach in the oil and gas sector places overly restrictive limits on collective agreements and union membership, and does not provide sufficient protection for women, ethnic groups, the disabled, and others facing discriminatory employment and pay practices.

Addressing unionization in such a global industry is difficult because of the great differences in the presence, power, and importance of unions across countries. In some countries, unions are strongly tied to government and politics, with the presence of labor parties. Further, the nature of labor relations in general varies greatly among countries, with some (e.g., Germany and Japan) generally having relatively collaborative relationships between unions and management, and others (e.g., the United States and Nigeria) historically having had a more combative relationship. The lack of any truly global unions, which would lead to a convergence of labor relations, further fosters the differences among countries. Thus, global firms may have very different labor relations policies around the world. Nevertheless, from a stakeholder and ethics perspective, multinational enterprises (MNEs) dealing with labor relations issues globally should strive to follow an approach that would be satisfactory to all stakeholders (including those in the home country) even if it goes beyond the requirements of the host country.[28] This is because the publicized poor treatment of workers and their unions is of concern to other stakeholders including governments, customers, communities, and NGOs.

Corporate Ethics and Accountability

Global business today is held accountable for its actions and impacts on society to a greater degree than ever seen before, and the past two decades have seen the growth in interest in corporate sustainability, CSR, and corporate ethics. Corporate sustainability tends to focus on environmental aspects, while CSR is broader and addresses how companies affect the overall welfare of society. A company's CSR performance appears to be an important factor in a firm's reputation and has a strong influence on the attitudes (and purchasing behaviors) of customers as well as society in general. Firms can have a strong positive CSR reputation even in industries with a generally poor reputation. For example, a 2013 survey of 21,000 business students rated Phillips 66 (a Houston-based downstream company recently spun off from ConocoPhillips) as the most ideal company to work for because of its strong CSR reputation.[29]

Similar, but somewhat broader than CSR, corporate (or business) ethics impose a moral responsibility on corporations and their managers to optimize the returns to all stakeholders. Companies generally signal the importance of ethics through a code of ethics or code of conduct. This is also true in the oil and gas industry, where a study found that the largest 10 oil and gas companies all had a code of ethics ranging from 9 (Valero Energy) to 84 (BP) pages. The overlap of ethics and stakeholder theory can be seen when looking at the content of the code of ethics across the companies. Topics that appeared in the content of codes of ethics in all the largest oil and gas companies include:

- Asking questions, raising concerns, and reporting unethical conduct
- Bribery and corruption
- Competition and antitrust
- Conflicts of interest
- Correct accounting and financial reporting
- Dealing with government officials
- Equal opportunity, nondiscrimination
- Gifts and hospitality
- Political activity
- Protection of company assets
- Safeguarding information[30]

Many of these topics relate to multiple stakeholders, and the stakeholder view proposes that satisfying all stakeholders will benefit the firm in the long run. This is not to say that all stakeholders should be treated equally or are equally important. Stakeholders that are viewed as more important are those that:

- Control key resources needed by the organization
- Are more powerful than the organization
- Are likely to take action for or against the organization
- Are likely to form coalitions with other stakeholders[31]

A key strategic management decision is how the firm allocates its resources to each particular stakeholder. Allocating too much effort toward stakeholders that will ultimately not benefit the firm much is seen as suboptimal. Certainly, primary stakeholders, those who are needed for the organization to survive in the short and long term, should be given attention first. In the oil and gas industry this includes shareholders, employees, unions, customers, suppliers, and governments. Other stakeholders, such as NGOs, communities, and other organizations are important to address, but are not likely to directly affect the survival of the company in the short term, but may considerably affect its performance and ultimately survival in the long term.[32]

Because of the visibility, proactiveness, and importance of many stakeholders in the oil and gas industry, these companies have been responsive to the stakeholder view. For example, in the mid-1990s, Shell was receiving considerable external criticism about its disposal of the Brent Spar oil storage facility in the North Sea as well as its relationship with the disreputable military regime in Nigeria. Shell responded with a management model that included the interests of diverse stakeholders, particularly nonprimary stakeholders. This led to the reframing of performance in terms of societal needs and expectations.[33] Many companies, like Shell, now consider the "triple bottom line," which looks at organizational performance as a combination of financial, environmental, and social performance.

Implementation of a stakeholder perspective goes beyond the creation of mechanisms used to manage specific stakeholders (e.g., investor relations for investors, and human resources [HR] for employees). It involves creating a culture where all stakeholders are considered, the business environment is monitored for new stakeholders, and relationships among stakeholders are considered. For example, employees tend to be more committed to their employers when they believe that the company is behaving in a socially responsible way.[34]

Human Resource Management from a Stakeholder Perspective

Clearly, human resource management (HRM) can have a large impact on all stakeholders. HR practices have been shown to affect a firm's financial performance, which is the primary interest of owners and investors. Specifically, HR practices have been shown to affect profitability, productivity, annual sales per employee, market value, and earnings-per-share growth of firms.[35] Obviously, HR practices can have a big impact on the firm's employees, such as their pay and benefits, quality of work life, and career security. Further, keeping employees satisfied seems to create a win-win situation, because it also relates to the firm's bottom line. For example, companies that appear on Fortune's "100 Best Companies to Work For" list outperform (as measured in stock gains) similar companies that do not.[36]

HR practices can have a great impact on all stakeholders through five main mechanisms. First, sound HR practices, including those related to staffing, training, performance management, and compensation, can improve the productivity of the workforce. Improving productivity improves the firm's financial resources, which can then be used to address the concerns of various stakeholders. Second, HR practices can shape employee behaviors. Companies can select, train, compensate, and manage the performance of employees in such a way as to focus on specific behaviors targeted at strategic goals that may relate to specific stakeholders. Third, HR practices can help reduce costs. Cost reductions can occur in several HR-related ways, including reducing turnover and absenteeism, promoting efficiency, staffing through cost-effective high-utility tools, and compensating employees in creative ways that reduce labor costs while keeping employees satisfied. As with productivity, lower costs improve the financial resources of the firm, which can then be used to address the concerns of various stakeholders. Fourth, HR practices can attract, motivate, and retain good employees, which helps productivity, lowers costs, and focuses behaviors on strategic goals. Fifth, the stakeholder view may result in firms creating new jobs or departments to address specific stakeholder concerns. Effective HR in those jobs will help achieve good performance targeted specifically toward stakeholders. We now discuss how oil and gas firms can address the concerns of their various stakeholders through the four HR functions: staffing, training, performance management, and compensation.

Staffing from a Stakeholder Perspective

Addressing all stakeholders rather than just shareholders has important implications for staffing. Staffing can be related to stakeholders in two distinct ways. First, whom we employ can directly and indirectly affect stakeholders. For example, multinational corporation (MNC) employers hiring locals rather than using expatriates directly affects the local community, the host government, and unions, and this practice may indirectly affect consumers, NGOs, and other companies, among others. Second, how a firm selects employees can affect stakeholders through the skills, knowledge, and abilities of their workforce. We now discuss this in greater detail using the framework of the key staffing decisions.

Determining labor markets from a stakeholder perspective

The determination of the labor market affects recruiting, which creates the pool from which firms then select potential employees. Recruiting from a specific labor market can strongly influence whom the company employs, which can subsequently directly and indirectly impact stakeholders. Targeting specific labor markets including local labor markets, underrepresented groups, and those with stakeholder specific skills can help firms address stakeholder concerns through the staffing function.

Targeting local labor markets. We discussed in detail in chapter 3 the benefits and possible downsides of MNEs hiring local employees rather than using expatriates. From a stakeholder perspective, using locals when possible can go a long way in addressing the concerns of several stakeholders, including the host government, the local community, local unions, and local NGOs.[37] Governments are so concerned about the hiring of locals that they frequently pass labor legislation or regulations requiring that a minimum level of locals compose the MNE's workforce in their country. These standards are often specified under "local content" laws. We discuss these in detail in the next chapter. Employing more local talent also increases the number of economic winners in the community, which in turn increases the benefits to the community and the benefit/cost ratio for the community. Providing economic gains to the community by hiring locals also helps address concerns of other local stakeholders such as local unions and local and national NGOs.

Targeting underrepresented groups. The increased hiring of underrepresented groups can help address some of the concerns of governments, communities, society, and certain NGOs. For example, greater workforce representation of women, those with disabilities, ethnic minorities, religious minorities, and indigenous populations is generally welcomed by many stakeholders. We showed in chapter 6 how women are underrepresented in the US oil and gas industry. The same could be said, although to a lesser degree, for racial minorities in the downstream and midstream sectors. African American and Hispanic employment in the US oil and gas industry is detailed in figure 7–5. Hispanic and African American minority employees make up 27% of the upstream, 17% of the downstream, and 23% of the petrochemicals workforce, while these two ethnic groups make up 27% of the total US workforce.[38] Increasing the hiring of underrepresented groups and thus increasing diversity has many benefits for an organization including:[39]

- Improving the reputation of the company favorably for women, minorities, and underrepresented groups, providing an edge in attracting talent
- Providing insights and cultural sensitivity of diverse and specialized markets, and thus generating an edge in marketing
- Providing more diverse perspectives and less conformity, leading to greater creativity in all facets of the business
- Providing more diverse perspectives in problem solving and decision making, leading to better critical analysis and better decisions
- Creating fewer standardized systems in the organization, providing more flexibility to adapt to changes in the firm's business environment
- Reducing the probability of lawsuits that can be financially costly as well as hurt the firm's reputation

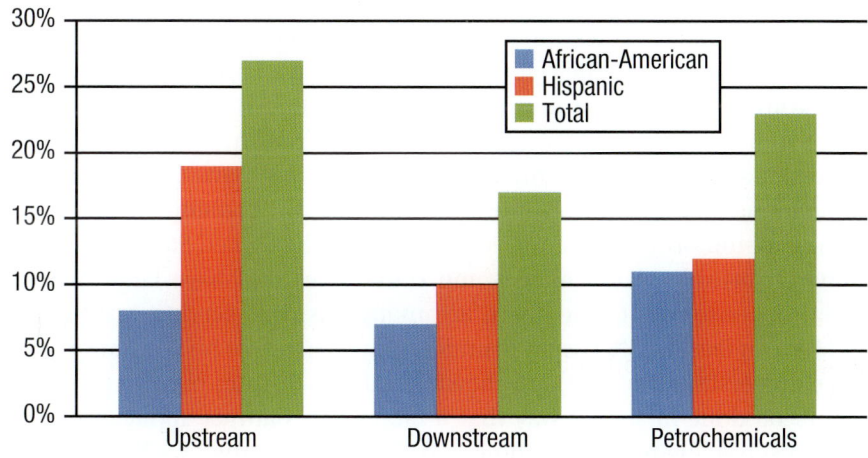

Fig. 7–5. Minority employment by sector (percent of minority workers in US oil and gas industry, 2019)
Source: IHS 2014 Report: *Minority and female employment in the oil & gas and petrochemical industries.*

Many of these positive outcomes benefit a number of different stakeholders. Thus, focusing on underrepresented groups in the labor market can increase the pool of applicants from these groups, satisfying a number of stakeholders and benefiting the organization in many ways.

Targeting those with stakeholder specific skills. Hiring those with skills that can help improve the firm's performance (as perceived by different stakeholders) can also help address stakeholder concerns. These skills may be task specific and relate only to one stakeholder, or they may be very general and apply to many stakeholders. We discuss how firm performance may be perceived by different stakeholders in detail below in the performance management section. By focusing on workers in the labor market who have skills that are applicable to different measures of performance, companies will help ensure that their employees have the skills needed to achieve those measures.

Recruitment from a stakeholder perspective

Targeting labor markets with stakeholders in mind requires new ways to think about recruiting—not only whom we focus on, as discussed above, but also how and where we recruit. In chapter 6 we discussed how firms can use recruiting methods specifically tailored to attracting female applicants. That discussion also applies here. Further, firms can use similar methods to target other underrepresented groups by using recruiting methods relevant for each group. Professional associations, university groups, and energy conferences focusing on minorities are a great place to recruit minorities. Some associations related to minorities in the oil and gas industry include the Society of Hispanic Professional Engineers (national.shpe.org), with 10,000 professional members; the National Society of Black Engineers (www.nsbe.org), with 31,000 student and professional members; and the American Association of Blacks in Energy (www.aabe.org), with 1,300 student and professional members. Further, some universities have large minority populations. Recruiting from these universities will certainly increase the number of minority applicants. For example, in the United States, there are many historically black colleges and universities that have large African American enrollment, and there are many Hispanic-serving institutions that have at least a 25% Hispanic undergraduate student body. Examples of these universities (by state) are shown in table 7–3. Thus, overall recruiting should be done in such a way that it either helps attract applicants who, in themselves, help address some stakeholder concerns or who have the skills to do so. Once we have attracted such a pool of applicants, we must determine who to actually hire.

Table 7–3. Examples of historically black universities and colleges, and Hispanic-serving institutions in the United States

State	Historically Black University and Colleges	State	Hispanic-Serving Institutions
Alabama	Alabama A&M University Alabama State University Stillman College	Arizona	Arizona Western College Northern Arizona U.- Yuma University of Arizona South
Florida	Bethune-Cookman University Florida A&M University Florida Memorial University	California	Cal State U, Fresno Cal State U, Los Angeles San Diego State University
Georgia	Albany State University Fort Valley State University Savannah State University	Colorado	Adams State University Colorado State U, Pueblo Emily Griffith Technical College
North Carolina	Fayetteville State University North Carolina Central University Winston-Salem State University	Florida	Florida International University Palm Beach State College St. Thomas University
South Carolina	Allen University Morris College South Carolina State University	Illinois	Northeastern Illinois University St. Augustine College Triton College
Tenn.	Knoxville College Lane College Tennessee State University	New Mexico	Eastern New Mexico University New Mexico State University University of New Mexico
Texas	Prairie View A&M University Texas College Texas Southern University	New York	Boricua College City College of New York New York City College of Tech.
Virginia	Norfolk State University Virginia State University Virginia University of Lynchburg	Texas	Texas State University University of Texas, El Paso University of Houston

Selection from a stakeholder perspective

Selection can be used to help address stakeholders in two ways. First, because the nature of employee performance can be fundamentally changed using a stakeholder view, naturally the selection tools used to predict a different type of performance would change. Second, as mentioned earlier, the demographics of the employees themselves may be of concern to some stakeholders (locals vs. expatriates, gender, race, etc.), and therefore these employee attributes could be a factor in selection.

Selecting for stakeholder relevant performance. Because the stakeholder view can fundamentally change how the organization's performance is measured (usually by adding additional performance measures related to stakeholders), it naturally follows that these changes in organizational goals would then be reflected at lower levels in division goals, department goals, team goals, and ultimately the performance goals of individual employees. Thus, for certain employees, the definition of what is good performance in their job can change when using the stakeholder view, particularly at higher levels in the company. For example, the performance of project managers historically was measured on three dimensions: Cost, timeliness, and quality. However, including the stakeholder view would include measures of safety performance, stakeholder satisfaction measures (e.g., customer satisfaction), efficient use of resources (e.g., costs of materials, tools, and equipment relative to productivity), reduced conflicts and disputes, and measures of management performance (e.g., absenteeism and turnover, employee performance measures, scheduling, team-building, and reporting). Because selection tools should be chosen that best predict job performance, changes in the definition of job performance should result in the use of new selection tools that best predict the new definition of performance. These selection tools may be tied to specific aspects of performance, (e.g., safety knowledge, certification, and relevant experience) or more general aspects such as ethics.

Selecting on employee attributes. As we mentioned earlier, hiring certain employees (such as locals in a multinational context) may alone address stakeholder concerns. So can we use these attributes as a factor in the selection decision? This is a tricky question, and the answer is largely driven by legal considerations. Defining the labor market as local and preferring local employees over others is defensible. However, using demographics such as gender, race, national origin, color, or religion as a hiring factor is almost always illegal in the United States (Title VII of the 1964 Civil Rights Act). It is even illegal if it is done as "reverse discrimination": for example, giving preferences to minorities or women (although there are exceptions for firms under a court order, government contractors, or those with bona fide affirmative action programs to overcome past hiring imbalances). Thus, depending on the country, a firm would put itself at considerable legal risk if were it to only hire women, for example. However, preferable treatment can be given to underrepresented groups when two candidates are equally qualified. Certainly, from a stakeholder perspective, firms should give preference to underrepresented groups in such cases. Further, increasing the representation of underrepresented groups in the applicant pool will allow for the hiring of more employees from underrepresented groups since the demographic makeup of the applicant pool is also considered in discrimination cases. This increases the importance of recruiting underrepresented groups, as specified earlier.

Combining selection tools from a stakeholder perspective

Addressing various stakeholder concerns leads to employees (particularly at higher levels of the organization) having many different dimensions of performance, which suggests selection should be based on more tools to tap into the qualifications of the applicant on these different dimensions. In cases where there are more selection tools used (particularly when the performance criteria being predicted are desirable, but not essential), the compensatory approach is generally preferable to the multiple hurdles approach. This is because as the number and difficulty of the hurdles (i.e., cutoff scores on the tests or number of years of applicable experience) increases, the applicant pool will become greatly reduced, making it less and less likely that even one applicant survives the multiple hurdles. That is because it is unlikely that many applicants (if any) will have the skills or experience to address all of the stakeholder-based performance dimensions, Thus, the compensatory approach will leave a bigger final pool that the company can choose from, providing more flexibility in the final choice.

Maximizing the effectiveness of staffing from a stakeholder perspective

The general principles of being legally compliant and treating applicants fairly are particularly relevant in the context of stakeholders, because these principles inherently recognize that the selection process itself can affect stakeholders. For example, discrimination in hiring is generally viewed negatively by governments, applicants, employees, customers, the local community, and society. Unfair treatment of applicants in the decision, process, or firm's interactions with the applicant may not only lead to lawsuits, but could also lead to negative perceptions of the company by the applicant, which may be widely communicated to others. Thus, monitoring, evaluating, and adjusting are particularly important relevant to stakeholders, so that any negative perceptions can be quickly addressed.

Training from a Stakeholder Perspective

The different stakeholders of the oil and gas industry have important implications for training. Each stakeholder's interests may require specific training. For example, focusing on sustainability, CSR, and ethics requires that employees be educated on these aspects. The same goes for diversity. Addressing customer concerns relates to more customer service training. Addressing employee and union concerns relates to more HR-related training. Further, because of the broad definition of performance, employees must have training on any area where they are deficient. We now discuss these differences in the framework of the key training decisions.

Who gets trained from a stakeholder perspective?

Because the stakeholder view fundamentally changes a firm's definition of good performance (e.g., the triple bottom line instead of just the bottom line), a firm's culture must reflect this new reality. Instilling this into the culture requires that every employee be educated on these organizational values. For example, employees have been shown to be more committed to corporate social responsibility when they are of aware of the firm's CSR guidelines, receive CSR training, and attend CSR conferences. However, because managerial support is critical to the success of any fundamental change, supervisors and managers must be well educated on the different stakeholders. For example, commitment from supervisors has been found to be an important predictor of employees being engaged in CSR. Even more specifically, when managers buy into CSR, employees are far more likely to develop and implement creative ideas that positively affect the environment, and they are far more likely to engage in CSR activities as part of a firm's core business.[40]

Because firms with a stakeholder view have a broader view of firm performance, more aspects of performance (than just profits, for example) should be captured in unit (and ultimately individual) performance. Because the individual employee's measure of performance is likely to be broader and encompass the interests of stakeholders, it is less likely that employees will be proficient in all dimensions of their performance. Thus, all employees should also be considered for training to overcome any deficiencies in specific dimensions of their performance.

Finally, training in itself may address some stakeholder concerns directly in three ways. First, when MNEs train local employees, this makes them more employable in the long term as well as in the short term, which helps to address some concerns of the community, the host government, local unions, and local NGOs. Second, training and developing employees in general is viewed positively by employees. Thus, developing employees to increase their skills for other jobs or to take on more responsibility in their current occupation will usually be viewed as beneficial by employee stakeholders. Third, stakeholders themselves may be trained to help satisfy themselves as well as other stakeholders. For example, training customers, suppliers, or employees of contractors or other organizations may foster behaviors that are beneficial for the stakeholders. Thus, training should be considered for every employee, as well as stakeholders, as a way to satisfy each of the different stakeholders.

What gets trained from a stakeholder perspective?

Clearly, what gets trained to address stakeholders is extremely broad and includes pretty much every type of training, since all training tends to benefit someone in the short or long term. This includes training in ethics, sustainability, CSR, health and safety, diversity, cultural sensitivity, customer service, job-specific skills, technology, people skills, leadership, and developmental skills. Companies could probably have employees spend most of their time in training, which of course would mean that not much work would get done. Thus, who gets what training and how much is a matter of prioritizing all the potential training opportunities to those that are really necessary. Nevertheless, a stakeholder view will increase the amount of training that will be deemed necessary. Firms with a stakeholder view will want all employees to understand the company's ethics and CSR codes, guidelines, and guiding principles. For example, Shell, Exxon, and ConocoPhillips require all their employees to regularly attend training courses on their ethics policies. More extensive courses are necessary for managers because of their broader discretion and the effect their attitudes have on the behaviors of the employees below them. For example, Total has developed a training program called "Our Ethical, Environmental, and Social Responsibilities" for corporate managers and "Ethics and Business," which is a 1.5-day seminar for senior executives, line managers, and corporate managers.[41] Here are some rules of thumb that may help in prioritizing what to train:

1. Training that ultimately benefits more stakeholders should be given priority over training that benefits only one or a few stakeholders.
2. Training that will impact outcomes of greater importance to stakeholders should be given priority over training that affects only minor issues.
3. Training that benefits critical stakeholders should be given priority over training that benefits secondary stakeholders.
4. Training that addresses the short-term and helps address long-term concerns of stakeholders be given priority over training that only addresses short-term concerns.

Table 7–4 lists some common courses that companies offer to specifically address stakeholder issues. Overall, training is clearly an important mechanism in HRM to address stakeholder concerns, but because of its cost (and thus its short-term financial impact), it needs to be used in ways that maximize its positive effects on stakeholders.

Table 7–4. Examples of stakeholder related training courses

Course	Content
Corporate social responsibility	Generally a broadly oriented program focusing on social and environmental issues that may contain elements of all other topics below
Sustainability	Training programs that generally focus on environmental issues. Some companies broaden the definition to include social issues
Sustainability reporting	Training programs that focus on effective reporting of sustainability performance
Human rights	Training programs that focus on human rights issues, including defining human rights, guidelines, and practical applications
Stakeholder management	Training programs that identify stakeholders, their concerns and how to address them
Anti-corruption	Training programs that focus on legal requirements regarding corruption and practical aspects of compliance
Ethics policies	Training programs that focus on the company's ethics policy, including interpretation and application of the policies

How do we train from a stakeholder perspective?

Addressing stakeholders of the oil and gas industry has several implications for the training methods that can be used. As is generally true, training can be classroom based or on the job. It can be personally delivered or online. Usually, the training method is affected by who and what is being trained. For example, online is a cost-effective and convenient method of training for stakeholders (customers, suppliers, or employees of contractors or other organizations) who are not located close to a company training facility. Shell uses e-learning courses to train contract partners and suppliers (as well as employees) on its ethics codes so that they understand and meet Shell's expectations. Further, online training also allows specialized courses that would not be cost-effective in person because of the limited demand of highly specialized topics.

One way to emphasize to employees the importance of company specific codes, guidelines, and guiding principles related to ethics, CSR, sustainability, health and safety, and diversity is to make this training part of all new employees' orientation. Socialization of the employees with respect to the company's culture in these areas is best done early in the training process to create accurate first impressions and in a centralized way to maintain consistency of the message across employees.

Related to the attraction and fair treatment of underrepresented groups, mentor programs (as mentioned in chapter 6) are generally viewed as a very positive long-term development method by employees and other stakeholders. For example, the majority of young female employees believe that having a senior female mentor is important to career success. A recent survey found that 63% of the respondents believed that mentorship programs led by women in the energy industry would have the greatest impact on attracting female engineers into the industry after graduating.[42] Thus, mentoring appears to positively affect recruiting as well as longer-term development. Overall, the optimal training and development method to use depends on who and what is being trained as well as the method's expected effectiveness relative to its cost.

Maximizing the effectiveness of training from a stakeholder perspective

As with training in general, maximizing the effectiveness of training related to addressing stakeholder concerns can be achieved by maximizing learning; facilitating training transfer; and monitoring, evaluating, and adjusting the system. Focusing training on stakeholders has several specific implications for the maximization of the effectiveness of training. Different targeted groups are likely to learn in different ways. For example, local workers may be much more willing to learn from a respected peer whom they can relate to, rather than an unknown, foreign trainer. Organizations should be mindful that training should be tailored to the individuals being trained. Second, training improves knowledge more with higher engagement and feedback. However, greater knowledge does not necessary result in better performance as perceived by stakeholders, unless that knowledge is transferred to the job. Facilitating transfer of knowledge can be increased by incorporating the training content into performance measures. Thus, feedback during and after training by monitoring and evaluating trainee performance can help maximize the effectiveness of the training.[43] Further, goal setting (another component of performance management) is also likely to improve performance as perceived by stakeholders. Finally, from a stakeholder perspective, the evaluation of training must take into consideration the perspectives of different stakeholders. What was the financial cost of the training? Did the trainees perceive the training to be worthwhile? How did the training impact stakeholders in the short term? In the long term? From a stakeholder perspective, the effectiveness of the training needs to consider all the costs and benefits to all of the stakeholders. We now discuss the role of performance management in addressing stakeholder concerns.

Performance Management from a Stakeholder Perspective

Satisfying stakeholders has a great impact on performance management because the stakeholder view fundamentally changes the definition of firm performance. This should have implications for the definition of performance of individuals. How stakeholders can be satisfied through performance management is straightforward: Make the satisfaction of each stakeholder an important part of performance. In doing so, assuming the firm has a good performance management system, goals will be set for different aspects of performance (each related to stakeholders), each aspect of performance will be measured, feedback will be given on them, employees will be developed relative to them, and meaningful rewards will be tied to overall performance. We now discuss addressing stakeholders through performance management relative to the key decision points in performance management.

How do we measure performance from a stakeholder perspective?

The stakeholder view discussed in this chapter can have a dramatic impact on the fundamental question: "What is performance?" At the firm level, the definition of good performance changes from only financial measures to many different measures that address the concerns of all the firm's stakeholders. Therefore, at the firm level, performance now includes not only financial performance measures such as profits, return on assets, return on equity, return on sales, and stock appreciation, but also measures such as environmental performance, social performance, customer satisfaction, employee satisfaction, community satisfaction, and supplier satisfaction. These measures may apply directly as measures of individual performance for upper level management who presumably have influence over all these aspects of unit performance. For example, at Total, unit performance is measured on 87 criteria related to many stakeholders including shareholders, employees, customers, suppliers, contractors, business partners, and host governments.[44] However, unit-level performance measures are generally not good measures of individual-level performance at lower levels of the organization since individual workers have little influence on any of them directly (or perhaps even indirectly). Nevertheless, changing the definition of firm performance should have implications for the definition of performance of individual jobs, even at lower levels.

The stakeholder view results in a broader view of performance even for lower-level jobs. This broader view is fundamentally different in two ways.

First, task performance should now include not only how the employee performs the tasks but also the impact of the job on stakeholders. How was the performance from the customer's perspective? How was the performance from the community's perspective? (And so forth.) Second, relatedly, this broader perspective must now go beyond the achievements or outcomes of the performance, to include the means used to achieve those outcomes. From a stakeholder perspective (and an ethical perspective), the ends do not justify the means, since unethical means are likely to negatively affect other stakeholders. Thus, large profits at the expense of the environment, high sales by misleading customers, meeting deadlines by behaving unsafely, or large cost reductions by paying workers unfairly should not be scored as high levels of performance. Thus, when using a stakeholder perspective, performance should be measured as a multidimensional construct that includes not only the outcomes but also the means by which they were achieved.

So what should the dimensions of performance be? They should generally follow the aspects of task performance as specified in the job description, keeping in mind the impact of the job on the various stakeholders. Ideally, job descriptions should capture the stakeholder view of performance in their description of the job, and they may need to be rewritten if they do not. Capturing the impact of the job on stakeholders should include customer-related performance, safety performance, supplier-related performance, and so on, beyond the standard measures of task performance, whenever it is relevant. Supervisory or managerial jobs would include performance measures of employee satisfaction, the hiring of underrepresented groups, union relations, and so forth.

The actual measures of performance may be objective, subjective, or a combination of both. Again, objective measures tend to be seen as fairer, but it is frequently difficult to find objective measures that are largely due to only one individual's performance. Frequently, the evaluation of the means used to achieve outcomes must be subjective. Although subjective measures may suffer more from biases than objective measures, they can be more accurate, because an informed judgment can take into consideration context and circumstances that objective measures cannot.

Who is involved in performance management from a stakeholder perspective?

Following the broader definition of performance under the stakeholder view, the number of people involved in performance evaluation is likely to be greater because it should include all stakeholders relevant to the job. Although the employee's supervisor is generally responsible for the performance evaluation, it should be performed with the input of many other sources. Input should be sought from stakeholders who can make an accurate assessment of any relevant aspect of performance. This input may be formal or informal, and confidential or nonconfidential. Input from co-workers may also help with the evaluation, particularly with respect to the means used to achieve the outcomes. Co-worker evaluations should be confidential or anonymous to get more honest feedback. Other sources of input may include customers, suppliers, relevant NGOs, union representatives, community representatives, or any other relevant stakeholder. More input from more relevant sources is likely to lead to a more accurate assessment of the employee's performance. However, consideration must be given to the cost of obtaining the information, the time it will take, the reliability of the information, and the political implications of collecting the information before casting too wide a net to collect feedback from every stakeholder possible.[45]

What format should be used from a stakeholder perspective?

Because performance from a stakeholder perspective is generally multidimensional, a multidimensional format is needed. As mentioned in chapter 5, the most common multidimensional format today is the balanced scorecard approach, which considers not only the typical financial measures but also customer-oriented, internal business, innovation, and learning measures.[46] These dimensions may need to be broadened to include other stakeholders that are relevant to that particular job. The scores on all the subdimensions must then be combined to come up with an overall score. However, this raises another issue: How should the scores on subdimensions be combined?

There are two basic approaches to combining scores on subdimensions of performance to get an overall score:

1. Treat all subdimensions equally, and simply total all scores to get an overall score.
2. Weigh subdimensions differently, multiplying the score on each subdimension by its weight to get an overall score.

Although the first approach is simpler and more straightforward, the second one allows the overall score to better reflect the importance of each different aspect of performance. The weights are important since they also signal to employees which aspects of performance the organization considers the most important. One reasonable way of determining the relative weights is to have them reflect the importance of the stakeholders (as described earlier in the chapter) that they apply to the most.

Maximizing the effectiveness of performance management from a stakeholder perspective. As is generally the case, maximizing the effectiveness of performance management from a stakeholder perspective requires feedback; accurate evaluations, including goal setting and rewards; and monitoring, evaluating, and adjusting the system. The stakeholder view has some implications for these aspects of performance management:

- The timing of feedback may need to be more flexible to better correspond with the timing of different aspects of performance (e.g., the results of an employee survey). With multiple aspects of performance being measured, it may be necessary to have many feedback sessions throughout the year.
- The accuracy of all aspects of the evaluation may be more difficult due to the likely variance in the quality of the feedback from a wide range of stakeholders. When the quality of feedback is questionable from certain stakeholders, managers should consider gathering additional data for that aspect of performance.
- Setting goals tied specifically to each stakeholder-based performance dimension can lead to goals that are in conflict with (or contradict) each other. In such cases, consider setting goals that are more holistic, which can be achieved while still achieving the other goals.
- Multiple rewards tied to multiple measures can also cause complications. Employees are likely to make calculated decisions regarding where to put their efforts, depending on the perceived difficulty of achieving the specific goal and the value of the reward tied to it. This increases the importance of making the reward commensurate with the effort required to achieve it.
- Monitoring and evaluation of the system are probably more important when using the stakeholder perspective because of the fluidness of the definition of performance. New stakeholders and changes in stakeholder importance should drive changes in the definition of performance and thus must be frequently monitored.

Therefore, because of the added complexity of multiple performance measures, maximizing the effectiveness of performance management from the stakeholder perspective requires greater consideration of the complications that could arise.

Compensation from a Stakeholder Perspective

A stakeholder perspective has important implications for compensation in the oil and gas sector. Pay level impacts performance as perceived by shareholders, employees, and the community. Pay for performance can be used to pass some financial risk from shareholders to employees. It can also be used to help motivate employees to prioritize their behaviors toward specific stakeholder-focused outcomes. Benefits can affect shareholders and employees through their costs and benefits. Benefits can also affect NGOs and the community by motivating employee giving and volunteering. We now discuss these differences in the framework of the key compensation decisions.

Pay level from a stakeholder perspective

Pay level has important implications for several stakeholders. First, because high pay levels (relative to competitors) can greatly increase labor costs, pay level impacts a company's financial performance, thus affecting shareholders. The high financial costs of high pay may be offset by lower administrative costs and greater productivity because of the following:

- Higher pay levels tend to increase the recruiting capabilities of a company by increasing the attractiveness of working there.
- Higher pay levels are motivational to employees because employees believe that they are being fairly treated, and coupled with a good performance management system, they will perform well in order not to jeopardize their high-paying job.
- Higher pay levels increase retention of employees because they are less likely to be lured away by other offers (which will tend to be lower). This reduces turnover costs.

Naturally, high pay levels are perceived positively by employees. High pay levels are also likely to be seen positively by the community because of the positive economic impact on the workers and subsequently on community businesses. Further, in foreign locations, paying local employees relatively high pay levels has an added benefit. Research has shown that locals tend to perceive their pay from MNCs as unfair, because they know that expatriates are earning considerably more than they are; however, these negative feelings tend to be neutralized when the local employees are getting paid more from the MNC than they would from other local employers.[47] So overall, from a stakeholder perspective, higher pay levels tend to be seen as positive by more stakeholders, particularly if the increased labor costs can be offset through increased productivity and lower administrative costs.

Pay mix from a stakeholder perspective

The pay mix can have an effect on stakeholders in two ways. First, when bonus pay is a greater percentage of total pay, companies can shift some financial risk from shareholders to employees. This is because when bonuses are tied to company or unit performance, then during tougher times, companies will pay less, making it easier for the company to reduce costs during economic downturns. Of course, during these times, employees will also earn less. Variable pay does not shift risk if bonuses are based on individual performance because there is not a close relationship between the company's financial performance and the performance of one individual employee. Second, when bonus pay is a greater percentage of total pay, the bonuses can be used to motivate employees to focus on certain aspects of the job, which may be associated with specific stakeholders. This will be discussed in greater detail in the next section.

Performance-based pay from a stakeholder perspective

As we discussed relative to performance management, because the stakeholder perspective has implications for the definition of performance and subsequently performance measures, it has implications for the measures of performance-based pay. Tying rewards to the employee's achievements as measured by the performance scores will increase the motivation of the employee to score high. More specifically, tying rewards to reaching specific performance goals will motivate employees to focus on those goals. Thus, employees can be motivated to focus on specific stakeholders by linking rewards (such as bonuses or recognition awards) to the measures of performance related to that stakeholder. However, from a stakeholder perspective, there are likely to be many different goals related to different stakeholders. Thus, tying rewards to each would result in the availability of many rewards tied to many different performance goals. In such cases, it is difficult for employees to focus on all goals, and they are likely to focus on the ones that have the greatest chance for success or those that provide the most valued rewards. Prioritizing rewards by providing greater rewards for the most important goals will help employees focus on the goals the organization deems the most desirable.

Benefits from a stakeholder perspective

Benefits have at least three important implications for stakeholders. First, benefits can be very costly, affecting the company's financial performance. Because benefits average about 30% of total labor costs, increasing benefits can greatly increase a firm's labor costs. This, of course, affects shareholders.

Further, although benefits such as vacation days can have implications for retaining and attracting workers (as mentioned in chapter 6), the gains in increased productivity and the savings in administrative costs are likely to be less for providing higher benefits than for higher pay levels. Nevertheless, employees view more benefits very positively. Thus, increasing benefits is likely to increase the satisfaction of employees, while decreasing the satisfaction of shareholders. Thus, optimizing this trade-off by focusing on benefits that are relatively low cost while highly valued by employees (e.g., flexible work hours, employee assistance programs, wellness programs, and child care services) creates more of a win-win situation.

Second, many benefits provide a safety net to employees, and thus are viewed positively by society and the community (as well as employees). Retirement plans and most types of insurance (including health insurance, dental insurance, life insurance, disability insurance, and supplemental unemployment insurance) take care of employees when they would otherwise face financial hardships. Thus, organizations that take a more paternalistic view of employees and provide many safety nets for employees help society and governments by reducing the need for safety nets at the community and national level.

Third, benefits may directly benefit some stakeholders. Organizations may match the charitable contributions of employees dollar for dollar, directly benefiting local NGOs and local communities. For example, in Oklahoma, Phillips 66 not only donated $1 million to tornado relief efforts, but they also matched all employee relief fund donations.[48] Companies may also provide employees with free time to volunteer at charities, again to the benefit of NGOs and local communities.

Maximizing the effectiveness of compensation from a stakeholder perspective

As with compensation in general, compensation that addresses stakeholder concerns will be more effective if employees accept the system. This should actually be easier using the stakeholder approach since employees generally favor this perspective. It does suggest that companies should be vigilant in communicating and educating employees on the system and its drivers to get the most bang for buck. As mentioned earlier, monitoring, evaluating, and adjusting the system are critical because of the possibility of unintended consequences when changing the fundamental notion of performance and the rewards tied to it. Further, the added complexity of the system because of the multiple dimensions of performance requires a careful evaluation of the value of the reward tied to each goal and the difficulty of achieving each.

Conclusion

Like few others, the global oil and gas industry is subject to many influential, vocal, and visible stakeholders. Addressing the concerns of these stakeholders has implications for how to manage HR in the industry. Specifically, staffing affects stakeholders directly by who the firm hires and indirectly through the skills, knowledge, and abilities of the workforce. Training can help educate workers about stakeholder interests such as sustainability, CSR, diversity, customer service, and ethics. Performance management is greatly changed with a stakeholder perspective because the definition of performance at the organizational level fundamentally changes. With respect to compensation, pay levels, pay for performance, and benefits can all affect how the firm addresses the concerns of different stakeholders. Overall, satisfying all stakeholders in the oil and gas industry creates many challenges; however, many of these can be overcome through sound HRM practices.

References

1. Smil, V. (2008). *Oil: A beginner's guide*. Oxford, UK: Oneworld.
2. Ministry of Finance, Planning, and Economic Development, Uganda. (2012). *Oil and gas revenue management policy*. Retrieved from http://www.acode-u.org/documents/oildocs/Oil_Revenue_Mgt_Policy.pdf.
   Amoako-Tuffor, J. (2011). *Public Participation in the making of Ghana's petroleum revenue management law*. Retrieved from http://naturalresourcecharter.org/sites/default/files/Ghana%20Public%20Participation.pdf.
3. Boohene, R., & Peprah, J. A. (2011). Women, livelihood and oil and gas discovery in Ghana: An exploratory study of Cape Three Points and surrounding communities. *Journal of sustainable development*, 4(3), 185.
4. Galbrath, K. (2012, July 13). In oil boom, a housing shortage and other issues. *New York Times*. http://www.nytimes.com/2012/07/13/us/west-texas-oil-boom-creates-housing-shortage-and-other-issues.html?_r=0.
5. EBI. (2010). Negative secondary impacts from oil and gas development. Retrieved from http://www.theebi.org/pdfs/impacts.pdf.
6. API. (2013). *Environmental expenditures by the US oil and natural gas industry, 1990–2012*. Retrieved from http://www.api.org/statistics/~/media/Files/Publications/Environmental-Expenditures-2013.pdf.
7. API. (2013). *Environmental expenditures by the US oil and natural gas industry, 1990–2012*. Retrieved from http://www.api.org/statistics/~/media/Files/Publications/Environmental-Expenditures-2013.pdf.

8. API. (2013). *Environmental expenditures by the US oil and natural gas industry, 1990–2012*. Retrieved from http://www.api.org/statistics/~/media/Files/Publications/Environmental-Expenditures-2013.pdf.
9. IPIECA/OGP. (2002). *Key questions in managing social issues in oil and gas projects* (Report #2.85/332). Retrieved from http://www.ogp.org.uk/pubs/332.pdf.
10. European Commission. (2013). *Oil and gas sector guide on implementing the UN Guiding Principles on Business and Human Rights*. Retrieved from http://www.ihrb.org/pdf/eu-sector-guidance/EC-Guides/O&G/EC-Guide_O&G.pdf. IPIECA. (2010). *Oil and gas industry guidance on voluntary sustainability reporting*. Retrieved from http://www.ipieca.org/sites/default/files/publications/Reporting_Guidance-12_October_2012.pdf.
11. Ladd, A. E. (2013). Stakeholder perceptions of socioenvironmental impacts from unconventional natural gas development and hydraulic fracturing in the Haynesville Shale. *Journal of Rural Social Sciences, 28*(2), 56–89.
12. IPIECA/OGP. (2002). *Key questions in managing social issues in oil and gas projects* (Report #2.85/332). Retrieved from http://www.ogp.org.uk/pubs/332.pdf.
13. Hahn, J. (2013). *4 Oil and gas companies that are killing it in social media*. Retrieved from http://info.drillinginfo.com/4-oil-and-gas-companies-that-are-killing-it-in-social-media/.
14. BP. (n.d.). *Supporting development in societies where we work*. Retrieved from http://www.bp.com/en/global/corporate/sustainability/society/Supporting-development-in-societies-where-we-work.html.
15. IPIECA/OGP. (2012). *Social development: Stakeholder communities prosper from local economic growth*. Retrieved from http://www.ipieca.org/sites/default/files/publications/ipieca_ogp_fact_sheet_social_development.pdf.
16. IPIECA. (2010). *Oil and gas industry guidance on voluntary sustainability reporting*. Retrieved from http://www.ipieca.org/sites/default/files/publications/Reporting_Guidance-12_October_2012.pdf.
17. IPIECA/OGP. (2002). *Key questions in managing social issues in oil and gas projects* (Report #2.85/332). Retrieved from http://www.ogp.org.uk/pubs/332.pdf.
18. Hahn, J. (2013). *4 Oil and gas companies that are killing it in social media*. Retrieved from http://info.drillinginfo.com/4-oil-and-gas-companies-that-are-killing-it-in-social-media/.
19. Jackson, S. E., Schuler, R. S., & Werner, S. (2011). *Managing human resources*. CengageBrain.com.
20. Non-governmental organization. (2014, August 10). In Wikipedia. Retrieved from http://en.wikipedia.org/w/index.php?title=Non-governmental_organization&oldid=620605042.
21. Pew Charitable Trusts. (2013). *Arctic standards*. Retrieved from http://www.pewtrusts.org/~/media/legacy/oceans_north_legacy/page_attachments/PEWArcticStandards092313.pdf?la=en.
22. Mufson, S. (2014, January 30). Shell says it won't drill in Alaska in 2014, cites court challenge. *The Washington Post*. Retrieved from http://www.washingtonpost.com/business/economy/shell-says-it-wont-drill-in-alaska-in-2014-cites-court-challenge/2014/01/30/72dd06f8-89ab-11e3-916e-e01534b1e132_story.html.

23. Snow, N. (2014). Full set of tools needed for US Arctic spill response, NRC finds. *Oil and Gas Journal*. Retrieved from http://www.ogj.com/articles/2014/04/full-set-of-tools-needed-for-us-arctic-spill-response-nrc-finds.html.
24. Bureau of Labor Statistics. (2014). *Union affiliation of employed wage and salary workers by occupation and industry 2012-2013 annual averages*. Retrieved from http://www.bls.gov/news.release/pdf/union2.pdf.
25. United Steelworkers. (2014). *Union solidarity in oil levels the playing field with the global oil giants*. Retrieved from http://www.usw.org/union/mission/industries/oil/bargaining/lorem-ipsum-dolor-sit-amet-consectetur-adipiscing-elit.
26. ILO. (2002). The promotion of good industrial relations in oil and gas production and oil refining. Retrieved from http://s3.amazonaws.com/zanran_storage/www.ilo.org/ContentPages/4110886.pdf.
27. ILO. (2002). The promotion of good industrial relations in oil and gas production and oil refining. Retrieved from http://s3.amazonaws.com/zanran_storage/www.ilo.org/ContentPages/4110886.pdf.
28. Eweje, G. (2009). Labour relations and ethical dilemmas of extractive MNEs in Nigeria, South Africa and Zambia: 1950-2000. *Journal of Business Ethics, 86*(2), 207-223.
29. Smith, J. (2013, August 8). Student's name oil and gas firm as the best corporate citizen. *Forbes*. Retrieved from http://www.forbes.com/sites/jacquelynsmith/2013/08/08/students-name-oil-and-gas-firm-as-the-best-corporate-citizen/.
30. Szegedi, K. (2011). *Ethics codes and ethics management in the oil and gas industry*. MOL group. Retrieved from www.mol.hu/repository/664893.pdf.
31. Savage, G. T., Nix, T. W., Whitehead, C. J., & Blair, J. D. (1991). Strategies for assessing and managing organizational stakeholders. *The Academy of Management Executive, 5*(2), 61-75.
32. Clement, R. W. (2005). The lessons from stakeholder theory for US business leaders. *Business Horizons, 48*(3), 255-264.
    Harrison, J. S., & Bosse, D. A. (2013). How much is too much? The limits to generous treatment of stakeholders. *Business Horizons, 56*(3), 313-322.
33. Sachs, S. (2002). Managing the extended enterprise: The new stakeholder view. *California management review, 45*(1), 6-28.
34. Shen, J., & Jiuhua Zhu, C. (2011). Effects of socially responsible human resource management on employee organizational commitment. *The International Journal of Human Resource Management, 22*(15), 3020-3035.
35. Jackson, S. E., Schuler, R. S., & Werner, S. (2011). *Managing human resources*. CengageBrain.com.
36. Jackson, S. E., Schuler, R. S., & Werner, S. (2011). *Managing human resources*. CengageBrain.com.
37. Forstenlechner, I., & Mellahi, K. (2011). Gaining legitimacy through hiring local workforce at a premium: the case of MNEs in the United Arab Emirates. *Journal of World Business, 46*(4), 455-461.
    Banai, M., & Sama, L. M. (2000). Ethical dilemmas in MNCs' international staffing policies a conceptual framework. *Journal of Business Ethics, 25*(3), 221-235.
    Banai, M. (1992). The ethnocentric staffing policy in multinational corporations a self-fulfilling prophecy. *International Journal of Human Resource Management, 3*(3), 451-472.

38. API. (2014). *Minority and female employment in the oil & gas and petrochemical industries* (IHS Report). Retrieved from http://www.api.org/~/media/Files/Policy/Jobs/IHS-Minority-and-Female-Employment-Report.pdf.
39. Cox, T. H., & Blake, S. (1991). Managing cultural diversity: Implications for organizational competitiveness. *Academy of Management Executive*, 45–56.
40. Aguinis, H., & Glavas, A. (2012). What we know and don't know about corporate social responsibility a review and research agenda. *Journal of Management, 38*(4), 932–968.
41. Szegedi, K. (2011). *Ethics codes and ethics management in the oil and gas industry*. MOL group. Retrieved from www.mol.hu/repository/664893.pdf.
42. Gulf Intelligence. (2014, February). *Women in energy—On the right track*. White paper presented at the International Petroleum Technology Conference, Kuala Lumpur.
43. Komaki, J., Heinzmann, A. T., & Lawson, L. (1980). Effect of training and feedback: Component analysis of a behavioral safety program. *Journal of applied psychology, 65*(3), 261.
44. Szegedi, K. (2011). *Ethics codes and ethics management in the oil and gas industry*. MOL group. Retrieved from www.mol.hu/repository/664893.pdf
45. Martin, V. (2006). *Managing projects in human resources, training and development*. London, UK: Kogan Page.
46. Kaplan, R. S., & Norton, D. P. (1996). Using the balanced scorecard as a strategic management system. *Harvard business review, 74*(1), 75–85.
47. Chen, C.C., Kraemer, J., & Gathii, J. (2011). Understanding locals' compensation fairness vis-à-vis foreign expatriates: The role of perceived equity. *The International Journal of Human Resource Management, 22*(17): 3582–3600.
48. http://investor.phillips66.com/investors/news/news-release-details/2013/Phillips-66-Contributes-to-Oklahoma-Tornado-Relief-Efforts/default.aspx.

8

Government Involvement in the Oil and Gas Industry

The Influence of Governments in the Oil and Gas Industry

The government is involved in nearly every facet of the oil and gas industry; there is probably no other industry where government is such an important stakeholder. Chapter 7 mentioned that governments are usually considered a secondary stakeholder in most industries—not critical to the short-term survival of the company, but important in the long term. In the oil and gas industry, governments are closely involved in all aspects of the value chain and have an impact on every aspect of the business. In most countries the government owns oil and gas resources, even those below privately held land. Thus, in the oil and gas industry, government could be considered a primary stakeholder. One of the reasons governments are so involved in the industry is because of the economic, social, and national security aspects of oil and gas.

The importance of oil and gas to governments

Oil and gas are critically important to governments for many reasons. First, oil and gas resources have important economic implications for a country. Second, beyond their economic impact, oil and gas have important implications for a country's social well-being, including quality-of-life issues. Third, oil and gas specifically, and energy in general, have important implications for a country's national security. These implications are now briefly discussed.

Economic influence. As discussed in the last chapter, oil and gas discoveries and the subsequent entry of oil and gas companies can positively impact a country through their powerful role in economic development—including the improvement of transportation, the creation of new infrastructure, the creation of new jobs, the contribution of income, and the improved availability of energy. The oil and gas industry produces the largest, or at least one of the largest, sources of export revenues and hard-currency earnings for many large oil- and gas-exporting countries. For example, in Angola, Libya, and Venezuela, more than 50% of the government's budget revenues come from oil.[1] This dependence on oil causes difficulties when the price drops significantly, as it did in 2008 and in 2014.

Quality-of-life issues. Of course, the quality of life is strongly related to economic well-being, with positive economic outcomes, such as high income and low unemployment, leading to higher standards of living. Benefits such as improved infrastructure, better education facilities, and training programs are likely to follow from oil discoveries both at the community and national level. These enhancements can lead to higher quality of life when properly managed. However, government neglect may greatly lower the impact of oil and gas investments.[2] Nevertheless, oil and gas investments and development provide governments with many opportunities to improve the quality of life of their citizens.

National security issues. The economic impact and quality-of-life effects of oil also influence national security. Prosperous, satisfied citizens are more likely to foster a stable government. However, oil also has a direct effect on national security. One example is how national security can be affected by an energy crisis, which increases country instability. Thus, many countries stockpile strategic petroleum reserves to reduce the negative effects of an energy crisis. The largest strategic reserves, in the United States, have a capacity of 727 billion barrels (the current inventory is available at www.spr.doe.gov/dir/dir.html). China is expanding its reserves, with the goal of a capacity of 476 billion barrels by 2020.

Another example of oil's effect on national security is the important role oil plays in national defense and in modern armies. Not surprisingly, the US Department of Defense is the single largest consumer of energy in the nation, consuming 5 billion gallons of gasoline and diesel annually.[3] Although many factors are usually involved, the pursuit or protection of oil resources has been a factor in a number of wars, the most obvious example being Saddam Hussain's occupation of Kuwait in 1990 and the resulting Gulf War.[4] The bottom line is that as long as oil and gas are the dominant transportation fuel and energy sources worldwide, relative stability in availability and prices will be an important factor in global security.[5]

Finally, the vast wealth that comes with oil and gas can be used for self-preservation by governments. Rather than using income from oil-based taxes, leases, or royalties for broad societal benefits, those in power may use it for rewarding supporters and maintaining patronage networks. Thus, oil money can be an important source of income for governments to maintain both broad public support and private self-interested support.

In summary, governments have become integral to the oil and gas business in many ways:

1. Governments own many of the largest oil and gas companies.
2. Governments own most of the oil and gas resources.
3. Governments control the regulations that oil and gas companies must comply with.
4. Governments and national oil companies (NOCs) frequently partner with private oil and gas companies.

Thus, governments not only have the ability to be heavily involved in the industry because of their regulatory and ownership power, but they are highly motivated to do so because of the importance and value of oil and gas to governments. Each of these areas is now discussed in detail.

Government ownership of companies

Because of the impact of oil and gas on governments and their citizens as described above, many governments have nationalized their country's oil and gas industry. The first NOC occurred in Austria-Hungary in 1908. NOCs were created in the UK in 1914 and in France in 1924. Early NOCs were largely created to ensure that fuel would be available for the military (and civilian) fleets. Beginning in the mid-1920s in Latin America, NOCs were created to capture wealth for the state. NOCs were created in Argentina, Chile, Uruguay, Mexico, Peru, Venezuela, and Bolivia. The takeover of Standard Oil of New Jersey's operations by the Bolivian government in 1937 was the first case of nationalization of the operations of a private corporation. This was followed by the nationalization of a foreign company in Mexico in 1938 and lead to widespread nationalization in the Middle East in the 1940s. To reduce risk, companies formed consortiums to develop oil in the Middle East. The nationalization of these consortiums was usually gradual. For example, Aramco (Arabian-American Oil Company), which was half-owned by Texaco and half-owned by Standard Oil of California, agreed to share half of its profits with the Saudi Arabian government in 1950. In 1973, the government took a 24% ownership share, which was expanded to 60% in 1974. That share became 100% in 1980. Even today nationalization is still a major concern for international, integrated oil company (IOCs) in countries with high political risk. In 2012, Argentine President Cristina Fernandez de Kirchner seized control of YPF (Yacimientos Petrolíferos Fiscales, translated as Treasury Petroleum Fields) from the Spanish company Repsol YPF by replacing the CEO with two top aides and expropriating assets and 51% of YPF shares. A year later Repsol was compensated for the expropriations.[6]

NOCs in resource-rich countries tend to be quite different from those in resource-poor countries. In resource-rich countries, NOCs exist to control and manage a country's oil and gas resources. These NOCs can vary in their level of investment and their level of activity. Some NOCs are small enterprises with little capital and few assets other than state-owned oil and gas. They tend to collect taxes and royalties from private upstream oil and gas companies that manage the oil operations. NOCs in Trinidad and Tobago, Brunei, and Gabon are examples of those with the lowest levels of investment and activity. At the opposite end are NOCs with large workforces and a presence around the globe. These NOCs have expanded outside their home countries and operate as IOC-type firms. NOCs in Brazil (Petrobras), Malaysia (Petronas), and Norway (Statoil) are examples of those with the highest levels of investment and activity.

Statoil, one of the best managed NOCs, is publicly traded, with the state retaining the golden share—in this case, a 69% controlling interest. Norway has largely allowed Statoil to be run as a private firm, with profitability as an explicit goal. Outside of some specific strategic interest and obligations, such as the continuing exploration and development of the Norwegian Continental Shelf, Statoil is largely free to run itself according to private-sector interests. As such, it continually strives to cut costs, including limiting employment and employment costs. The state collects its dividends and, outside of some special oil and gas taxes that many NOCs are subject to, leaves the company alone to run its own business. Statoil's governance model has been called the Norwegian model of governance. This model is based on the strict separation of the institutional roles responsible for regulatory policy and commercial interests.

In resource-poor countries such as India and China, NOCs exist to manage the energy security needs of their country. The NOCs of China (Sinopec, CNPC, and CNOOC) are unique because of the communist government. Although they are publicly listed, their leadership is closely tied to the government (the majority owner), and thus the energy security of China is a clear goal of the companies. Based on the large employment of these companies, it appears that maintaining employment is also an important goal. These goals affect how the Chinese NOCs operate in the global markets. For example, Chinese NOCs are willing to bid much higher for access to oil and reserves than their IOC and independent competitors, indicating a lower priority for economic returns than their competitors.[7]

The United States is the largest oil producer without an NOC. The establishment of a US NOC is occasionally mentioned in Congress, but very unlikely to happen. The benefits to the United States of having an NOC would be a more direct linkage between oil issues and US government interests. However, a US NOC would be more vulnerable in other countries during conflicts, would likely be much less efficient and productive than the IOCs, and would possibly weaken the position of private companies, threatening US energy security. Thus, the establishment of a US NOC is highly unlikely in the foreseeable future.[8]

NOC performance. It is hard to compare the general performance of NOCs versus non-government-controlled firms because of their different goals. For example, while shareholders would consider overemployment a wasteful inefficiency that negatively affects financial performance, a government may see overemployment as a successful strategy to maximize employment levels (and perhaps reward cronies for political support). Nevertheless, from a financial performance standpoint, NOCs usually have suboptimal financial performance compared to non-government-owned competitors. NOCs tend to be less profitable, less efficient, and more highly leveraged when compared to private enterprises.[9] Figure 8–1 shows the average revenue per employee (a measure of efficiency) for the largest NOCs and privately owned oil companies for 2001–2009.[10] The figure clearly shows the large differences between the two. One reason for this lackluster financial performance is the fact that other goals such as wealth distribution, energy security, goals related to foreign policy, economic development, job creation, and national security may take precedence over financial goals in NOCs. Table 8–1 shows the strategic interest of IOCs, resource-rich NOCs, and resource-poor NOCs.[11] Note how the IOC interests focus far more on financial returns. Another reason for lower NOC financial performance is that NOCs are insulated from standard markets and therefore are not subject to market controls. In light of this, governments sometimes want to improve efficiency and change the mission and strategy of state-owned companies, and they can do this through privatization.

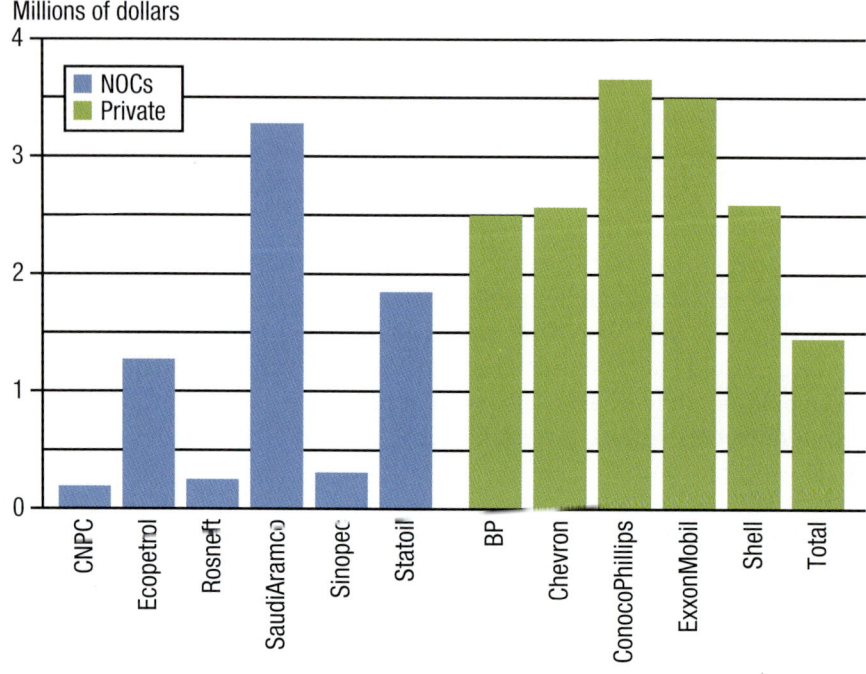

Fig. 8–1. Revenue per employee for the large NOCs and private oil and gas companies, 2001–2009
Source: Hartley & Medlock[12]

Table 8–1. Strategic interests of IOCs and NOCs

IOCs	NOCs (Resource Rich)	NOCs (Resource Poor)
Returns on invested capital that meet shareholder expectations	All	Control over their domestic reserves
Long-term growth in income	Control over their domestic reserves	Long-term domestic energy security
Access to reserves and reserve replacements	Domestic employment	Access to reserves globally
Partnerships with NOCs and other IOCs	**Low levels of investment and activity**	International expansion
Efficient and disciplined management	Sufficient cash flow to provide a large share of government revenues	Breakeven returns on invested capital
Stakeholder satisfaction	Social contributions for the nation	Proprietary technology
Proprietary technology	Improved efficiency and management know-how	Partnerships with IOCs in international projects
Talented employees	IOCs as operators/contractors	Improved efficiency and management know-how
Minimal political involvement	**High levels of investment and activity**	
	International expansion	
	Integration across value chain	
	Returns that meet stakeholder expectations	
	Proprietary technology	
	Efficient and disciplined management	
	Partnerships with IOCs	

Privatization. Selling government companies (in whole or in part) to private interests produces a number of benefits at the cost of the government losing some control of the firm. Privatization helps to raise capital. Access to capital is a crucial element for efficient operations in the domestic market as well as for international expansion. This is particularly true in the oil and gas industry, a highly capitalized industry. NOCs may have trouble raising capital because of their subpar financial performance, particularly during industry down cycles. Exceptions include the Chinese NOCs because of their close relationship with the Chinese government and Petrobras and Statoil because of their long-term record of operational efficiency. NOCs such as NIOC, NNPC, Pemex, and Petróleos de Venezuela S.A. (PDVSA) have limited access to capital and thus have been unable to build capacity to meet domestic demand, particularly downstream. Hence, Iran, Nigeria, Mexico, and Venezuela all import refined products.[13] For some background on PDVSA, see Current Challenge Box 8–1.

Current Challenge Box 8–1. Government Exploitation of its NOC: Hugo Chàvez and PDVSA

Stopping governments from exploiting their NOCs is a challenge for citizens when short-sighted politicians are elected. Petróleos de Venezuela, S.A.(PDVSA) was long considered one of the best managed NOCs. Founded in 1976 after the nationalization of the Venezuelan oil sector, PDVSA performed admirably in its task of managing Venezuela's vast oil reserves, often estimated as the largest in the world. In the early years, the company was responsible for more than 70% of Venezuela's export earnings and more than 80% of the government budget. Venezuela was a *petrostate*—and an increasingly successful one—as PDVSA managed the oil sector with skill.

PDVSA's successful run came to an end in 1998 with the election of President Hugo Chàvez, who proceeded to replace most of the company's leadership with political appointees who were beholden to Chàvez and not the Venezuelan people. Chàvez then turned to using PDVSA cash flows to fund a rapidly expanding socialist government. The siphoning of critical cash flows from the company resulted in underinvestment in many sectors of the oil industry, including refining. A country that had once been a major exporter of both crude oil and refined products now found itself importing refined products as the underinvestment in refining stifled output. A few years of failed leadership and investment resulted in the massive decline of a once successful NOC.

One of the most powerful policy tools used by Chàvez was the heavy subsidization of gasoline. The price of gasoline in Venezuela, at a government-set price of $0.18 per gallon, was the lowest in the world. Although a significant financial benefit to the citizens and industries of Venezuela, it had the unintentional result of increasing domestic consumption of oil and refined products, reducing the amount of oil available for export—and export was critical to the income of the country. With gasoline essentially free, consumption and imports of gasoline skyrocketed, adding to the industry's inefficiencies.

The rapid decline of the company and its prospects resulted in a massive PDVSA worker strike in 2001. The striking oil company workers demanded early elections, presumably to elect someone other than Chàvez. After a nearly two-month shutdown of company production, the government gradually reestablished control over PDVSA. It then quickly fired 18,000 of its workers, almost 40% of the company's workforce. This prompted a new wave of pro-Chàvez workers hired, many lacking the skills critically needed for the company's operations. By the time of Chàvez's death in 2013, Venezuela's oil exports had dropped nearly 50% since he took office in 1997 and export revenues were down nearly 25%.

Privatization also helps firms gain expertise (particularly when ownership is by an experienced company) and fosters efficiency through market controls. Interestingly, even the threat of privatization and increased competition appears to greatly increase efficiency and financial performance. For example, in Brazil, President Fernando Henrique Cardoso eliminated Petrobras's monopoly in 1995 through reforms that opened the sector to private firms and increased the threat of privatization. After the reforms, Petrobras's productivity doubled in six years.[14] Petrobras is now almost as capable as any private oil company[15] (although it did suffer from a corruption scandal in 2015). Privatization also tends to reduce overemployment, which results in layoffs for many government workers, who are then most likely to be employed in lower-paying jobs.[16] Thus, the increased efficiency and financial performance come at a social cost.

The general view regarding privatization is that as government ownership decreases and private ownership increases, efficiency and financial performance will improve, because such firms become more similar to private firms than NOCs. That has been the result for Statoil and Petrobras. A major downside from the government's perspective is loss of control. In 2014 the Mexican government decided that giving up some control over the industry was necessary to create change in how the industry is managed. (See Current Challenge Box 8–2.)

Current Challenge Box 8–2. Old Dogs Do Learn New Tricks: Mexico, Pemex, and the Renewal of the Mexican Oil Industry

Underperforming by NOCs is a current challenge for several countries. Mexico nationalized its entire oil industry—from exploration and development through service stations and electricity production—in 1938. The Mexican constitution was amended to require that only the country's NOC, Petróleos Mexicanos (Pemex), could own and profit from the production and sale of oil and gas assets. Within Mexico, Pemex and the oil industry became a symbol of the Mexican people's struggle against repression.

The next 75 years saw a Mexican oil industry that was regulated, owned, and operated by a single firm that became an institution unto itself. Mexico was, in the eyes of some, a *petrocracy*, a state whose leadership was increasingly disenfranchised from its own people by oil and its tool, Pemex. Many believe that this era of readily available oil revenue allowed Pemex to become complacent and inefficient. The Mexican government increasingly relied on Pemex for its income, as did a large part of the Mexican population.

Although part of a complex web of intergovernmental structures and relations, Pemex was officially regulated by the Ministry of Energy and held largely accountable to the Federal Income Tax Authority of Mexico—the *Hacienda*. Pemex paid all its income beyond current direct costs and general administrative expenses to the Hacienda. Without the ability to control its own financial resources, Pemex was unable to undertake investments it believed necessary for its future. For many years, investment was largely confined to production only—not exploration, refining, or petrochemicals. In the words of one analyst,

> Pemex's manager[s] are effectively stripped of much of the responsibility and accountability that comes with running such a large company—they are unable to pursue an optimal investment strategy because they have neither the capital resources nor the autonomy to make intrinsically risky investment decisions on commercial merits.[17]

Pemex was in many ways controlled by its workers' union, the Oil Workers Union of the Mexican Republic. The union had resisted significant organizational change, protecting the jobs of its 140,000 employees. The union lost one major employment fight in the 1980s, when Pemex employment dropped from 220,000 to 120,000, but 20,000 of that was regained in the following years. The union was also the subject of increasing debate over the corruption thought to be rampant throughout the organization. Three of the union's general directors ended their administrations in recent years under indictment. The union maintained its stranglehold on much of the Pemex organization through hiring, because the union controlled nearly all new hiring at Pemex (termed *filling spaces*).

In 2013 the Mexican oil industry experienced its 10th year of declining production and 12th year of declining reserves. Pemex was the source of 42% of the Mexican government's income, and that income was in serious decline. The company had made no new significant discoveries in years, had been largely unsuccessful in its exploration of the deepwater potential of the Gulf of Mexico, and was increasing its imports of refined products and natural gas from the United States. Pemex was failing, and it was largely considered inefficient and corrupt. It was time for change.

On Friday, December 13, 2013, President Enrique Peña Nieto signed into law changes to three key articles in the Mexican constitution that would allow private companies to share in oil and gas production and book reserves as expected cash flow (inferring ownership of the hydrocarbons). The changes to both Pemex and the Mexican oil industry as a whole are dramatic. Mexico's oil market will cease to be run by a single firm and Pemex will now have the ability to enter into joint ventures with foreign and domestic private companies, while also facing the possibility of new competitive entrants. Pemex will continue to be state-owned, but the company will gain control over its own budget and its own financial performance. Time will tell if old dogs can indeed learn new tricks.

Government ownership of oil

In almost all countries, the oil and gas beneath the surface of the earth is owned by the government. The United States is a notable exception, where private citizens may own natural resources. Because the state usually owns the resources, they are used in such a way to maximize the government's goals, rather than the goals of conventional shareholders and other stakeholders. The fact that the government owns the oil in the ground has tremendous implications for the oil majors and the other companies in the oil and gas industry. The future of any upstream oil and gas company depends on the reserves it has in the ground. The government ownership of that oil puts huge constraints on companies that want access to those reserves. As owners, governments wield their power by controlling the fiscal terms of access, which can include not allowing any access at all. In the United States many high-potential plays, such as the Alaskan National Wildlife Refuge, are not available for exploration because of the expected costs to the environment. Also, all offshore drilling in the Gulf of Mexico was halted for six months after the Macondo accident in 2010. Governments can also control access by targeting certain companies. For example, BP was suspended from bidding on gulf oil drilling rights until 2014 because of the Macondo accident.

Leases. A lease is an agreement for a company (the lessee) to gain access for exploration and development on a property. A specific time period is determined during which the lessee has the right to access the land. The lease usually specifies the rights of the lessee, including minimum drilling requirements, nonperformance penalties, and exploration techniques and drilling methods allowed. Leases also generally specify the up-front payments, bonus payments, and royalties (the percentage of production proceeds that the lessee will pay). Royalty payments are usually calculated as market price (a set percentage of the market price), although some are based on revenue received (a percentage of the actual sale price) or in-kind (a portion of the oil or gas produced).[18] Figure 8–2 shows the number of land leases that were in effect in 2013 compared to 2003 for some states in the United States.

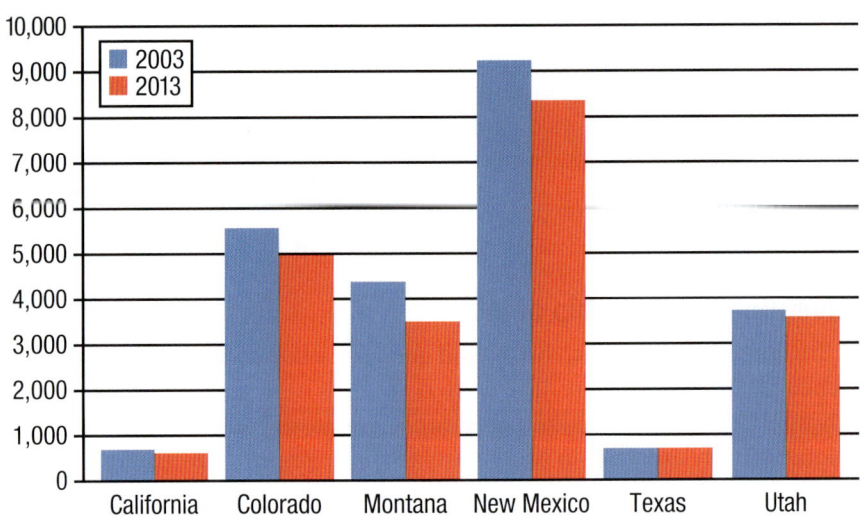

Fig. 8–2. Number of leases in effect in 2003 and 2013
Source: Bureau of Land Management.

Lease auctions. The right to access reserves is granted by the government on the basis of individual negotiations, relationships, political affiliations, personal preferences, or, most commonly, lease auctions. Auctions are the favored method of leasing because they allocate and price resources in an uncertain context. Further, auctions allocate the exploration and development rights in an efficient and transparent way, and help maximize lease revenues. Auctions differ on a number of aspects, such as whether blocks are valued collectively or individually, the basis for bidders' valuations, whether the rights are sold simultaneously (in bundles) or sequentially (individually), and whether the auction is static (single bid) or dynamic (rebidding allowed).

In the United States, the Bureau of Ocean Energy Management (for offshore) and the Bureau of Land Management (for onshore) identify the areas to be made available for exploration and drilling and divide it into blocks (or tracks) roughly 9 square miles in size. Historically, blocks were only put up for auction after they were nominated by companies interested in developing them. Today, nearly all blocks are available for auction on an ongoing basis. Companies wishing to bid on blocks are free to assess them. This is usually done through complex and expensive 3-D seismic analysis. Companies frequently bid as part of a consortium to reduce the risks, but since 1975 the largest oil companies are not allowed to bid jointly (although they may join with nonmajor partners). For onshore leases, auctions are required to be held at least quarterly, but may be more frequent in states where the number of nominated parcels is high. Most acres offered do not receive any bids, as can be seen in figure 8–3. Auction winners are determined through the first-price sealed bid method, although high bids are occasionally deemed insufficient and are rejected. Winners have exclusive exploration, drilling, and production rights of the block for the length of the lease, usually 10 years. At the end of the lease period, undeveloped land reverts back to the government; for developed land, the lease extends automatically as long as royalty payments continue.[19]

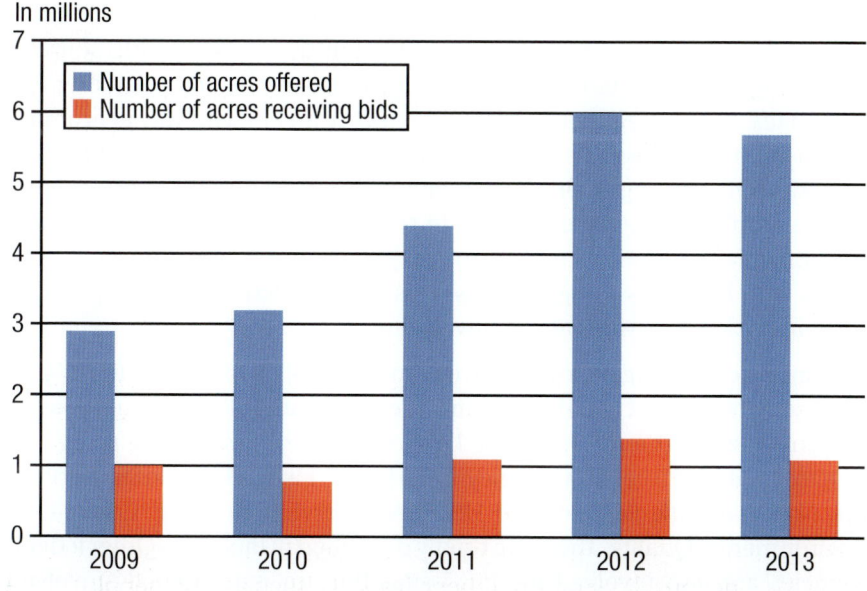

Fig. 8–3. Number of acres offered and number of acres receiving bids by year in the United States
Source: Bureau of Land Management.

Regulations

Governments can control many aspects of the oil and gas industry, including who has access to the resources (as described earlier). Much of this control is accomplished through regulations from government agencies and laws passed by their legislatures (for simplicity, the generic term *regulations* applies to both). Regulation is seen as one of the greatest challenges in the oil and gas industry, with one in five industry professionals believing that tighter industry regulation is a factor that will have the greatest impact on the industry in the short term. The same survey also found that 65.4% of the respondents agreed with the statement, "The regulatory outlook in our firm's key markets continues to get tougher," while only 5% disagreed.[20] Regulations that are particularly applicable to the oil and gas industry include health, safety, and environmental (HSE) regulations; drilling regulations; pricing and taxation regulations; labor regulations; and foreign practice regulations.

Health, safety, and environmental regulations. HSE regulations are particularly applicable to oil and gas companies because of the nature of the work, the working conditions, locations, and materials. Environmental regulations may be related to broad-sweeping energy policies that aim to diminish a nation's use of fossil fuels and increase the use of renewable sources. Increasing emission standards is one way to accomplish this. Another is a carbon tax. Some suggest a carbon tax would be an efficient way to promote a transition to renewable energy sources, while increasing government revenues. Others suggest that such regulations would hurt the economy and reduce employment.[21] Interestingly, in the United States the recent natural gas bonanza (and low natural gas prices) has led to substituting coal-fired plants with low-emission gas plants. This has resulted in large drops in US carbon dioxide emissions at a relatively low cost.[22]

Major federal agencies involved in general HSE regulations and enforcement in the United States are the Occupational Safety and Health Administration, the Centers for Disease Control and Prevention, the Environmental Protection Agency (EPA), the Fish and Wildlife Service, the National Park Service, The US Forest Service, and the Council on Environmental Quality. These are briefly described in table 8–2. Other federal agencies are also involved in HSE issues but are more directly involved specifically with the oil and gas industry. These include the Bureau of Ocean Energy Management, the Bureau of Safety and Environmental Enforcement, the Bureau of Land Management, and the US Geological Survey. These are briefly described in table 8–3.

All of these agencies may be directly involved with regulations relative to oil and gas companies. For example, the EPA's regulations on air and water quality, solid waste disposal, and greenhouse gas emissions have a direct effect on oil and gas companies. Specifically, in 2014 the EPA started requiring offshore oil and gas operations to report chemicals discharged in the ocean during fracking. Further, these agencies have indirect effects. The EPA's recent study of fracking and its potential impact on drinking water is considered an important factor for state-level fracking regulations.[23]

Table 8–2. Agencies concerned with health, safety, and environment (HSE) in the United States

Agency	Description
The Occupational Safety and Health Agency (OSHA) www.osha.gov	OSHA was established in 1970 and assures safe and healthy working conditions through regulations that include limits on chemical exposure, safety procedure requirements, and protective equipment requirements. It also conducts training, provides compliance assistance, provides or funds education and outreach programs, and recognizes exemplary organizations.
Centers for Disease Control and Prevention (CDC) www.cdc.gov	The CDC was started in 1942 and its mission is to protect America from health, safety, and security threats through research, the dissemination of information, and responses to threats.
The Environmental Protection Agency (EPA) www.epa.gov	The EPA was started in 1970 and protects health and the environment through regulations. It also conducts environmental assessments, research, and education.
Fish and Wildlife Service (FWS) www.fws.gov	Created in 1940, this agency is devoted to conserving, protecting, and enhancing fish, wildlife, plants, and their habitats through the enforcement of federal wildlife laws, and the management of environmental resources.
National Park Service (NPS) www.nps.gov	Founded in 1916, this agency manages the 401 parks of the US national park system, and is directed to help conserve natural and historic lands and their wildlife.
The United States Forest Service (USFS) www.usfs.gov	Established in 1905, this agency manages public lands in national forests, grasslands, and wilderness areas covering over 229 million acres.
Council on Environmental Quality (CEQ) http://www.whitehouse.gov/administration/eop/ceq	Established in 1969, this council coordinates the activities of the environmental efforts of other agencies and works with the White House to achieve its environmental agenda.

Table 8–3. Agencies concerned with oil and gas in the United States

Agency	Description
The Bureau of Land Management (BLM) www.blm.gov	This agency manages public lands and handles oil and gas lease sales on public lands and issues development permits.
Bureau of Ocean Energy Management (BOEM) www.boem.gov	Formerly the Minerals Management Service, this agency created in 2011, manages offshore energy and mineral resources by developing offshore leasing programs, handling offshore oil and gas lease sales, and conducting environmental reviews.
Bureau of Safety and Environmental Enforcement (BSEE) www.bsee.gov	Also created in 2011, as part of the reorganization of the Mineral Management Service, this agency is responsible for the safety and environmental oversight of oil and gas operations, including permitting and inspections (which used to be a joint responsibility of MMS and the Coast Guard).
Federal Energy Regulatory Commission (FERC) www.ferc.gov	Established in 1977, this agency regulates the interstate transmission of natural gas, oil, and electricity as well as natural gas and hydropower projects.
Pipeline and Hazardous Materials Safety Administration (PHMSA) www.phmsa.dot.gov	An agency with the US, Department of Transportation, it regulates the safety aspect of pipelines and other means of transportation of hazardous materials including oil, natural gas.
The United States Geological Survey (USGS) www.usgs.gov	Established in 1879, this agency collects, monitors, and analyzes data, researches, and provides information regarding natural resource conditions, issues, and problems.

Drilling regulations. Drilling regulations may be related to offshore, onshore, or both. Many of the drilling regulations are directly related to HSE, as described above. Offshore drilling regulations were considerably tightened in the United States after the Macondo accident. Drilling regulations may stem from many laws and be uncoordinated among and within agencies, or they may be highly integrated, depending on the country. For example, in the United States, regulations and many different laws apply to offshore drilling, including those related to offshore leasing, offshore health and safety, jurisdiction, pipelines, national security interests, rights of native peoples, marine mammals, endangered species, pollution, waste disposal, environmental impact, and liability. Not only are all these laws not integrated or harmonized, but they are all complicated by judicial decisions. The resulting regulations are highly prescriptive and may include inspection requirements for compliance and resulting sanctions for noncompliance. In contrast, in Norway the legal framework for offshore drilling is highly integrated, directing companies toward building self-regulation through internal controls.[24]

As in all areas of regulation, drilling regulations may occur at the state, county, and municipal level in the form of regulations, laws, and ordinances. At the state level, for example, Alaska has the Alaska Oil and Gas Conservation Commission, which regulates oil and gas drilling and production; the Alaska Department of Natural Resources Oil and Gas Division, which regulates leasing; and the Alaska Department of Environmental Conservation, which issues regulations related to human health and protection of the natural environment. At the municipal level, for example, the city of Houston has eight pages of ordinances covering oil and gas wells, including specifications for drilling units and blocks, number of wells and locations.[25] Not surprisingly, oil and gas companies tend to prefer less regulation rather than more, and prefer that federal regulators not get involved in regulating oil production when the process is already being managed by states. Also not surprisingly, state governors agree with this view.[26]

Pricing and taxation regulations. Taxes can be implemented for strict revenue purposes or to change firm behaviors, as mentioned previously regarding a carbon tax and environmental concerns. Opponents of a carbon tax view it as an antibusiness government intervention, while proponents see it as an efficient way to control pollution with the benefit of raising revenues. The EU and Australia currently have carbon taxes and China is considering one.[27] Although taxes of profits are imposed on all industries, profits have been taxed at higher levels for the oil and gas industry. For example, in the United States, Congress enacted the Crude Oil Windfall Profit Tax Act in 1980 to gain revenue from the expected high profits oil and gas companies would earn following the deregulation of oil prices in the United States while global prices were rising sharply. This tax, which was actually implemented as an excise tax, lasted until 1988. *Excise taxes*, which are taxes on the sale of specific goods sold within a country, are common on gasoline.

Governments can also affect firms through tax incentives, by providing tax credits for certain investments. Related to the oil and gas industry, one common tax incentive is for science, technology, engineering, and mathematics (STEM) research and development. For example, Australia, Canada, France, the UK, and the United States all offer tax credits for research and development (R&D) investments. Governments can also encourage R&D though grants, business incubators, and technology parks.[28]

This category also includes subsidies. Energy-related subsidies are common, particularly in energy-producing countries. Energy subsidies may be used to lower costs for citizens, to encourage the use and development of specific energies (usually renewables), to alleviate market failures, or as political tools. In the United States, the most controversial energy subsidy is for corn-based ethanol, but many other energy subsidies are in place.[29] Many developing countries (about two-thirds) subsidize oil or gas to shield citizens from the pain of price increases. Subsidies lower prices, but also alter the markets and can have negative consequences such as crowding out other government spending, discouraging investment, reducing consumption efficiencies, and encouraging corruption and black markets. Further, removing subsidies is politically difficult. However, the most negative thing about subsidies is that they are expensive. For example, natural gas subsidies in Uzbekistan equaled a quarter of that country's gross domestic product (GDP) and a fifth of the GDP in Turkmenistan, while about 10% or more of the GDP of Saudi Arabia, Bahrain, and Ecuador go to petroleum product subsidies.[30] Looking at it another way, subsidies in many countries pay for more than half of the cost of the fuel. In Venezuela, subsidies pay for more than 80% of the fuel cost, while those in Saudi Arabia pay almost 80%; in Uzbekistan, 60%; and in Algeria, nearly 60%.[31]

Labor regulations. Labor regulations include any laws, regulations, and ordinances that apply to conditions of employment. These may deal with discrimination, termination or layoff criteria, wage determination, working hours, independent contractors, and working conditions, among other things. These laws apply to any employer and usually do not apply specifically to the oil and gas industry. The same is true for labor regulations for unions and labor relations. Laws and regulations (and subsequent court/commission/labor board interpretations) related to union formation, collective bargaining, work councils, board representation, and union and management rights can also have far reaching implications for the employment relationship. Regulations related to labor relations can have a big impact on the relationship among unions, employees, and management, including the level of conflict and cooperation among them. For example, the Industrial Relations Commission's rulings in Australia were found to have tremendous influence on the level of conflict and modes of conflict resolution between Australian refinery workers and the management of the Shell Clyde refinery in the short and long term.[32] Nevertheless, as with most labor laws, labor relations laws also tend to apply to all businesses. However, some labor laws are particularly applicable to the oil and gas industry. For example, in Ecuador, private companies are required to pay 15% of their profits to employees. Because of the high profits and high capitalization rate, the profit-sharing bonuses from oil and gas companies can be 10 times greater than that of other industries and many times more than the employees' average salary. Laws regarding working conditions and work permits may also be particularly applicable to oil and gas because of the generally difficult working conditions and the global nature of the industry.

Foreign practice regulations. Because most oil and gas firms are international, their home country government (as well as the governments where they do business) can have big effects on the international aspects of the business. For example, trade agreements and credit guarantees have an important impact on foreign business. Also, governments, through diplomatic, political, and economic strategies, can encourage the creation of cross-national joint ventures and the development of specific projects, as the Brazilian government has recently done in Africa.[33] Governments can also provide aid to other countries through overseas development agencies that can affect their firms abroad through social and health projects that affect local labor.

Home country governments may also have laws regarding interactions with host country governments. An example of this in the United States is the Foreign Corrupt Practices Act, which prohibits bribery (payments to attract or retain business) of foreign government officials since 1977. Relatedly, since 2013, Section 1504 of the Dodd-Frank Wall Street Reform and Consumer Protection Act requires oil, gas, and mineral companies to report all payments greater than $100,000 (by project or by government). Oil and gas companies must report such payments to governments (including US agencies) related to all commercial developmental activities, including taxes, royalties, fees, production entitlements, bonuses, dividends, or infrastructure improvements.[34]

Government and private firm partnerships

Governments may partner with firms to achieve goals that are in the best interest of both parties. A good example of this consists of collaborations between governments and oil and gas firms to foster and sustain STEM talent. These collaborations can include initiatives to raise public awareness of STEM, promote STEM education, and create favorable working environments to improve the appeal of STEM industries. Government actions to help promote STEM talent may also include establishing science weeks, funding science museums or centers, implementing STEM education initiatives, and promoting STEM careers.[35]

Historically, NOCs lacked skills in project management and integration, financial analysis, upstream technologies, contractor relationships, and logistical support in addition to their lack of access to capital. To acquire this knowledge (and access to capital), they partnered with IOCs that were willing to work with them to acquire an equity stake (allowing them to book reserves) in the project. These partnerships allowed IOCs access to the oil and upside potential in the risky capital investments in exchange for the IOCs' skills and knowledge. This relationship between NOCs and IOCs still exists today for NOCs with low levels of investment and activity. However, many NOCs now have high levels of investment and activity, so the relationship is changing for them.[36]

The capabilities of NOCs with high levels of investment and activity are approaching the capabilities of IOCs, reducing the need for partnerships. However, partnerships are still useful to both parties, for two reasons. First, as discussed earlier, because NOCs are owned and controlled by the state, they are less efficient than firms with a shareholder mentality. Second, the IOCs' managerial experience, technical experience, and access to capital are still useful to NOCs. Although the know-how of some NOCs is now nearly equal to that of IOCs in many areas of the business, IOCs still have an edge in their ability to manage large-scale projects, new technologies, and project integration. Although partnerships are still useful to both, the nature of the partnership changes when the NOC has high levels of investment and activity. Such NOCs are more interested in buying or renting the IOC's technical expertise rather than giving up some of their resources. This new type of relationship is counter to the historic role of IOCs, that is, booking reserves and participating in upside returns, and more closely resembles a contractor relationship. However, IOCs are finding themselves with less power in their quest for access to oil, given that they currently own less than 10% of the world's reserves, compared to 85% in the 1970s.[37]

In summary, governments are closely involved in all aspects of the value chain of the oil and gas industry because of oil's strong economic, social, and national security influences on countries. Governments own many of the largest oil and gas companies; governments own most of the oil and gas; governments control the regulations that oil and gas companies must comply with; and governments and government-owned companies frequently partner with other companies. The next section considers the implications these findings have on the management of human resources (HR) in oil and gas companies.

Improving Government Relations through Human Resources Management

The above discussion suggests that for firms to prosper in the upstream oil and gas business, they must learn how to work with governments in acquiring the rights to oil, they must comply with government regulations, and they must partner effectively with governments and NOCs. Further, an overall positive relationship with governments is likely to help in these endeavors. This is one reason why following the stakeholder approach (as discussed in the previous chapter) is also likely to help long-term relationships with governments. By addressing the concerns of communities and society, nongovernmental organization (NGOs), unions, and other stakeholders, the firm can show that it brings tangible economic, environmental, and social benefits to the country. Firms that provide such benefits are far more likely to be seen as long-term allies compared with other firms that also build relationships with high-level officials and meet recovery and production targets, but that neglect the other stakeholders that are important to the government. Positive long-term relations will then result in winning and keeping access to resources, increasing the probability of smooth operating conditions, fostering a positive company reputation in the country, and increasing the likelihood that the company will be the preferred partner for future projects.[38] We now discuss how managing HR effectively can help achieve positive long-term relations with governments. The term *government relations* is used to describe the fostering of a long-term relationship between the firm and the government, rather than the common, more narrow interpretation related specifically to lobbying. The discussion is organized around the key decisions in each HR area.

Improving Government Relations through Staffing

Selection is an important HR function relative to government relations because governments have a strong interest in whom firms (particularly foreign firms) hire. As mentioned in the previous chapter, most local stakeholders benefit from hiring local labor, as does the country and government in general. This is the reason many countries have laws regarding the minimum level of local employment. These laws are frequently part of "local content" laws or policies that also deal with the use of local suppliers, contractors, and resources. See Current Challenge Box 8-3 for more detail regarding local content laws. In general, hiring locals is not only preferred by the host government, but it also benefits the firm directly. As discussed in chapter 3, hiring locals helps the firm acquire deep local knowledge and results in employees who are much more likely to have a broad local network. This can be particularly important in government relations.

Current Challenge Box 8–3. Local Content Laws

One way governments can increase the likelihood that their citizens benefit from the presence of foreign companies is through local content laws. These laws can specify minimum levels of local workforce employment and development, requirements regarding the use and development of local suppliers, and requirements regarding the use of local resources. Local content in actuality frequently means national content and can be extremely constraining. For example, in 2013, the Indonesian government passed a decree that for both upstream and downstream firms in the Indonesian oil and gas industry, the local content of employees in certain jobs must be 100%. Specifically, employment of expatriates is prohibited for jobs in HR, legal, HSE, supply chain management, quality control, and exploration and exploitation functions below superintendent level (with limited exceptions).[39] Governments introduce local content laws to create jobs, to promote economic development, and to ensure that their citizens and firms acquire new skills and technologies.[40] However, requirements that are too stringent can backfire by negatively affecting the development of reserves and slowing foreign investment when local capabilities are inadequate. For example, some believe that Brazil's vast reserves are greatly underdeveloped because of the 55% local content requirement for exploration projects.[41] Nevertheless, local content laws are generally believed to be beneficial to developing countries.[42]

The benefits of local content

As should be clear from this chapter and from chapter 7, more local content can be beneficial to the firm as well as to local stakeholders. More local content helps meet regulatory requirements, is likely cheaper in the long term, helps protect commercial interests, and helps meet the goals of various stakeholders. In the long term, a reliable local supply chain will be more cost effective and have greater efficiencies than other alternatives. Long-term labor cost savings will result from reductions in expatriate employee costs. Thus, overall and consistent with this chapter, greater local content will improve a company's relationship with the government, while improving the firm's local supply capabilities, reducing long-term labor costs while addressing industry labor shortages, improving the firm's reputation, and helping address the concerns of many stakeholders.[43]

The costs of local content

Local content may create inefficiencies in the short term since the company may be using suboptimal suppliers, and the labor market may be inadequate. This can negatively affect company performance, lowering revenues and company competitiveness. There are large development costs for both workers and suppliers in the short term. Further, these investments may benefit other firms if they poach these employees or use the same suppliers. Thus, in the short term, developing local content may be quite costly, result in suboptimal performance, and limit growth opportunities.

Achieving local content goals

Country requirements should only be a starting point for local content goals given the long-term benefits of increased reliable and qualified local content. When contracting work, companies may also require contractors to hire and train the national workforce. Identifying the benefits as well as costs can help create internal support for the long-term process of developing local content. Greater local content can be achieved by the following:

- Thoroughly understand the local context before starting.
- Start the planning phase early to allow plenty of lead time for local development projects.
- Gain stakeholder support and working with stakeholders including national and local governments, NOCs, communities, NGOs, international organizations, other oil and gas companies, lead contractors, and financial institutions to help achieve mutually beneficial outcomes.
- Make training and development an important part of the local content plan.
- Take a long-term perspective.
- Be transparent in information flows.[44]

Clearly, the management of HR is critical in meeting local content laws, particularly with respect to the workforce. The staffing function recruits and selects qualified local employees. In cases where there are not enough qualified local employees, the training function kicks in. Training ensures that locals have the capabilities needed to do the jobs, in the short and long term. This training can take many forms. For example, Total has opened a training center (the Pazflor Centre) in Luanda, Angola, that trains locals in risk analysis; geological systems and structures; industrial drawing; rotation equipment; and an introduction to drilling, valves, and tubing.[45] Schlumberger works with local universities through its Ambassador Program to create long-term relationships with faculty and administrators. ExxonMobil requires that its expatriates share their expertise with, train, and mentor local employees for long-term placement in higher-level operational and leadership roles. The performance management function ensures that the local workforce performs at acceptable levels. The compensation function ensures that the locals are motivated to perform and stay with the company. Overall, from a stakeholder perspective (and from a government relations perspective), establishing reliable and qualified local content should be a top priority for foreign firms in developing countries.

Determining labor markets from a government relations perspective

As we have previously discussed in this chapter, as well as in chapters 3 and 7, targeting local labor has many positive benefits for the firm, including the increased satisfaction of local stakeholders and governments. Based on all these positives, firms should exert a great deal of effort to meet their labor needs locally. Of course, this is not always possible and in some cases absolutely impossible (e.g., when the location is so remote that there are no local inhabitants). From a national government perspective, widening the targeting from local to regional to national recruits is still largely preferred over a foreign company bringing in expatriates or third-country nationals.

Government relations may also be improved by targeting employees who have experience in working with or for governments. This may include those who have worked for governments, in government relations departments, and for NOCs, as well as those who have worked for oil and gas alliances or associations. Table 8–4 shows some of the larger US oil and gas alliances and associations that represent firms in federal legislatures and regulatory agencies. Employees who have experience working with or for governments will have a better understanding of the government's perspective, are more likely to have established relationships with government officials, and are more likely to be competent in their interactions with governments. Having such employees throughout the organization rather than just in the government relations department will help foster better government relations.

Table 8–4. Major US oil and gas associations and alliances involved in government relations

Association/Alliance	Description
American Fuel and Petrochemical Manufacturers (AFPM) www.afpm.org	Represents its 450 member firms in the US House and Senate and in US state and federal regulatory agencies
American Gas Association (AGA) www.aga.org	Represents its 200 local natural gas delivery member firms in US and State legislatures and regulatory agencies
American Petroleum Institute (API) www.api.org	Represents its 600 member firms who are involved in all aspects of petroleum in US and State legislatures and regulatory agencies
America's Natural Gas Alliance (ANGA) www.anga.us	Represents its 22 independent natural gas exploration and production member firms in US and State legislatures and regulatory agencies
Independent Petroleum Association of America (IPAA) www.ipaa.org	Represents its 10,000 plus independent crude oil and natural gas explorer/producer member firms in the US House, Senate, and executive branches and in US federal regulatory agencie
Interstate Natural Gas Association of America (INGAA) www.ingaa.org	Represents its 25 interstate natural gas transmission pipeline member companies in dealing with the US legislatures and regulatory agencies, primarily the Federal Energy Regulatory Commission
Texas Alliance of Energy Producers (TEAP) www.texasalliance.org/government-relations	Represents its 3,300 member firms in the Texas legislature, the US House and Senate and in US state and federal regulatory agencies
Western Energy Alliance (WEA) www.westernenergyalliance.org	Represents its 480 member firms in the US House and Senate in addressing any federal regulations that affect western exploration and production

Recruiting for better government relations

As mentioned in chapter 3, when recruiting local employees, recruiting methods should be tailored to the local culture. The use of local recruiters and local recruiting agencies should be seriously considered. Using local government (or regional or national government) employment agencies could also help to establish a positive relationship with the host government.

The recruiting of those with government experience is likely to require methods commonly used with passive recruits (those currently not actively looking for a job), such as referrals, targeted marketing, and headhunters. Several cautions regarding recruiting with a focus on government relations should be noted. First, closer relations with governments and particularly government employment agencies may lead to attempts at cronyism, where politically connected applicants are recommended. Such situations are difficult, particularly when the recommended applicant is not particularly qualified. Although such hiring is tempting from a relationship-building perspective, ethical issues, possible reputational effects, precedence setting, and the negative job performance implications all suggest that the relationship-building benefits would come at large costs that would most likely to outweigh the benefits. Second, recruiting and hiring current government (or NOC) employees could have a negative effect on the relationship if the government views the poaching of their employees negatively. Thus, in some cases hiring former or retired government employees may be preferable. Finally, the hiring of (current or former) government employees may be viewed negatively by some as reciprocity for past favors. This perception can be addressed by proactively clarifying the employees' qualifications and how their skills will benefit the company.

Selection for better government relations

Selection can be used to address government relationship issues in three ways. First, the demographics of employees (e.g., national versus foreign, indigenous ethnicity, or underrepresented groups) hired is likely to be important to governments. As described in chapter 7, in cases where candidates are equally qualified, preference should be given to those that would help improve firm–government relations, including locals, underrepresented groups, and indigenous natives. Further, firms should be extremely proactive in recruiting preferable candidates so that they are represented in the applicant pool. This may take long-term strategies tied to training and government partnerships, which is discussed later in the chapter. Second, selection tools can be used that will result in a workforce that is more adept in dealing with governments. Selecting based on experience with and knowledge of the government and its historical context will increase the number of employees with government relationship-building skills. Third, selection tools can be used to determine applicant's knowledge of and familiarity with applicable regulations. Employees with a deep knowledge of applicable government regulations will be more likely to help the firm in its quest for compliance.

Combining selection tools for better government relations

Adding selection criteria that tap into government relationship-building skills to higher-level jobs outside of the government relations department will add another requirement to applicants beyond the skills already needed specific to the job. Because these (government relationship) skills are desirable but not essential, the compensatory approach is generally preferable to the multiple hurdles approach. As the number and difficulty of the hurdles (i.e., cutoff scores on tests or number of years of applicable experience) increase, the applicant pool will be reduced, making it less and less likely that even one applicant can survive all the multiple hurdles. Thus, the compensatory approach, at least with respect to these particular skills, will leave a bigger final pool that the company can choose from, providing more flexibility in the final choice.

Maximizing the effectiveness of staffing for better government relations

The general principles of being legally compliant and treating applicants fairly are particularly relevant in the context of governments, because these principles inherently recognize that the selection process itself can affect them. For example, discrimination in hiring is generally viewed negatively by governments, as well as other stakeholders. Many laws and regulations are likely to apply to the hiring process. Thus, monitoring, evaluating, and adjusting are particularly important relevant from the government perspective, so that any possible violations can be quickly addressed. Further, the process and criteria regarding hiring decisions should be well justified and well documented.

Training for Better Government Relations

The importance of governments in the oil and gas industry has major implications for training. First, it is necessary to train employees regarding government regulations to ensure that employees comply with them. Second, training in and of itself may be a government requirement (e.g., certain safety training). Third, extensive training may be necessary to be able to hire a workforce that is consistent with government goals (e.g., local workers, underrepresented groups, or indigenous inhabitants). This training may require long-term programs focused on potential workers and may include partnerships with other firms, governments, and universities. See Current Challenge Box 8–3 for more details regarding this aspect. Fourth, organizations should consider training that includes teaching proper protocol and interpersonal skills to employees who interact with government officials. Finally, the training discussed in chapter 7 to address the goals of other stakeholders (e.g., NGOs and the community) is applicable to governments, given that these stakeholders share many of the same goals as governments.

Who gets trained for better government relations?

There are a number of issues that have implications for who gets trained when using the government relations perspective. First, because of the similarity of some of the goals of government with the goals of other stakeholders, all employees should be trained regarding the broader stakeholder approach and its implications for performance, as discussed in chapter 7. Second, all employees should be considered for training to overcome any deficiencies in specific dimensions of their (stakeholder-related) performance. Third, any employee directly or indirectly involved in the firm's compliance of regulations should be trained on those regulations. Fourth, training and developing local workers (or potential workers) will allow the company to increase the percentage of local workers hired. This may include programs in cooperation with local elementary, middle, and high schools; local vocational schools; local colleges; and local universities. It may also include programs in cooperation with the government or with other firms through associations or alliances. Fifth, employees of NOC partners may receive technical or other training related to a joint project, increasing the value of the partnership to the NOC. Finally, anyone interacting with government officials should receive training to maximize the effectiveness of those interactions. Clearly, the importance of governments in the oil and gas industry has implications for the training of all employees.

What gets trained for better government relations?

As is generally the case, what gets trained is largely related to the "who gets trained?" issue discussed above. Table 8–5 provides some examples of what each of the employee groups could receive training on. This includes stakeholder-related training (e.g., ethics, CSR, and safety and health) and task-specific training related to any performance deficiencies for all employees. Those directly or indirectly related to regulation compliance should receive training on the regulations, how to comply with the regulations, as well as any training specified by the regulations. Locals should receive basic education (when necessary), language training (if the operating language differs from the local language), and specific training on any job-related tasks, followed by practical experience that would result in their becoming qualified employees of the company.[46] Employees of NOCs the firm is partnering with for a joint project may receive technical or job-related training. Such training is highly valued by governments because it transfers knowledge to the NOC and increases the human capital of its citizens.[47] Of course, such transfers may lessen the dependence of the NOC on the company in the future.[48] Finally, those who may personally interact with governments should be trained on soft skills, such as communication, negotiation, and conflict resolution, as well as a deep understanding of local history, customs, culture, and protocols to maximize the positive outcomes of those interactions and increase their network of government officials.

Table 8–5. Training content related to government relations

Who Gets Trained?	What Gets Trained?
All employees	Regulations applicable to all employees Stakeholder management Corporate social responsibility Anti-corruption Ethics Human Rights Sustainability
Local employees and potential employees	Basic skills STEM knowledge Job-related skills, knowledge, and abilities
Employees interacting with governments	Local history, customs, and culture Local laws and regulations Protocols Interpersonal skills Negotiation skills Communication skills Conflict management skills Networking skills
Employees directly or indirectly related to regulations	Content of applicable regulations Regulation compliance policies and procedures Training mandated by regulations
Employees of NOC partners	Job related skills, knowledge, and abilities Technical knowledge Process knowledge

How to train for better government relations

There are several implications regarding the training methods that can be used for addressing government relations. As is generally true, training can be classroom based or on the job, or personally delivered or online. In addition, the training method is affected by who and what is being trained. As discussed in chapter 3, when training local employees, the training program must be created by someone who knows the local laws, local practices, the level of employee skills and knowledge, local learning styles, and the country's culture.[49] When training on regulations and policies, which may be seen as uninteresting to many employees, choosing a training method that is high in engagement is very important. In cases where knowledge of the material is critical (e.g., regulations and safety), the cost of the method should not be as important as its effectiveness. Overall, the optimal training and development method to use depends on who and what is being trained, as well as the method's expected effectiveness.

Maximizing the effectiveness of training from a government relations perspective

As with training in general, maximizing the effectiveness of training related to government relations can be achieved by maximizing learning; easing training transfer; and monitoring, evaluating, and adjusting the system. Training related to government relations has a number of specific implications for the maximization of the effectiveness of training. First, training local workers requires using methods consistent with local cultural expectations. Local workers are generally much more willing to learn from a respected peer to whom they can relate than an unknown, foreign, company trainer. Second, learning can be maximized by emphasizing the importance of the material (e.g. regulation compliance) and emphasizing the long-term, costly repercussions of noncompliance to both the individuals and the organization. Third, transfer of training is particularly important regarding compliance with regulations. Training regarding regulation compliance will improve the knowledge of the regulation and how to comply with it, but this greater knowledge does not necessarily result in applying this knowledge on the job. Improving the transfer of this knowledge can be increased by incorporating the training content into performance measures directly related to complying with regulations. Thus, feedback during and after training by monitoring and evaluating performance can help maximize the effectiveness of the training.[50] Finally, because of the importance of much of this content (e.g., regulations), monitoring, evaluating, and adjusting the system are particularly important. In the event of any compliance failures, the role of training should be immediately investigated.

Performance Management for Better Government Relations

The importance of governments in the oil and gas industry also has major implications for performance management. First, compliance with government regulations is an important aspect of organizational performance and thus has implications for performance management through the incorporation of regulatory compliance into performance measures. Second, regulations may directly require the measurement of certain outcomes or behaviors that could be considered performance. Third, the quality of the firm's relationship with the government should be considered an aspect of organizational performance. Finally, governments are likely to have the same goals has many other stakeholders, further changing the nature of performance as described in chapter 7. Thus, the discussions of performance management in that chapter are also applicable here. These differences in how organizational performance is viewed should have implications for the definition of performance of individuals. How government relationships can be improved through performance management is straightforward: Make a good government relationship (which includes regulatory compliance) an important part of performance by setting goals related to it, measuring, providing feedback on it, and tying meaningful rewards to goal achievement.

How to measure performance for better government relations

From the government's perspective, the fundamental question "what is good firm performance?" is different from the historical standard of increasing shareholder value. The definition of good performance has changed from considering financial measures only to looking at different measures that address the concerns of the government, some of which are shared by other stakeholders such as NGOs, customers, suppliers, employees, and the community. Therefore, at the firm level, performance now includes not only financial performance measures but also those directly related to positive government relations. Such measures may include measures of regulatory compliance (e.g., number of regulation violations, citations, fines, lawsuits, environmental performance such as emissions, and safety performance), measures of practices designed to improve government relations (e.g., percent of local workers hired, training expenditures on local workers, percent of goods and services from local suppliers, and measures of technology transfer), and measures of outcomes of positive government relations (e.g., access to resources, smooth operating conditions, the company's reputation in the country, and standing as a preferred partner for projects).[51] These measures may also include measures of performance important to the government as well as other stakeholders such as environmental performance, social performance, customer satisfaction, employee satisfaction, community satisfaction, supplier satisfaction, and so forth. These measures may apply directly as measures of individual performance for the highest levels of management who have influence over various aspects of unit performance. However, these performance measures are generally not good measures of individual-level performance at lower levels of the organization since individual workers have little influence on any of them directly (or perhaps even indirectly). Nevertheless, changing the definition of firm performance as described above should have implications for the definition of performance of individual jobs, even at lower levels.

The consideration of government relations can result in a broader view of performance even for some lower-level jobs. Specifically, task performance should include measures that directly relate to regulation compliance. Thus, as mentioned in chapter 7, this broader perspective must now go beyond the outcomes of the performance to include the means used to achieve those outcomes, since regulations can apply to both outcomes and processes. That is, performance should be measured as a multidimensional construct that includes regulatory compliance regarding not only the outcomes but also the means (e.g., following all relevant policies and procedures diligently) by which they were achieved.

Who is involved in performance management for better government relations?

Supervisors are generally responsible for the performance management of their subordinates, using input from other sources. Consistent with the stakeholder view (since the goals of stakeholders and governments largely overlap), appraisals should be performed with the input of many other sources, including those that can shed light on the employee's performance related to regulatory compliance (for most employees) and government relations (for higher-level employees) when applicable. Data regarding compliance are more likely to be objective (citations, warnings, etc.), but may include more subjective assessments by any relevant source that has knowledge that could lead to a more accurate assessment of the employee's performance. This may include co-workers, inspectors, contractors, or any others who have observed compliance violations. Given the importance of regulatory compliance, it is up to the supervisor to determine if performance deficiencies are so great that disciplinary actions are necessary. Disciplinary actions should meet the following criteria:

- Company policies and procedures should be strictly followed.
- The level of disciplinary action should "fit the crime," with only egregious transgressions warranting termination, the most severe penalty possible at the workplace.
- The level of disciplinary action should be consistent with that given to others who have committed the same or a similar offense.
- Repeated performance issues warrant stronger disciplinary actions.
- Disciplinary actions should occur within a reasonable time frame.

- Unions should be involved as specified in the collective bargaining agreement, when applicable.

For less serious performance transgressions, additional training may be helpful in fostering improved performance of the worker. Table 8-6 shows types of progressive discipline and specifies the appropriate disciplinary actions for various types of performance problems.

Table 8–6. Progressive discipline

Disciplinary Action	Description
1. Verbal warning	Common disciplinary action for first time relatively minor offenses such as unexcused absences, dress code violations, minor rule violations, tardiness, etc. The warning should include the requirements (goals and time tables) needed to avoid the next disciplinary action.
2. Written warning	This disciplinary action entails a formal warning that is part of the employee's employment record. Can result from a repeated offense in violation of the ground rules specified in a previous verbal warning. Can also result from a first time more serious policy or regulation violation. The warning should specify the consequences of further violations in detail.
3. Suspension	Suspending employees (generally without pay) can result from further violations as specified in a written warning. May also result from first time serious violations including sexual harassment, insubordination, or regulatory non-compliance.
4. Termination	The most severe disciplinary action of the workplace. Should be used only for the most serious of first time violations including theft, drug use at work, physical assault, and sabotage of operations. Can also be used as a last resort for continued minor violations or poor performance that has not been corrected through training.

What format should be used for better government relations?

Again, following chapter 7, the additional criteria related to performance from a government relations perspective can best be incorporated into the appraisal process using something like a balanced scorecard approach. Critical compliance criteria may be separated out as essential performance criteria that are not used to calculate an overall score. Rather, in such cases the separate criteria are requirements that must be met or disciplinary actions will result.

Maximizing the effectiveness of performance management for better government relations

As is generally the case, maximizing the effectiveness of performance management from a government relations perspective requires providing feedback; having accurate evaluations; including goal setting and rewards; and monitoring, evaluating, and adjusting the system. The government relations perspective has some implications for these aspects of performance management. First, the timing of feedback needs to be flexible to correspond to feedback regarding regulation compliance. Because regulation noncompliance needs to be addressed seriously and quickly, feedback regarding noncompliance and possible disciplinary actions should be immediate. Adding measures related to government relations can cause complications regarding where employees are likely to put their efforts. Thus, critical aspects of performance, such as regulatory compliance, would require larger rewards to ensure that they are given the importance they deserve. This could lead to the need for larger bonuses or other rewards. Alternatively, as described above, critical criteria can be specified as so important that failure to perform satisfactorily on them warrants immediate disciplinary action, including (in order of level of severity) verbal warnings, written warnings, suspension, or termination (recall table 8–6). Although employee relations are likely to be better with the use of positive reinforcement (such as recognition or bonuses) rather than punishment, prompt disciplinary action may be warranted and appropriate in cases of willful or neglectful regulatory noncompliance. Finally, monitoring and evaluation of the system are probably more important when using the government relations perspective because of the importance of these performance measures. Thus, given the added complexity of additional performance measures and the importance of these measures, maximizing the effectiveness of performance management from the government relations perspective requires diligent monitoring and timely responses to nonperformance.

Compensation for Better Government Relations

The importance of governments to the oil and gas industry has several implications for compensation and benefits. As with the other areas, because many of the goals of other stakeholders are similar to the goals of governments, the implications discussed in chapter 7 regarding stakeholders also apply here. This includes paying for performance to motivate employees toward stakeholder-focused outcomes, using benefits to provide a safety net for employees, and using benefits to motivate employee giving and volunteering. Additionally, pay for performance and recognition can be used to motivate employees to comply with regulations and maintain positive relationships with governments.

Pay level for better government relations

Chapter 7 discussed how higher pay levels tend be seen positively by employees and communities. This is also true of governments, which would view the increased economic welfare of their citizens positively. Governments would also benefit from the increased tax revenues from higher wages (which would most likely be greater than the tax revenue from higher corporate profits). Higher pay level would also help with recruitment of employees with government-related experience. Conversely, paying below market for local employees would increase the perception that the foreign company is exploiting the local workers. The foreign government would most certainly see this negatively. Thus, higher pay levels would tend to be seen more positively by governments and other stakeholders, with the exception of shareholders (if the higher labor costs are not offset by increased productivity and lower administrative costs).

Pay mix for better government relations

Pay mix can have an effect on government relations in two ways. First, a greater proportion of benefits relative to salary and variable pay provides a greater safety net for employees. Benefits such as retirement pay, health insurance, life insurance, dental insurance, vision insurance, disability insurance, and supplemental unemployment insurance provide financial assistance to employees when they would otherwise face financial hardships. Thus, organizations where benefits are a larger part of the pay mix provide many safety nets for employees. This helps governments by reducing the need for safety nets provided by the government while giving its citizens a sense of security. Second, a greater proportion of variable pay allows for more bonuses that may be tied to government-related performance such as meeting regulatory standards.

Performance-based pay for better government relations

Including government-related measures of performance in employees' performance evaluations can affect pay for performance because new criteria are added. As discussed in the performance management section, these new criteria can be incorporated into a total performance score, which can be tied to rewards such as bonuses or can be looked at separately with an independent reward tied to it. Of course, it could also have no rewards tied to it. The recognition program is one type of pay-for-performance program that would be well-suited for meeting important compliance requirements. Recognition is relatively inexpensive, but is valued by employees and has good motivational effects. Providing small rewards such as plaques, gift cards, or company merchandise to recognize high levels of regulatory compliance is a relatively inexpensive way to create positive reinforcement. It may also help offset any negative effects (about the perceptions of the organization's culture) of tying disciplinary actions to instances of noncompliance. Some options regarding recognition awards include the standards that must be met. They could be given to all employees who have met compliance standards over a specific time period (such as annually) or given only to those who have performed at a certain level above the minimum level of compliance. Finally, pay for performance in itself may be part of a government's labor regulations, as is the case in Ecuador's requirements that 15% of profits be given to employees.

Benefits for better government relations

As mentioned in our discussion of pay mix above, many benefits can be offered by firms that would be viewed positively by governments. Table 8–7 provides a list of benefits that protect employees from financial hardships and thus would be seen positively by governments. Table 8–7 also includes the percentage of firms in the United States that offer that benefit. The most popular of these are prescription drug insurance programs, dental insurance, retirement plans, health insurance, and life insurance.[52] It should be noted that benefits can be quite expensive, adding up to more than a third of the total labor costs. Thus, the goodwill of employees and government toward the employer may come at considerable cost. Also, benefits themselves are likely to be part of any government's regulatory framework. Firms must make sure they are vigilant in providing all government-mandated benefits and follow all applicable regulations in their administration.

Table 8–7. Benefits protecting employees from financial hardship

Benefit	Percent Offering
Prescription drug insurance	98%
Dental insurance	96%
Defined contribution retirement savings plan	92%
Mental health insurance	89%
Preferred provider health insurance	86%
Life insurance	86%
Accidental death and dismemberment insurance	83%
Vision insurance	82%
Chiropractic insurance	80%
Employee assistance program	77%
Long-term disability insurance	77%
Employer match for defined contribution retirement plan	73%
Short-term disability insurance	68%
Wellness programs	64%
Supplemental accident insurance	50%
Paid sick leave	34%
Health maintenance health insurance	33%
Long-term care insurance	31%
Retiree health care coverage	23%
Intensive care insurance	20%
Defined benefit retirement plan	19%
Loans to employees for emergency/disaster assistance	14%
Cash balance pension plan	6%
Subsidized cost of elder care	3%

Source: Society of Human Resource Management 2013 Employee Benefits Report.

Maximizing the effectiveness of compensation from a government relations perspective

Legal and regulatory compliance is important aspect of the effectiveness of compensation. Compliance is much more complicated in the oil and gas industry because of its global nature (as described in chapter 3). National labor laws cover compensation issues such as minimum wage, overtime pay rate, night shift differentials, hazardous pay differentials (e.g., 30% in Brazil), differentials for work that is dangerous to health, discrimination, vacation pay, Christmas bonuses, profit-sharing requirements, payroll taxes, pension requirements, and health care requirements for workers. However, home country laws may also apply, particularly to expatriates. The reconciling of home country and host country regulations that may appear to be in conflict is a common difficulty that arises in regulatory compliance in the oil and gas industry. The use of local experts (as well as home country experts) can help multinational enterprises (MNEs) navigate through the difficult terrain of varying country laws and conflicts between home country and host country laws. Finally, as is true with all aspects of HR, the effectiveness of compensation from a government relations perspective is likely to be enhanced through careful and diligent record-keeping. Documentation to support all compensation actions and decisions will help verify compliance.

Conclusion

Companies in the oil and gas industry have an important stakeholder in the government. Governments own many of the largest oil and gas companies; governments own most of the oil and gas; governments control the regulations that oil and gas companies must comply with; and governments and government-owned companies frequently partner with other companies. Addressing government relations has implications for how to manage HR in the industry. Specifically, staffing affects government relations indirectly by whom the firm hires and indirectly through the skills, knowledge, and abilities of the workforce. Training can help educate workers about regulations. Performance management needs to adapt to new government-related performance criteria. Pay levels, pay for performance, and benefits can all directly or indirectly affect the relationship between the firm and government, with respect to compensation. Overall, having a positive long-term relationship with governments is critical to the performance and the very survival of oil and gas companies: we have attempted to show how human resource management practices can help in this endeavor.

References

1. Inkpen, A. C., & Moffett, M. H. (2011). *The global oil & gas industry: Management, strategy & finance.* Tulsa, OK: PennWell Books.
2. Idemudia, U., & Ite, U. E. (2006). Corporate-community relations in Nigeria's oil industry: Challenges and imperatives. *Corporate Social Responsibility and Environmental Management, 13*(4), 194–206.
3. Lyle, A. (2012). DOD must have petroleum fuel alternatives, official says. *DOD News.* Retrieved from http://www.defense.gov/news/newsarticle.aspx?id=117084.
4. Smil, V. (2008). *Oil: A beginner's guide.* Oxford, UK: Oneworld.
5. Goldwyn, D. L. (2012, November 13). Making an energy boom work for the US *New York Times.* Retrieved from http://www.nytimes.com/2012/11/13/business/energy-environment/making-an-energy-boom-work-for-us.html?pagewanted=all&_r=0.
6. Morris, S., & Bronstein, H. (2013, November 25). Spain's Repsol has initial deal with Argentina on YPF. *Reuters.* Retrieved from http://www.reuters.com/article/2013/11/25/us-repsol-ypf-idUSBRE9AO0UB20131125.
   Romero, S., & Minder, R. (2012, April 17). Argentina to seize control of oil company. *New York Times.* Retrieved from http://www.nytimes.com/2012/04/17/business/global/argentine-president-to-nationalize-oil-company.html?pagewanted=all.
7. Canucks, meet CNOOC: China wants expertise even more than oil. (2012, July 28). *The Economist*, p. 56.
8. Pirog, R. (2007). *The role of national oil companies in the international oil market* (CRS Report for Congress). Retrieved from http://fas.org/sgp/crs/misc/RL34137.pdf.
9. Inkpen, A. C., & Moffett, M. H. (2011). *The global oil & gas industry: Management, strategy & finance.* Tulsa, OK: PennWell Books.
10. Hartley, P.R., & Medlock, K. B., III. (2013). Changes in the operational efficiency of national oil companies. *The Energy Journal, 34*(2), 27–57.
11. Inkpen, A. C., & Moffett, M. H. (2011). *The global oil & gas industry: Management, strategy & finance.* Tulsa, OK: PennWell Books.
12. Hartley, P.R., & Medlock, K. B., III. (2013). Changes in the operational efficiency of national oil companies. *The Energy Journal, 34*(2), 27–57.
13. Inkpen, A. C., & Moffett, M. H. (2011). *The global oil & gas industry: Management, strategy & finance.* Tulsa, OK: PennWell Books.
14. Bridgman, B., Gomes, V., & Teixeira, A. (2011). Threatening to increase productivity: Evidence from Brazil's oil industry. *World development, 39*(8), 1372–1385.
15. Victor, D. G., Hults, D. R., & Thurber, M. C. (Eds.). (2011). *Oil and governance: State-owned enterprises and the world energy supply.* Cambridge, UK: Cambridge University Press.
    Oil in Brazil: The perils of Petrobras. (2012, November 17). *The Economist*, pp. 32–33.
16. Galiani, S., & Sturzenegger, F. (2008). The impact of privatization on the earnings of restructured workers: Evidence from the oil industry. *Journal of Labor Research, 29*(2), 162–176.

17. Stojanovski, O. (2008, April). *The void of governance: An assessment of Pemex's performance and strategy* (Working Paper #74). Palo Alto, CA: Stanford University, Program on Energy and Sustainable Development.
18. Inkpen, A. C., & Moffett, M. H. (2011). *The global oil & gas industry: Management, strategy & finance.* Tulsa, OK: PennWell Books.
19. Inkpen, A. C., & Moffett, M. H. (2011). *The global oil & gas industry: Management, strategy & finance.* Tulsa, OK: PennWell Books.
20. GL Noble Denton. (2013). *Seismic shifts: The outlook for the oil and gas industry in 2013.* Retrieved from http://www.longituderesearch.com/wp-content/uploads/2013/01/FINAL-Seismic-Shifts.pdf.
21. Wieners, B. (2013, June 25). Obama's climate plan to ditch coal will be good for business, really. *Bloomberg Businessweek.* Retrieved from http://www.businessweek.com/articles/2013-06-25/the-business-argument-obama-should-have-made-for-ditching-fossil-fuels.
22. Goldwyn, D. L. (2012, November 13). Making an energy boom work for the US *New York Times.* Retrieved from http://www.nytimes.com/2012/11/13/business/energy-environment/making-an-energy-boom-work-for-us.html?pagewanted=all&_r=0
23. Fracking in the west: Big reserves, big reservations (2013, February 16). *The Economist*, p. 32
24. Baram, M. S. (2010). *Preventing accidents in offshore oil and gas operations: The US approach and some contrasting features of the Norwegian approach* (Boston University School of Law Working Paper No. 09-43). Retrieved from http://ssrn.com/abstract=1705812.
25. Fan, L., & Cevallos, N. E. (2008). Profit-sharing as a catalyst for recognition of oil company employees in Ecuador. *2008 IEEE International Conference on Computer Science and Software Engineering, 5,* 718–720.
26. Handley, M. (2013, February 25). Energy execs: More drilling, less regulation. *US News.* Retrieved from http://www.usnews.com/news/articles/2013/02/25/energy-execs-more-drilling-less-regulation.
    Niquette, M., & Oldham, J. (2014, February 19). US Governors want states to take lead on drilling regulations. *Bloomberg Businessweek.* Retrieved from http://www.bloomberg.com/news/2014-02-19/u-s-governors-want-states-to-take-lead-on-drilling-regulations.html
27. Esty, D. C., & Charnovitz, S. (2012). Green rules to drive innovation. *Harvard Business Review, 90*(3), 120ff.
28. International Gas Union. (2012). *Nurturing the future generations for oil and gas industry* (Triennium work report). http://agnatural.pt/documentos/ver/nurturing-the-future-generations-for-oil-gas-industry_4400092093c413c4f-fa44d246d461fe2ff0cc194.pdf.
29. Esty, D. C., & Charnovitz, S. (2012). Green rules to drive innovation. *Harvard Business Review, 90*(3), 120–123.
    Energy policy: Biofuelery. (2013, March 2). *The Economist,* p. 28.
    Blown away: Wind power is doing well, but it still relies on irregular and short-term subsidies. (2013, June 8). *The Economist,* p. 36.
30. Greeley, B. (2014, March 13). Why fuel subsidies in developing nations are an economic addiction. *Bloomberg Businessweek.* Retrieved from http://www.businessweek.com/articles/2014-03-13/why-fuel-subsidies-in-developing- nations-are-an-economic-addiction.

31. Cheaper oil: Winners and losers. (2014, October 25). *The Economist*. Retrieved from http://www.economist.com/news/international/21627642-america-and-its-friends-benefit-falling-oil-prices-its-most-strident-critics.
32. Westcott, M. (2012). "Worker control" in the Australian oil industry. *Labour & Industry: A Journal of the Social and Economic Relations of Work, 22*(4), 399–414.
33. Direct Government Involvement. (December, 10, 2012). *African Business Magazine*. Retrieved from http://africanbusinessmagazine.com/special-reports/direct-government-involvement/.
34. Ernst and Young. (2013). *Understanding the effects and challenges of Section 1504*. Retrieved from http://www.ey.com/Publication/vwLUAssets/Dodd_Frank_section_1504/$FILE/Understanding_the_effects_and_challenges_of_Section_1504.pdf.
35. International Gas Union. (2012). *Nurturing the future generations for oil and gas industry* (Triennium work report).
36. Inkpen, A. C., & Moffett, M. H. (2011). *The Global Oil & Gas Industry: Management, Strategy & Finance*. Tulsa, OK: PennWell Books.
37. Inkpen, A. C., & Moffett, M. H. (2011). *The Global Oil & Gas Industry: Management, Strategy & Finance*. Tulsa, OK: PennWell Books.
38. Tideman, D., Kombargi, R., Oushoorn, R., Rizzi, C., & Landau, R. (2012). *Government-facing strategy for oil and gas companies: Developing a productive relationship with host governments*. Booz. Retrieved from http://www.strategyand.pwc.com/media/file/Strategyand_Government-Facing-Strategy-for-Oil-and-Gas.pdf.
39. Dawborn, D., Tang, V., & Harto, N. 2014. *Indonesia: Further restriction on employment of expatriates in the oil and gas industry* (White Paper, Herbert Smith Freehills, LLP). Retrieved from http://www.lexology.com/library/detail.aspx?g=45cd0f4a-8f39-48be-b22f-b92682361760.
40. IPIECA. (2011). *Local content strategy* (IPIECA Document). Retrieved from http://www.ipieca.org/sites/default/files/publications/Local_Content.pdf.
41. Asher, B. 2012. *Brazil local content policy inhibits oil development*. Retrieved from Rigzone website: www.rigzone.com/news/oil_gas/a/11941/Brazil_Local_Content_Policy_Inhibits_Oil_Development.
42. Ado, R. (2013). Local content policy and the WTO rules of trade-related investment measures (TRIMs): The pros and cons. *International Journal of Business and Management Studies, 2*(1), 137–146.
43. IPIECA. (2011). *Local content strategy* (IPIECA Document). Retrieved from http://www.ipieca.org/sites/default/files/publications/Local_Content.pdf.
44. IPIECA. (2011). *Local content strategy* (IPIECA Document). Retrieved from http://www.ipieca.org/sites/default/files/publications/Local_Content.pdf.
45. ILO. (2012). *Current and future skills, human resources development and safety training for contractors in the oil and gas industry*. Geneva, Switzerland: ILO Sectoral Activities Department.
46. IPIECA. 2011. *Local content strategy* (IPIECA Document). Retrieved from http://www.ipieca.org/sites/default/files/publications/Local_Content.pdf.
47. Aroge, S. T., & Hassan, M. A. (2011). The responsibility of human resource management and development professionals in the development of low-skilled workers in the Nigeria public sector. *International Journal of Business and Management, 6*(11), 227.

48. Hickey, W. (2012). The oil PSA and its inverse effect on human resource development (HRD). *Procedia-Social and Behavioral Sciences, 65,* 1060–1065.
49. Perkins, S. J., & Shortland, S. M. (2006). *Strategic international human resource management: Choices and consequences in multinational people management.* Philadelphia, PA: Kogan Page.
50. Komaki, J., Heinzmann, A. T., & Lawson, L. (1980). Effect of training and feedback: Component analysis of a behavioral safety program. *Journal of applied psychology, 65*(3), 261.
51. Tideman, D., Kombargi, R., Oushoorn, R., Rizzi, C., & Landau, R. (2012). *Government-facing strategy for oil and gas companies: Developing a productive relationship with host governments.* Booz. Retrieved from http://www.strategyand.pwc.com/media/file/Strategyand_Government-Facing-Strategy-for-Oil-and-Gas.pdf.
52. Society for Human Resource Management. 2013. *2013 Employee benefits.* Retrieved from www.shrmstore.shrm.org.

Case 3

The Bakken Boom: Unconventional Oil and Employment

"If you wish for this oil, be careful what you wish for, because life as you know it is done," said Ken Norgaard, road department supervisor for Roosevelt County, the vast and sparsely populated county of rolling farmland that includes Bainville. County jobs were once coveted for their solid benefits and retirement plan, Norgaard said. Now, he has trouble finding workers. Norgaard advertised a road grader job as far away as Wyoming. In six months, he received two applications.

One of the most dramatic changes in the oil and gas industry in the past decade has been the development of *unconventional sources* of oil and gas. One of the largest such areas is the Bakken, an area extending from western North Dakota into eastern Montana in the United States. The rapid growth in oil drilling and development in the Bakken has had significant social, economic, and employment impacts on the communities in and around Williston, North Dakota. By 2013 the Bakken was the source of more than 10% of all US petroleum production, and North Dakota's unemployment rate, 1.8%, was the lowest in the nation. With job growth in the Bakken oil sector itself rising from essentially zero to 60,000 in less than a decade, the human resource challenges have been dramatic.

Unconventional Oil

Oil was first discovered in the Bakken in 1951, but was largely left undeveloped due to the technical complexity and high cost of its production. Conventional oil production occurred sporadically over the next 50 years, generally expanding during steep oil price increases. The Bakken Shale Formation, at an average depth of 10,000 feet, is made up of three layers: the Upper Bakken Shale, the Middle layer (which is not itself a shale), and the Lower Bakken Shale (figure C3-1). The US Geological Survey estimates that there are approximately 4.6 billion barrels of recoverable oil in the Bakken.

Current production expansion in the Bakken began in 2001 with the first horizontal well drilled in 2004.

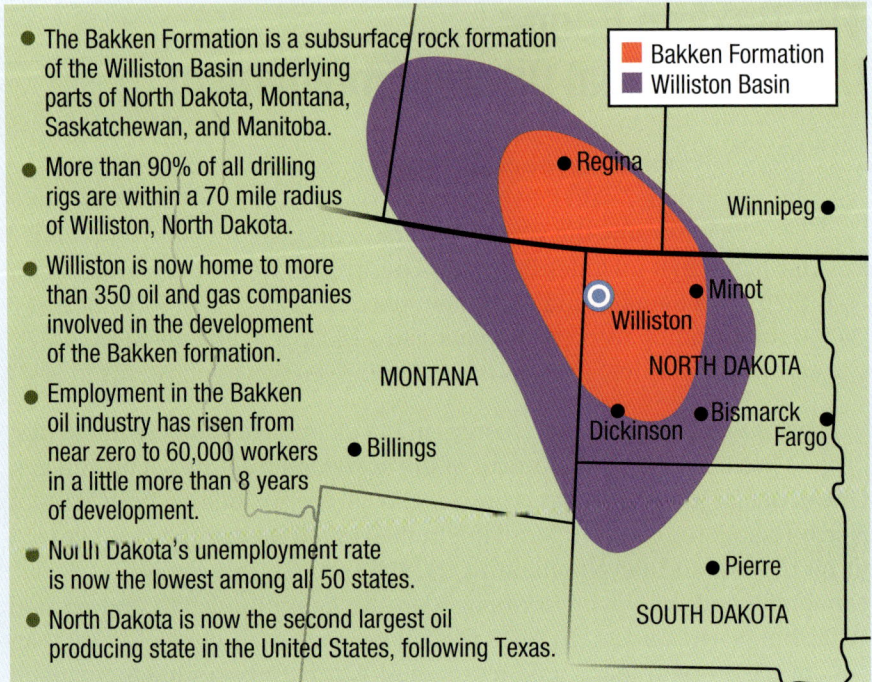

Fig. C3–1. The Williston Basin and Bakken Formation
Source: "The Good, the Bad, and the Ugly," Bakken Development Impacts, North Dakota State Legislative Assembly Special Session, November 7, 2011.

Unconventional oil in Bakken

The Bakken Formation holds what is termed *unconventional oil*, a category that includes oil found in shale, oil sands, heavy oil formations, and other difficult to produce oil sources. As opposed to conventional oil fields or reservoirs often characterized as "sticking a straw into the ground," unconventional oil does not flow freely upon penetration. Bakken oil formations, sometimes referred to as *shale oil* or *tight oil*, require that the shale formation be broken down to allow the oil to flow and be gathered. The US Department of Energy divides unconventional oil into four distinct categories: heavy oil, extra-heavy oil, bitumen, and shale oil. Bakken oil is the fourth: shale oil. (Some in the oil and gas industry may also include various gas-to-liquids processes for converting natural gas to oil, as well as coal-to-liquids, in which coal is converted to oil, as unconventional.)

The introduction of horizontal drilling and hydraulic fracturing technologies ("fracking") to tight formations like the Bakken, combined with the rising oil prices (making higher-cost sources of oil like the Bakken economic), has resulted in rapid oil industry development in an area that had previously been predominantly agricultural. This has posed an enormous array of challenges for the communities impacted, including lack of oil industry regulation related to permitting; groundwater protection; environmental impacts related to drilling, transportation, and spillage; inadequate infrastructure to support heavy industry development on roads and public utilities; and inadequate transportation capacity by trucking, railroad, and pipeline of the oil produced.

As illustrated in figure C3–2, crude oil production in the Bakken has risen very rapidly over the past decade, from less than 100,000 barrels of oil per day to more than 1.2 million. This rapid industrial development has placed enormous pressures on employment; skills shortages; wages; housing; public services like roads, schools, and utilities; and the general cost of living in the region.

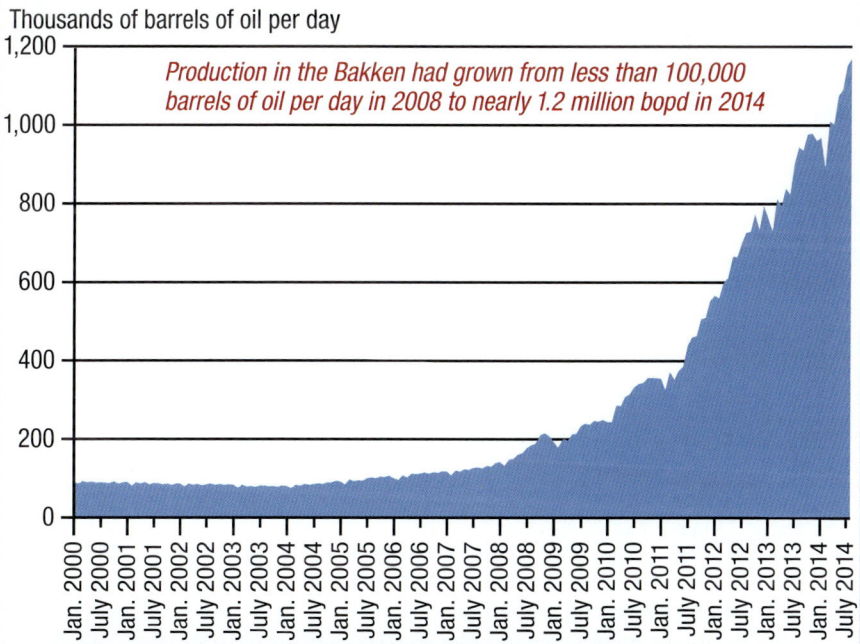

Fig. C3–2. North Dakota (Bakken) production of crude oil
Source: Energy Information Administration, US Department of Energy.

Unconventional oil and employment

Fracking is an oil or gas production process that pumps a mixture of water, sand, and sometimes chemicals into a fractured well at high pressure.

The high pressure of the mix fractures the rock surrounding the drill pipe, and sand is then injected into the well to act as a blockage to prop (proppant) the fractures open to allow the oil to flow to the pipe. In the Bakken, the horizontal drill passes through the Middle Bakken layer, and then fractures radiate outwards into the Upper and Lower Bakken Shales.

The employment impacts of an unconventional development differ from traditional oil field developments. An unconventional well will suffer an extremely rapid fall in output, often in only the second year of production.

Figure C3-3 highlights the significance of employment gains to North Dakota and Montana from the Bakken. Within a four-year period beginning in 2010, oil-industry-related employment in the Bakken surged from just over 50,000 to more than 85,000, all while employment levels in both states were largely flat or declining.

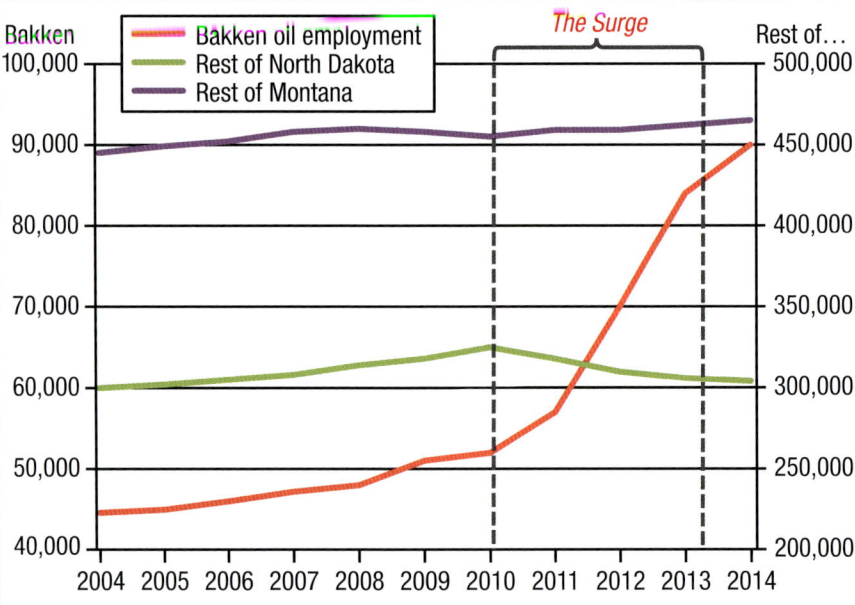

Fig. C3–3. Bakken's booming employment
Source: US Bureau of Labor Statistics, *Local Area Unemployment Statistics*.

Figure C3-4 details the employment by category impacts of unconventional oil's drilling cycle approach. A typical Bakken horizontal oil well will produce only 55% of the oil it produced in its first full year of production, falling to 45% or less in the third year.

Oil drillers and developers will therefore continue to drill and frack new wells on a regular near-cycle basis for years to produce more and more of the oil-in-place. This means that drilling, fracking, and production employment is higher on a per-barrel basis and lasts much longer over the life of an oil field or production area. Because new wells must be continually drilled and fracked, employment in drilling stays at a higher level than is typically seen in conventional oil field developments.

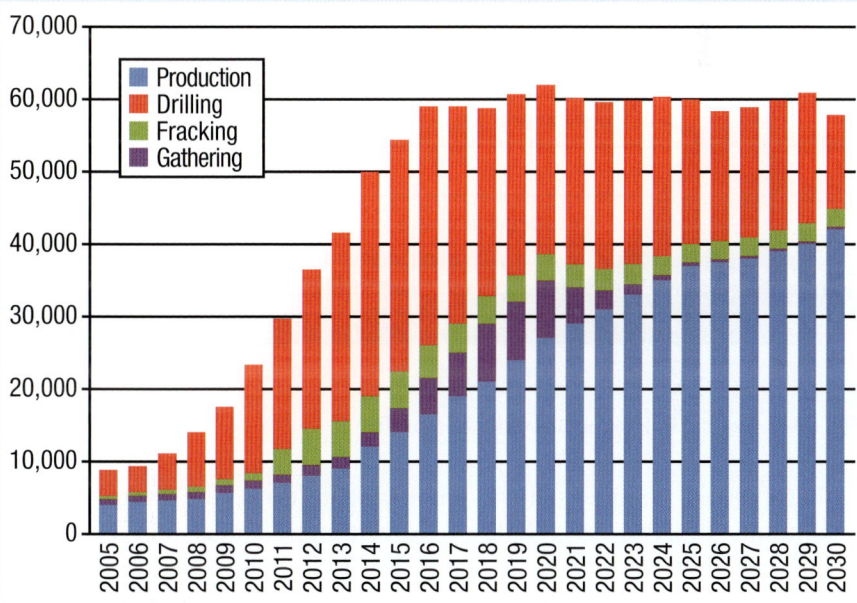

Fig. C3–4. North Dakota's oil industry jobs—actual and projected
Source: North Dakota Department of Mineral Resources.

Creating Boomtowns

Although horizontal drilling and hydraulic fracturing are described as a technology, it is an exceedingly labor-intensive technology that is only justified by higher oil prices—those prices seen since the early to mid-2000s—and constant improvement in lowering the cost of new well drilling and development. The continuous drilling cycle is required because of what is seen in figure C3–5, the extremely rapid decline in production per well expected over the life of a Bakken well.

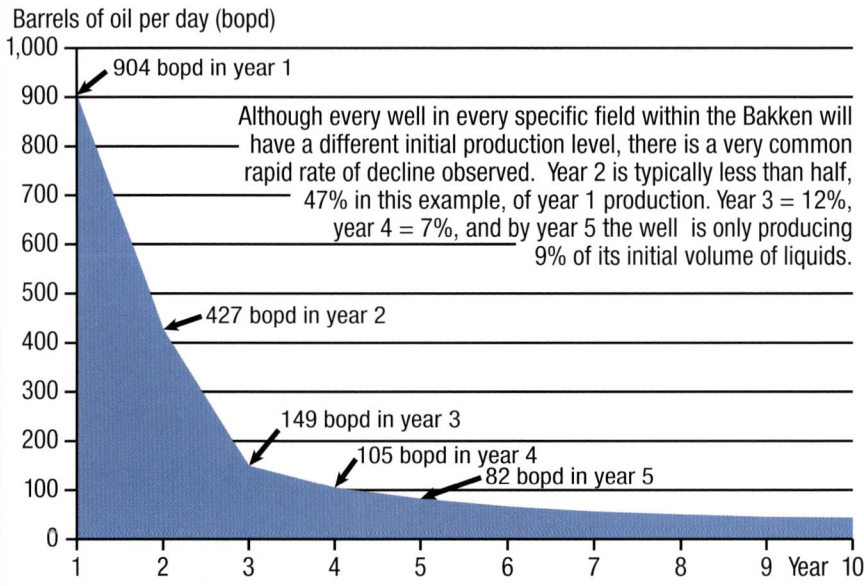

Fig. C3–5. Typical Bakken well production profile

With higher semipermanent employment levels, the industrial and community impacts from Bakken development are greater and more continuous than impacts from conventional oil fields. This actually helps financially support and fiscally justify greater community and public services and utility development to support the greater boomtown demands.

Boomtown wages and prices

The economics of boomtowns are well known and have been observed many times in many places around the world. A rapid buildup in a local population, often in an isolated geographic location with little preexisting infrastructure or population, places enormous stress on every dimension of social and economic society.

Figure C3-6 provides some insight into one of the major attractions, and at the same time distortions, related to the Bakken development: the escalation of local area wages. According to the US Bureau of Labor Statistics, by the second quarter of 2012 average weekly wages in the Bakken were approaching $1,200, while the US average was roughly $960, and the non-Bakken portions of North Dakota and Montana averaging $757 and $707, respectively.

This same oil and gas industry wage inflation has been observed in the past everywhere from Norway to Nigeria. The result is escalating costs for every activity and business in the region, as well as absolute shortages of workers for any nonoil activity as wage earners migrate to the industry in ever-growing numbers.

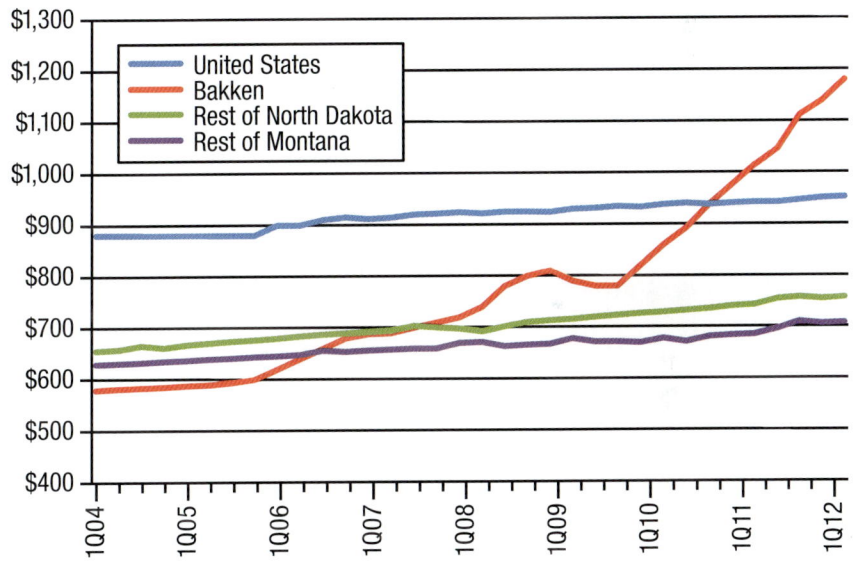

Fig. C3–6. Average monthly wages in the Bakken
Source: US Bureau of Labor Statistics, *Quarterly Census of Employment and Wages.*

Jobs and wages had combined to draw workers in from not only every corner of the United States but also from customary occupations in the North Dakota–Montana labor market in general.

Much like the oil and gas booms seen in countries like Nigeria and Angola, the Bakken had disrupted traditional labor markets in the impact region.

Figure C3-7 illustrates the change in employment in the Bakken during the employment surge, the 2007–2011 period. The most notable changes in employment by industry are the increase in mining and oil industry extraction and the relative decrease in health care and social employment, areas of relatively lower wage histories and profiles. With wages 60% or higher than those of typical of the region, the exodus from all other employment sectors had left them with dire shortages as well. Companies like Walmart in 2014 were offering prospective workers wages that were 2.5 times the minimum wage, roughly $17.60 per hour, to attract workers away from the multitude of oil industry opportunities.

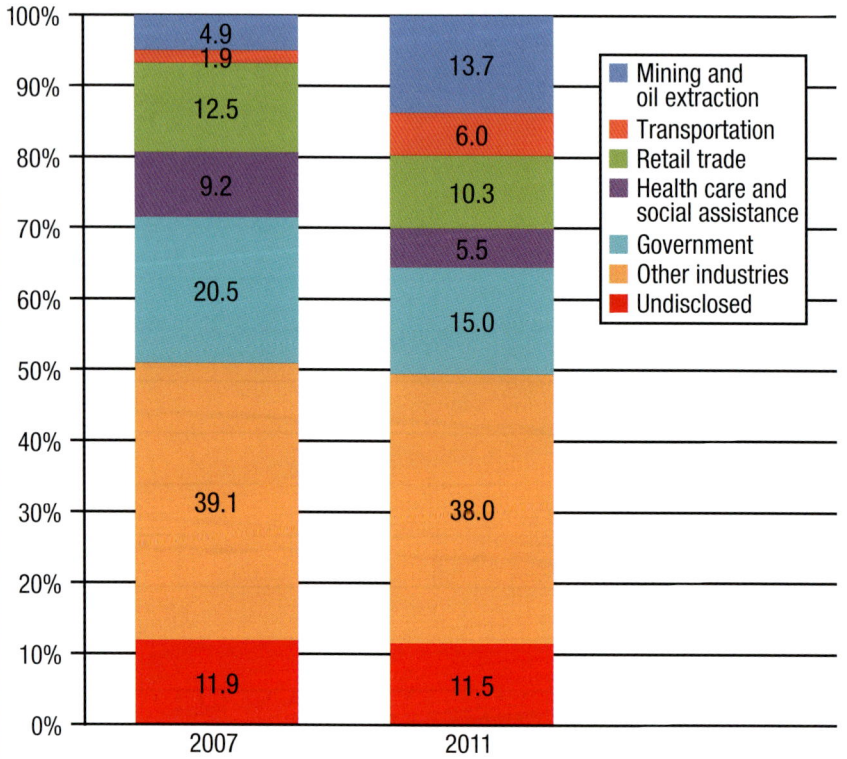

Fig. C3–7. Industry share of employment in Bakken oil counties (percentage of total country employment by industry classification)
Source: "Employment and wage changes in oil-producing counties in the Bakken Formation, 2007–2011," by Paul Ferree, and Peter W. Smith, *Beyond the Numbers*, US Bureau of Labor Statistics, April 2013, Volume 2, Number 11.

Boomtowns and societal impact

The unconventional oil boom in the Bakken is obviously not the first experience the Western United States have had with boomtown development. Most recently a number of communities in Wyoming and Montana and Utah had similar developments in the coal and copper industries, which has also resulted in a number of socioeconomic and demographic studies.

The impact on communities is significant. Housing shortages arise rapidly, as the influx of workers rapidly inundates the housing capacity (and its expansion ability) quickly. Oil industry workers quickly bid up the price of housing, often crowding out residents, particularly those on nonoil related incomes and fixed incomes in retirement. Housing expansion capacity itself is cropped by the diversion of all construction related jobs and capabilities to the oil industry itself. This diversion capacity has been much greater in the Bakken's unconventional oil fields because of the cycle of repetitive drilling and fracking which must be maintained to continue oil production.

The impacts do not stop with housing. Increasing instances of crime, truancy, alcoholism and drug use, traffic congestion, air and water quality, essentially every dimension of community life is impacted.

A brief overview of some of the more common findings of rapid energy-related growth's impact on communities includes the reduction of the density of acquaintanceship, the reduction of community satisfaction, and the increase of employment opportunities.

Density of acquaintanceship. The proportion of the community that any one individual knows personally is reduced rapidly, reducing the sense of identity, solidarity, and community. This is often linked with perceived social impacts such as less control over deviant behavior, reduced respect for law and order, increased drug and alcohol use, and lower satisfaction with the community. Although there may be no real increase in crime rates on a per capita basis, there is often a general perception that crime initiated by new emigrants increases.

Reduced community satisfaction. A reduced sense of community results from residents' perceptions of infrastructure problems, such as housing shortages, increased costs for goods and services, inadequate health care services, overcrowded schools, and deteriorating streets and roads.

Increased employment opportunities. Increasing employment opportunities has a number of immediate impacts such as redistribution and influx. The increased employment opportunities do not appear to increase the ability of the community to retain its youth, and youth appear to be less happy than their counterparts in rural communities not experiencing boom conditions. The increase in workers is often accompanied by family members, placing new and growing demands on public services (e.g., such as schools and hospitals) which the community finds difficult to absorb and cope with. In postboom periods these same increased populations are often slow to disperse, placing continuing pressures on public services after the income benefits of the boom have largely played out.

Conclusion

In the end, the core attributes of unconventional oil development may also contribute to better than traditional boomtown employment and impact. Unconventional oil requires a repetitive, cyclic investment, employment, and production process which not only increases the ultimate recovery of oil, but also provides a more sustainable employment opportunity for those involved in and around the industry's continued development.

An oil boomtown could stop booming for a variety of reasons: The oil runs out, the demand for oil drops, or the price of oil falls below the cost of production. During 2014 the price of oil fell from over $100 to less than $60 a barrel, putting pressure on the highest-cost producers. Bakken oil, relative to most conventional oil plays, is high cost oil. A sustained drop in oil prices below $50–$60 will likely mean cutbacks in production for the Bakken's highest cost producers and a slowdown or even end to the region's high growth. A rapid contraction in activity would create new economic and social stresses such as layoffs and unemployment, falling house prices, and reduced revenue sources for state and local governments.

9

Final Thoughts on Managing Human Resources in the Oil and Gas Industry

Over the next several decades, the demand for energy is expected to increase significantly, especially in the large emerging markets of Brazil, China, and India. As it has for the past 150 years, the oil and gas industry will play a key role in supplying the energy that supports the world's quality of life. The thousands of oil and gas industry companies and their human resources (HR) can expect to be part of a dynamic and evolving industry. This chapter begins with an overview of the major challenges facing the industry in the coming 5–10 years. This section is followed by a list of the key competencies that will be necessary to deal with the challenges. The chapter then summarizes the main topic areas covered in the book.

Major Challenges Facing the Industry

Throughout the book we have noted the differences of the oil and gas industry from most other industries. These differences and other factors in the oil and gas business environment have led to a number of large challenges facing the industry today. These include price volatility, resource access, technological changes, changes in the legal and regulatory environment, cost pressure, the growth of national oil companies (NOCs), shrinking talent pools, and the growth of social media. We now briefly discuss each.

Continued crude oil and natural gas price volatility

Crude oil and natural gas are both commodities and subject to the vagaries of international supply and demand forces. The companies that have survived and prospered over the years are those that invest and manage their assets for the long term. As the world possibly enters a period of low crude prices, there will be winners and losers. Inevitably, the firms with a "get rich quick" mentality will see their performance suffer. The best managed firms, which include all of the supermajors, will survive the down cycles, as they have for over a century.

Commodity price volatility has a direct impact on investment and hiring by oil and gas firms. For example, as oil prices fell in late 2014, many companies began implementing restructuring programs. BP, which had already begun downsizing, announced that it would accelerate its layoffs because "the fall in oil prices has added to the importance of making the organisation more efficient."[1] In contrast, during the period 2010–2013 when oil prices were steadily rising, BP added thousands of upstream jobs, as did its major competitors.

Resource access challenges for the majors and other exploration and production (E&P) firms

In the upstream, resource access is essential. Firms competing in the upstream must be able to engage in the primary upstream activities of exploration, development, and production. With most of the world's oil and gas resources controlled by states through their NOCs, access has become increasingly difficult. Although the development of shale oil and gas in the United States has resulted in new resources outside of the control of NOCs, these resources will not last forever.

Technological change

Since the earliest days of the oil and gas industry, technology change has been a constant. Advances in areas such as drilling have allowed exploration in ultra-deep water and innovations in shipping have been critical to the development of the liquefied natural gas (LNG) industry. Changing technology is both a challenge and an opportunity. There have always been industry opportunities for the companies that can drive the change and exploit new technologies first. For the technology laggards, life can become very difficult. The huge growth in U.S. oil and gas production, which is driven by a combination of technology, productivity, and innovation across many value chain activities, shows how an entire new industry segment can be created through technology and innovation.

The rapid advancement of oil and gas technology and innovation above and below the ground has created a more complex web of firms, activities, and skills in the industry. This expansion is geometric in size and staffing and poses both challenges and opportunities for those working in adjacent industries and markets.

Climate change action and legislation: Doing business in a greener world

Although some oil and gas firms may see climate change and the related policy and environmental consequences as harmful to the industry, that would be ignoring the reality of an interdependent and interconnected world. Climate change is an important global political issue, and the oil and gas industry is in the center of the debate. The climate change debate has an indirect impact on many aspects of oil and gas operations, such as pipeline development, natural gas flaring, refinery emissions, and oil sands development. Oil and gas firms must ensure that they build the necessary competencies in climate change or risk being excluded from the debate.

Continued cost pressure

In commodity-based industries where product prices are set by the market forces of supply and demand, cost control is critical for achieving competitive advantage. Over the past decade the industry has seen a sharp increase in the cost of many industry inputs. In particular, labor costs have risen significantly in areas such as the Alberta oil sands, Australia, and the shale areas in the United States. Much of the increase in costs has occurred because mega-projects have grown in size and complexity. It is likely that the next decade will see aggressive efforts by oil and gas firms to rein in costs, increase productivity, and possibly reduce their involvement in complex mega-projects.

NOCs growing and evolving as global competitors

As discussed in chapter 8, many NOCs have expanded beyond their original home country markets. NOCs like Statoil and Petrobras have developed technical skills that are close to those of the integrated oil companies (IOCs). The Chinese NOCs have expanded into many countries to satisfy the energy demand of the huge Chinese market. Going forward into the next decade, it is likely that other NOCs will expand beyond their domestic markets. For some NOCs expansion will be driven by declining home country resources. Other NOCs will see international expansion as an opportunity to build international legitimacy and increase their market share.

For the oil majors and independent exploration and production (E&P) firms, the growth of NOCs is a major challenge and is linked to the challenge of resource access. For NOCs looking to expand globally there is a need to increase the quality and productivity of their HR.

Shrinking talent pools in the industrialized countries

Global challenges are occurring within a traditional industrial country talent pool. This talent pool, used predominantly by the world's major oil and gas firms for decades, is aging and continues to suffer scarcity of interest in the oil and gas sector, especially from the millennial generation. This lack of interest found in the industrialized countries offers great opportunity for new talent development in the nonindustrial world. The raw and inexperienced talent that is resident in countries like Mozambique and Papua New Guinea, to which the global oil and gas industry is now migrating, may be its future—with proper recruitment and development.

Risks and challenges enhanced by social media and information access

The oil and gas industry impacts everyone on a daily basis, from the energy consumed to the everyday products made from chemical product derivatives. Oil and gas firms should expect that the future consumers of these products will be better informed and better connected through social media. Social media will be how people learn about climate change and other industry issues. Embracing social media as a platform for advocacy will help oil and gas firms manage the evolving expectations of customers and stakeholders.

Skills Required to Deal with the Industry Challenges

The most successful oil and gas firms in the coming few decades will be those that can adapt to change and effectively manage the challenges discussed above. To do so, the people in these firms will also have to change, adapt, and develop new skills. Some critical skills that will be necessary are as follows:

1. Building and sustaining complex interfirm relationships; diplomacy and interpersonal skills
2. Managing global diversity and inclusion
3. Managing ambiguity in relationships
4. Technical expertise

5. Adaptability and flexibility
6. Risk management
7. Maintaining commitment to safety and integrity
8. Developing and exploiting new communication and learning platforms for the industry's next generation of leaders

Developing these skills in the industry workforce is one important way for firms to address these challenges. The entire HR function can help address these challenges as well. The entire HR function can also help address the stark differences between the oil and gas industry and other industries.

Managing Human Resources in the Oil and Gas Industry

Throughout this book we have emphasized the important differences between the oil and gas industry and other industries. The differences include the extremely global nature of the industry, the importance of health and safety, the project-based nature of the industry, the unique workforce and working conditions of the industry, and the importance of stakeholders and the government. After looking at these differences, we have shown how these differences relate to the staffing, training, performance management, and compensation of employees in the oil and gas industry. We now briefly summarize these implications and conclude with some final thoughts.

Staffing in the oil and gas industry

The above mentioned characteristics of the oil and gas industry can affect staffing, specifically the determination of the labor market, the type of recruiting, the choice of selection devices, and how those selection devices are used in conjunction with each other. In choosing a labor market, local markets should be the first targeted and long-term programs should be in place to develop local content. See Current Challenge Box 9-1—MOL of Hungary and Strategic HR for an example. When local markets are not sufficient, expanding the labor market to regional or even worldwide may be necessary. Targeting specific labor markets including women, under-represented groups, indigenous groups, and those high in safety conscientiousness, stakeholder-specific skills, or government experience will help address some of the challenges the oil and gas industry currently faces. Recruitment of internal candidates should include career development implications as well as employee expertise and experience. Recruiting methods should be employed that help target the groups mentioned above. To attract applicants to high-demand positions, oil and gas companies should also consider targeting employees from other industries as well as competitors, contractors, and suppliers. Recruiting methods should be consistent with local expectations for firms abroad, and they should include methods to attract passive applicants.

Current Challenge Box 9–1. MOL of Hungary and Strategic HR

MOL Group of Hungary is a company recognized for its commitment to future talent needs. MOL is a publicly traded integrated oil and gas group headquartered in Budapest. Employing more than 30,000 people across 40 countries, MOL is active in exploration, production, and the downstream with four refineries and has a growing retail presence with more than 1,700 gasoline stations. As a fast-growing company MOL is well aware of how critical talent and skill will be in the coming years.

MOL, in partnership with PetroSkills (a competency-based training provider for the oil and gas industry), is heavily invested in strategic workforce planning. The company has its eyes on the strategic horizon, stressing that sector specific sustainability and autonomy requires a minimum of five to seven years in the oil and gas industry. In this process MOL uses an inside-out approach to skill development that emphasizes leadership development focusing on the impacts of leadership and management on the outside world, without the internal blinders so commonly observed in many technical industries.

MOL received the European HCM Excellence Award for recruitment in 2011, recognized for its unique, integrated strategy for attracting talent. The company has used online games for secondary school students, as well as competitive simulations for undergraduate and graduate student selection and recruitment. MOL Group aggressively and massively mobilizes internal resources to ensure intellectual capital and knowhow transfer from internal experts to new hires. Technical competency mapping and gap analysis is performed to translate strategic goals into concrete people and learning actions. Technical training is delivered by internal resources to ensure knowledge transfer from the matured professional and to keep the intellectual capital of MOL internal to the group. Strategic workforce planning is used to create a methodology for dynamic modeling of midterm needs.

Selection tools tend to be conventional in the oil and gas industry, but these tools should be consistent with local norms when selecting abroad. Selection tools may include some that address the issues that make the oil and gas industry unique. Examples of this include:

- Safety knowledge tests, safety-related interview questions, personality tests, and past safety records to increase the probability of selecting safety conscientious applicants
- Assessing team skills for team-oriented jobs
- Certifications for certain jobs such as project manager
- Assessments of compatibility with rotator schedules when applicable
- Assessments of skills or characteristics that will affect stakeholder-related performance

Further, when selecting for international assignments, firms should assess factors beyond the ability to do the job and include factors to assess the applicant's cultural adaptability. The combination of selection tools in the oil and gas industry should follow the general principle of using multiple hurdles for measures that predict essential performance aspects of the job and use the compensatory model for measures that predict desirable (but not essential) performance aspects of the job.

Training in the oil and gas industry

We have shown throughout the book how the differences between the oil and gas industry and other industries can affect training, specifically who, what, and how the company trains its employees. Who gets trained typically depends on what is getting trained. Table 9-1 shows different types of general training content and the employee groups that typically receive each type. All employees should be trained in general health and safety, regulations, stakeholder management issues (ethics, anticorruption, etc.), organizational socialization issues, and general policies. Employees in specific locations should be trained on cross-cultural training, health and safety issues tied directly to that location, security training, and political risk management. Employees in certain jobs should be trained on job-related skills, knowledge, and abilities, job-related health and safety topics, technical and process knowledge, and team building, customer service, interpersonal skills, job-specific regulation compliance, project management, and leadership skills when applicable. Finally, contractor training is particularly important in the oil and gas industry because of the prevalence of contractors.

Table 9–1. General training content in oil and gas companies

Who Gets Trained?	What Gets Trained?
All employees	General health and safety General regulations Stakeholder management (ethics, CSR, anti-corruption, etc.) Organizational socialization (culture, history, mission, etc.) General policies (guidelines, rules, procedures, sexual harassment, etc.)
Employees at Certain Locations	Cross-cultural training (history, customs, culture, laws, etc.) Location specific health and safety Language training Security training Basic skills Political risk management
Employees in Certain Jobs	Job-related skills, knowledge, and abilities Job-related health and safety Technical knowledge Process knowledge Team building Regulation compliance policies and procedures Global leadership Project management Customer service Interpersonal skills

The type of training method also depends on who gets trained as well as what gets trained. Although computer-based methods are more commonly used now because of their lower long-term costs, easier accessibility, and greater customizability, they may be low on engagement depending on their level of interactivity. For critical training, behavioral modeling, simulations, and hands-on training should be used whenever possible because it creates higher levels of engagement which improves training results. Training abroad should use methods consistent with the country's culture. Further, mentoring has been shown to be particularly effective in furthering employees' career success.

Performance management in the oil and gas industry

The differences between the oil and gas industry and other industries can affect performance management, specifically how performance is measured, who is involved, and the format used. Performance measures in most jobs in the oil and gas industry should include essential and desirable task performance behaviors and outcomes, safety-related behaviors and outcomes, team contribution when applicable, and developmental activities. Measures should not just focus on the outcomes, but also on the means for achieving the results. For managers and some other higher level employees, performance is likely to include more outcome-based measures, and should include meeting stakeholder-related goals including measures of performance related to financial, environmental, customer, subordinate, community, social, union, and government concerns.

Research has shown that feedback from many different sources is better than performance feedback from only one source. Thus, viable sources for performance feedback in the oil and gas industry include supervisors, coworkers, and the employees themselves. They may also include other managers; health, safety, and environment (HSE) officers; HR professionals; team members; subordinates; trainers; customers; nongovernmental organizations (NGOs); suppliers; union representatives; government officials; inspectors; contractors; and community representatives when applicable. The format of the performance evaluation should be based on the behaviorally anchored rating scale (BARS) or behavioral observation scale when rating performance behaviors and relevant, justifiable outcome measures when rating results. Usually multidimensional formats will be needed to capture all relevant aspects of performance.

Compensation in the oil and gas industry

The differences between the oil and gas industry and other industries can affect compensation, specifically pay level, the pay mix, pay for performance, and benefits. Absolute pay levels and pay level strategy (paying at, above, or below the market) can be affected by country differences for multinational enterprises (MNEs). A strategy of paying above the market will increase the firm's ability to attract, retain, and motivate talent toward good performance (however it is measured), but at a higher cost of labor. Paying above the market will also reduce some of the difficulties that can occur due to pay levels in project-oriented companies and will help attract applicants for jobs with current talent shortages. Finally, paying above the market tends to be viewed positively by most stakeholders (employees, governments, the community, etc.), with shareholders being the exception because of its high cost and possible detrimental effect on the bottom line.

As with pay level, pay mix may also be affected by country differences (e.g., legally required benefits). In MNEs, the pay mix of international assignees will have a large portion of the pay mix be allowances and premiums. Making performance-based incentives a greater percentage of the pay mix will generally motivate employees toward greater performance (particularly toward the aspects that are actually measured). Project-based employees are likely to have their pay mix be greater in bonuses and benefits because of the nature of project-based work. The pay mix of contractors and rotators are likely to be more bonus heavy than the pay mix of permanent employees. Bonus heavy pay mixes can help shift some of the risk during industry downturns from the shareholders to the employees.

With respect to pay for performance, bonuses are likely to improve performance on the specific aspect of performance that the bonus is tied to. In the oil and gas industry bonuses may be tied to safety performance, productivity, sales, profits, project completion, location hardships, contractor "loadings," refinery turnarounds, or any one of many stakeholder-related goals. Recognition awards may also be effective in recognizing outstanding performance, including performance related to safety, achieving health specific goals, achieving project milestones, or meeting regulatory standards.

As stated above, benefits can vary tremendously across countries for MNEs due to prevailing norms, legal requirements, and tax implications. Health-related benefits such as health care, employee assistance programs, and wellness programs can improve workers' health and safety. Contractors tend to get fewer benefits than permanent employees, while rotators tend to get more. Benefits can be of value to many stakeholders (e.g., by providing safety nets to employees, or motivating volunteerism), but are likely to come at a high cost at the expense of shareholders.

Conclusion

The oil and gas industry has many challenges, as described in the beginning of the chapter, and will continue to evolve over the coming years. Many of the challenges stem from the key differences between this industry and most other industries. This book shows how many of these challenges can be addressed through the thoughtful and creative management of HR. We urge managers in the oil and gas industry to consider HR-based solutions in the future as one important tool in their toolbox of solutions, when undoubtedly new difficult challenges such as volatile demand, plunging prices, or new energy sources will emerge.[2]

References

1. BP quickens job cuts due to oil price fall. (2014, December 7). Retrieved from http://www.bbc.com/news/business-30372363.
2. The future of oil: Yesterday's fuel. (2013, August 3). *The Economist*, p. 12.

A

accidents, 127–128
    costs and losses of, 136
    investigation of, 133–136
    Macondo in, 133–136, 319, 324
accountability, 285–286
accuracy, 25–26
acquaintanceship, 359
acquisitions, 68–69
acres, 321
activities, 58–64, 123, 222–223
ad hoc approach, 107–108
African Americans, 289–291
African project, 119–126
agencies, federal, 322–324
agreements, 59, 264–265
Alaska Department of Environmental Conservation, 325
Alaska Department of Natural Resources Oil and Gas Division, 325
Alaska Oil and Gas Conservation Commission, 325
Alaskan National Wildlife Refuge, 319
allowances, 111
alternative fuels, 70
alternative international travelers, 92–93
American Petroleum Institute (API), 131, 152
analysis, 59, 260–261
analysis phase, 172
annual sustainability reports, 275
applicants, 16–17, 143–144
approaches, 15, 107–108, 144
Arab oil embargo, 42
arctic drilling, 279

Arctic Standards: Recommendations on Oil Spill Prevention, Response, and Safety in the US Arctic Ocean, 278
artificial intelligence, 236
assessments, 85
    future risk, 143–144
    Hofstede's dimensions of national culture for, 86–87, 110
    psychological, 87
assignees, short-term, 92
assignments
    foreign, 84–85
    international, 87–90, 92–93, 218
    recruiting and nature of, 84–85
    rotators and causes of failed, 220–221
assistance, mutual, 280
Association for Project Management (APM), 181
associations, female recruiting, 236
A.T. Kearney, 62
auctions, 320–321
Australia
    employment and gas development in, 258–259
    liquefied natural gas development case of, 253–265
    mentoring and, 241
    project locations and, 254–256
Australian Department of Immigration and Border Protection, 264–265
Australian Institute of Project Management, 152
Australian LNG industry, 253–261
    Chevron and, 253, 259–260, 263
    developments, 254–256
    employment structure and strategies for, 261–265

growth of, 253–254, 265
identified project improvements for, 260–261
integrated project development and, 263–264
of liquefied natural gas, 253–265
project benefits for, 264–265
staffing and skill requirements of, 257–259
strategic promise and challenge of, 260–261
awards, 32, 160, 203, 366

B

Baker Hughes, 236
Baker Report, 133–134, 160
Bakken Boom
average monthly wages in, 356–357
drilling growth and, 351–360
employment of, 354–355
industry share of employment in, 357–358
shale formation and, 351–352
unconventional oil, employment and, 351–360
unconventional oil of, 352–353
well production profile of, 355–356
Bakken formation, 351–352
balanced scorecard, 195, 300
bargaining, 103
behavior
evaluations for safety performance, 153
HR strategies and employee, 8
pay-for-performance for health-related, 160
behavioral observation scale (BOS), 24
behaviorally anchored rating scale (BARS), 24, 155–156
benefits. *See also* compensation and benefits; wages
Australian LNG industry project, 264–265
for better government relations, 344–345
compensation and percentage cost of, 108–109
danger pay, 111
firms providing rotator, 245–246
flexible plans for, 33
global context and, 110–111
global oil and gas' prevalent, 33
of increasing diversity, 289
legally required, 32–33
of local content, 330
oil discovery costs and, 270

project-oriented company travel-related, 203–204
for protection of financial hardship, 344–345
safety performance, 161
from stakeholder perspective, 303–304
in workforce, 245–247
biodiversity, 271–272
biofuel, 63
blacks. *See* African Americans
bonuses, 157, 244
caution to take with, 158
Christmas, 346
criteria for, 200–201
determining size of, 201–202
performance-based pay as safety, 160
boomtowns
creating, 355–360
density of acquaintanceship and, 359
increased employment and, 359–360
reduced community satisfaction and, 359
societal impact and, 358–360
wages and prices in, 356–358
break-even price, 60
British Petroleum (BP), 13, 16, 51–52, 67, 74, 133–134, 138–139
build, operate, transfer (BOT), 214
build, own, operate (BOO), 214
build, own, operate, transfer (BOOT), 214
Bureau of Labor Statistics, US, 152, 356
Bureau of Land Management, 320–321
Bureau of Ocean Energy Management, Regulation and Enforcement, 135, 320–321
Bureau of Safety and Environmental Enforcement, US, 279
business, 4–5

C

cafeteria approach, 108
capital expense (capex), 168–169, 262
capital resources, 1
Cardoso, Fernando Henrique, 317
career development, 22–27
centralized bargaining processes, 103
certifications, 181–182
Chad-Cameroon Petroleum Development Project
companies investing in, 119–126
countries in, 119–120
difficulties operating, 122–123
employment and, 119–126
ExxonMobil and, 119–126
host and parent-country nationals employed at, 121–122

incident safety performance and, 150–151
reinvestment and, 121–123
revenue management plan, 269–270
challenge box
  arctic drilling, 279
  cultural differences, 86–87
  expatriate localization, 108
  gender and sexual harassment, 226–227, 239
  government exploitation, 316–317
  local content laws, 90, 330–332
  long-term revenue management and, 269–270
  Mexican oil industry, 317–319
  MOL of Hungary and strategic HR, 365–366
  performance metrics for projects, 194–195
  retaining talent, 219–221, 245
  terrorism, piracy, and kidnapping, 138–139
  underreporting safety incidents, 159
Chandler, Alfred, 42
charities, 275, 304
Chavez, Hugo, 55, 316–317
Chemical Safety Board, 134–135
Chevron, 16, 79, 119–126, 253, 259–260, 263
China, 69, 315–316
Chinese National Offshore Oil Company (CNOOC), 95
Christmas bonuses, 346
climate, 271
coal bed methane, 69
coal seam gas (CSG), 253
Coast Guard, US, 135
code of ethics, 285–286
collective bargaining, 280, 282–283, 341
combined cycle gas turbine (CCGT), 66
communities
  addressing societal concerns of, 274–275
  boomtowns and reduced satisfaction of, 359
  corporate social responsibility and concerns of, 271–274
  residential, 260
  stakeholders, society and, 268–271
companies. *See also specific companies*
  in Chad-Cameroon Petroleum Development Project, 119–126
  global, 75–76, 364
  government ownership of, 311–319
  international oil, 223
  job skill gaps anticipated by, 227–228
  pipeline, 57
  recruiting employees from other, 234–235
  training content in project-oriented, 186–189
company-sponsored "upskilling," 263
comparisons, 111–112, 320
compensation, 273
  allowances, 111
  approaches for, 107–108
  average oil and gas workers', 103–104
  benefit cost percentages and total, 108–109
  for better government relations, 342–346
  comparisons of, 111–112
  danger pay, 111
  differences between countries for, 102–103
  in global environment, 102–113
  government relations' effectiveness and perspective of, 345–346
  hardship pay, 111
  maximizing effectiveness of, 33–35, 111–113, 162, 204, 247
  multinational enterprises and complications of, 103
  multinational enterprises and legal compliance in, 112
  oil and gas' imported labor, 105–106
  in oil and gas industry, 369–370
  parent-country nationals and, 107, 110
  plan types of international assignee, 105
  production workers hourly costs for, 103–104
  in project-oriented companies, 199–204
  safety through, 156–162
  from stakeholder perspective, 302–305
  third-country nationals and, 107
  in unconventional workforce, 243–248
compensation and benefits
  employees', 27–35
  HR function ties of, 27–28
  key decisions in, 30–33
  legal compliance importance in, 34
  monitoring and evaluating, 34–35, 112
  strategic implications of, 28
compensatory approach, 15
competencies, 28
competition, 4–6, 63–64, 260–261, 363–364
complexity, 167–169, 195
concerns, 129–130
  addressing community and societal, 274–275

corporate social responsibility
community, 271–274
social and economic, 273
conflicts, 193
construction, 216
contextual performance, 23
contract workers, 142, 211–212
contractors
firms for, 212–213
management, 173
prevalence of, 211–214
reasons for prevalence of, 212–214
safety and, 216–217
contracts, 142
costs and construction, 215
engineering, procurement, and construction, 215
engineering, procurement, construction, and management (EPCM), 215
financial risks of, 215
management based construction, 215
oil and gas project construction, 214–215
ownership based construction, 214
corporate ethics, 285–286
corporate social responsibility (CSR), 271–275, 285, 293–296, 305
corruption, 319, 327
cost overruns, 169, 215
costs, 108–109, 194–195
accident losses and, 136
compensation and production workers hourly, 103–104
construction contracts and, 215
foreign worker rotation, 265
Gorgon project blowout on, 259–260
labor, 265
local content, 330–331
oil discovery benefits and, 270
parent-country nationals and high, 123–124
countries, 42, 47–49
in Chad-Cameroon Petroleum Development Project, 119–120
compensation differences between, 102–103
fairness and perceived culture of, 90–91
resource-poor, 312–313, 315
rotators used in top, 232
shrinking talent pools in industrialized, 364
courses, training, 295
crew resource management training, 148
crime, 124
criteria, for bonuses, 200–201

crude oil, 42–43
continued price volatility of, 361–362
North Dakota production of, 353
in value chain, 60–61
value of, 60
Crude Oil Windfall Profit Tax Act, 325
culture
challenge of assessing differences in, 85–87
creating safety, 140–141, 162
fairness and perceived country, 90–91
feedback and differences of, 101
goals and differences in, 101–102
Hofstede's dimensions and differences in, 86–87, 110
performance management differences and, 98–99
resources and, 283
rewards and different, 101–102
training in cross-, 94
curse, resource, 49
customers, 268, 277–278

D

danger pay, 111
data, 3-D seismic, 68
days away from work case rate (DAWC), 150, 152
Declaration on Fundamental Principles and Rights at Work, 280
Deepwater Horizon, 135–136
Department of Education, Employment and Workplace Relations (DEEWR), 258–259
Department of Energy, US, 352
Department of Labor, US, 11
Department of State, US, 111
design, build, finance, operate (DBFO), 214
developmental purposes, 183, 193
developments, Australian LNG industry, 254–256
disciplinary actions, 340
discipline, 341–342
discrimination, 90, 112, 183, 239, 326, 335, 345–346
females and, 15–16
reverse, 292
stakeholders and, 280, 285, 293
workforce and, 224, 226
disease, 127, 139–140
disparate impact, 15–16
disparate treatment, 15–16
diversity, 271–272
benefits of increasing, 289
ExxonMobil and, 83
integrated oil companies,' 82–83
Shell and, 83

Dodd-Frank Wall Street Reform and Consumer Protection Act, 327
dominance, male, 222–227, 239, 248
downstream activities, 61–64
Drake, Edwin, 41
drilling
    Bakken Boom and growth in, 351–360
    challenges of arctic, 279
    offshore, 319
    regulations, 324–325
    top oil field services for, 57
    wells, 59
drive-in and drive-out (DIDO), 262

E

economy, 44–49, 273, 310
ecosystems, 271–272, 279
effectiveness
    compensation's, 33–35, 111–113, 162, 204, 247
    global context and, 90–91, 97, 111–113
    government relations and staffing, 335
    government relations' perspective of compensation, 345–346
    performance management, 25–27, 101–102, 243
    performance management for government relations, 338–342
    project-oriented companies', 183–184, 191–192, 204
    safety and health, 149
    safety's, 144–145, 149, 156, 162
    staffing, 15–17, 90–91, 144–145, 183–184, 237–238, 293
    stakeholder perspective's, 297, 301
    stakeholders and maximizing staffing, 293
    training and government relations, 335–338
    training's, 20–22, 97, 240–241, 248, 297
    workforces', 237–238, 240–241, 243, 247–248
employees, 8, 268
    BP's operating, 74
    compensation and benefits of, 27–35
    developmental purposes and, 183, 193
    focusing on training and development of, 17–22
    foreign workers as, 265
    gaining acceptance of, 34
    global context selection methods of, 87–90
    global context training and, 91–93
    national oil companies' revenue of, 313–314
    permanent nationals as, 124
    recruiting other companies, 235
    recruiting other industry, 234–235
    recruiting safety and health valuing, 142–143
    rotators as, 217–219
    Royal Dutch Shell's, 75
    selecting attributes of, 292
    selecting training and development, 19
    short-term assignees as, 92
    special-project nationals as, 124
    as stakeholder group, 276–277
    Total's operating, 74–75
    training and development satisfying, 18
    value proposition of, 27
    young, 235–236, 238, 240–241, 244–245, 297
employment
    Australian gas development and estimated, 258–259
    Bakken and booming, 354–355
    Bakken Boom, unconventional oil and, 351–360
    Bakken Boom's industry share of, 357–358
    boomtowns and increased, 359–360
    Chad-Cameroon Petroleum Development Project and, 119–126
    contract workers' relationship, 211–212
    host-country nationals' proportion of project, 125
    issues, 273
    nontraditional sources of, 263
    oil and gas industry's minority, 289–290
    project sponsor investments in, 263
    quality of life and, 277–278
    rotators' current challenges in, 219
    sector and region of female, 222–223
    structure and categories of, 123–125
    structure and strategies for Australian LNG industry, 261–265
    unconventional oil and, 353–355
    in upstream African project, 119–126
energy project, Gorgon, 253
engagement, 148–149
engineering, procurement, and construction (EPC), 215, 219
engineering, procurement, construction, and management (EPCM), 215
Enron, 202
Enterprise Migration Agreement (EMA), 264–265
environment, 129–130, 271
environmental expenditures, 271–272
Environmental Protection Agency (EPA), 323
errors, 25–26

ethics, corporate, 285–286
"Ethics and Business," 295
ethnocentric staffing philosophy, 82–83
European HCM Excellence Award, 366
evaluations, 186, 197–198
    compensation and benefits' monitoring and, 34–35, 112
    global context performance, 99–100
    performance management's monitoring and, 26–27
    project personnel's performance, 193
    reducing errors and improving accuracy in, 25–26
    safety performance behavior, 153
    stakeholder perspective and, 301
    training and development monitoring and, 21–22, 97
evolution, 68–70
execution phase, 173–174
expatriates. *See* parent-country nationals
expenses, travel, 247
exploration, 362
    leases and development in, 319–320
    rights, 59
    upstream activities and, 58–60
exploration and production (E&P) firms, 362
explosion
    BP's Baker Report on, 133–134
    Chemical Safety Board findings on, 134–135
    Texas City Refinery, 133–135, 160
ExxonMobil, 52, 61, 73–74, 83, 259, 279, 331
    Chad-Cameroon Petroleum Development Project and, 119–126
ExxonMobil Chemical, 67

F

fairness, 90–91, 111–112
families, 92–93, 108
fatalities, 141, 152–153
fatigue, 141
feasibility phase, 172
federal agencies, 322–324
Federal Income Tax Authority of Mexico, 318
Federal Mediation and Conciliation Service, 283
feedback, 24–25, 101, 198, 242, 301
females, 245
    company type and occupations of working, 223–225
    discrimination and, 15–16
    oil and gas industry and, 222–227, 239
    reasons for underrepresentation of, 223–224
    recruiting, 236
    recruiting associations for, 236
    as taking less risks than males, 144
final investment decisions (FIDs), 173, 253
finances, 215
financial analysis, 59
financial hardship, 344–345
financial performance, 46–47
firms, 56–57, 212, 245–246
fiscal regimes, 59
flexible benefit plans, 33
floating natural gas liquefaction, regasification, and storage unit (FLRSU), 58, 65
floating storage and offloading vessel (FSO), 119
flown-in and flown-out (FIFO), 262
Foreign Corrupt Practices Act, 327
foreign practice, regulations, 327
foreign workers, 265
formats
    global context, 100–101
    government relations, 341
    multidimensional, 300
    performance management, 24–25
    safety performance measuring, 155–156
    stakeholder perspective for usage of, 300–301
    to use in project-oriented companies, 197–198
    workforce involvement and, 242
Fort McMurray, 43
fracking. *See* hydraulic fracturing
Frade Japao Petroleo Limitada (FJPL), 79
Frade Project, 79
fuels, 63–64, 70
funds, 269
future, 63, 143–144, 370
Future Generations Fund, 269

G

gaps, 227–228
gas
    Australia's employment for developing, 258–259
    Australia's liquefied natural gas development case for, 253–265
    coal seam, 253
    combined cycle turbine technology, 66–67
    floating natural gas liquefaction, regasification, storage unit and, 58, 65
    growth of unconventional oil and, 69–70, 351–360
    liquefied natural, 56, 64, 167, 214, 253–265, 362

nations as major producers of, 66
natural, 64–67, 361–362
natural gas to liquid, 65
regions of selected company reserves for, 95–96
shale, 65–66, 69
tight, 69
utilities, 57
gasoline, 64
gender, 226–227, 239, 292
geocentric staffing philosophy, 83–84
global companies, 75–76, 364
global competitors, 363–364
global context. *See also* world
benefits in, 110–111
effectiveness of, 90–91, 97, 111–113
employee selection method for, 87–90
format to use in, 100–101
labor market in, 80–84
pay levels in, 103–108
pay mix in, 108–109
performance in, 99
performance management in, 101–102
performance-based pay in, 109–110
recruiting method for, 84–87
skills to train in, 92–93
training in, 91–93
ways to train in, 94–96
who evaluates performance in, 99–100
global economy, 44–49
global environment
compensation in, 102–113
human resources and, 79–80, 84, 113, 126
performance management in, 98–102
staffing in, 80–91
training in, 91–97
global industry
of oil and gas industry, 73–79
oil and gas industry as world's most, 73, 113
potential leaders in, 92–93
unionization in, 284
global leaders, 92–93
global oil and gas industry
background of, 41–43
China and India in, 69
companies by stock market capitalization, 51–52
compensation for workers in, 103–104
complexities and issues of, 126
evolution of, 68–70
financial performance of, 46–47
global economy and, 44–49
impacts and events involving, 41

industry players and competitors in, 49–57
innovation and technology in, 68
integrated oil companies in, 51–52, 63
mergers and acquisitions in, 68–69
nations producing in, 44–45
predictions for, 70
production of, 47–48
reserves, 47–48
resource curse as paradox of, 49
rotators and expected growth of, 233–234
safety performance of, 151–152
work hour trends in, 212–213, 280
global operations, 73–74
global public trading, 53–54
global supermajors, top 10, 49–50
goals, 198, 201, 278
different cultures and, 101–102
local content for achieving, 331–332
rewards and setting, 26
stakeholder perspective, 301
Gorgon energy project, 253, 259–260
government, 253, 264
company ownerships by, 311–319
economic influences of, 310
exploitation of national oil companies by, 316–317
human resource management and improving, 328–329
importance of oil and gas to, 309, 346
as integral to oil and gas industry, 311, 346
integrated oil companies and, 312–313, 315–316
involvement in value chain, 328, 346
national oil companies and, 311–319, 327–328
national oil companies suffering intervention from, 55, 71
national security issues and, 310–311
oil and gas industry and influence of, 309–328, 346
oil and gas industry regulations and, 322–328
organizations as non-, 268, 275, 278–279, 288–289, 294–295
ownership of oil by, 319–321
performance of national oil companies and, 313–315
private firm partnerships with, 327–328
privatization and, 315–317
quality-of-life issues and, 310
science, technology, engineering, mathematics and, 325, 327, 337

staffing and improving relations with, 329–335
United States' influence and, 310–327
government relations
  benefits for better, 344–345
  combining selection tools for better, 334–335
  compensation for better, 342–346
  compensation's effectiveness from perspective of, 345–346
  content for training, 336–337
  effectiveness of staffing and, 335
  effectiveness of training from perspective of, 338
  formats for, 341
  human resource management for improving, 328–329
  involvement in performance management for, 340–341
  labor market perspective and, 332–333
  measuring performance for better, 339–340
  oil and gas associations involved in, 332–333
  pay level for better, 343
  pay mix for better, 343
  performance management for better, 338–342
  performance-based pay for better, 343–344
  recruiting for better, 333–334
  selection for better, 334
  staffing and improving, 329–335
  training candidates for, 336
  training for better, 335–338
Great Recession, 46
greenfield project, 258
gross domestic products (GDPs), 168–169, 268, 326
growth, 42–43, 69–70, 253–254, 265, 351–360
A Guide to the Project Management Body of Knowledge (PMBOK Guide), 187–189

H

harassment, 226–227, 239
hardship, financial, 344–345
hardship pay, 111
health, safety, and environment (HSE), 128, 147, 216, 271, 322–324
health and safety, 2, 124, 127–162
Health insurance, 32
Hispanics, 289–291
HIV/AIDS, 139
Hofstede, G. H., 86–87, 110
Hofstede's dimensions of national culture, 86–87, 110

home-based approach, 107
host-based approach, 107
host-country nationals (HCNs), 81–82, 92–93, 121–122, 125, 217
HQ-based approach, 107
HR practices, 7
HR professionals, 100
HR strategies, 7, 8
human resource management (HRM), 1
  challenges of project, 170–205, 305, 351
  four functional areas of, 8–35
  in global environment, 79–80, 126
  improving government relations through, 328–329
  safety through, 140–141
  stakeholder perspective of, 287, 295–296, 305
  strategic implications of, 3–6, 35
  surveys, 24, 26, 34–35
human resources (HR), 140–141
  function ties of compensation and benefits in, 27–28
  global environment and, 79–80, 84, 113, 126
  managing oil and gas industry, 365–370
  in project-oriented companies, 174–184
  project-oriented companies' aspects related to, 175
  project-oriented companies' differences in, 175–176
  strategies and, 365–366
hydraulic fracturing, 270–271, 274, 323, 352–355

I

incentives, 160–161
incidents, 150–151, 159
independents, 56
India, 69
indigenous skills, 261–262
Industrial Relations Commission, 326
industrial unions, 283
industries, 1–3, 234–235, 281, 370
industry players and competitors
  global oil and gas, 49–57
  independents in, 56
  integrated oil companies, 51–52
  national oil companies in, 53–56
  oil field services and firms in, 57
innovation, 68
integrated oil companies (IOCs), 4–5, 74–75, 95, 363
  differences in, 51–52
  diversity and, 82–83
  global oil and gas industry's, 51–52, 63

governments and, 312–313, 315–316
industry players and competitors, 51–52
Seven Sisters, 51
strategic interests of, 313, 315
integrated project development, 263–264
intelligence, artificial, 236
Internal Revenue Service (IRS), 246–247
international assignees, 105, 109–110
international assignments, 87–90, 92–93, 218
International Association of Oil and Gas Producers (IOGP), 151–152, 155, 275
International Labour Organization (ILO), 129, 280, 284
international oil companies (IOCs), 223
International Petroleum Industry Environmental Conservation Association (IPIECA), 155, 275
International Project Management Association, 182
Internet job boards, 13
internships, 235–236
interviews, 14, 87–88
inventions, 42
investigations, 133–138
investment decision phase, 173, 253
investments, 121–123, 263
final investment decisions and, 173, 253
investors, 268, 275–276
iron triangle, 194–195

J

jobs
applicants' fair treatment for, 16–17
considerations for growth in, 9–10
deciding on methods of screening for, 14–15
descriptions of, 199
Internet job boards for, 13
North Dakota oil and gas industry, 354–355
oil and gas industry growth comparison of, 9–10
oil and gas industry's common, 11–12
oil and gas industry's projected openings for, 224–225
rotations of, 190
screenings and, 14–15
security of, 277
skill gaps anticipated for company, 227–228
training for performance on, 21
training specific to, 147–148

K

Kellogg Joint Venture Gorgon (KJV), 260
kidnapping, 138–139
Kirchner, Cristina Fernandez de, 312
knowledge, 187–189, 191
knowledge, skills, and abilities (KSAs), 19

L

labor, 11, 152, 283, 356
Australian LNG industry challenges and, 253–261
availability, 9
compensation for oil and gas imported, 105–106
costs, 265
laws, 90
multinational enterprise sources for, 80–81, 84
oil and gas compensation for imported, 105–106
proportion of categorized skilled, 124
regulations, 326
shortages, 220, 263–265
skilled, 261
labor markets
expanding, 231–232
global context choices for, 80–84
government relations perspective on, 332–333
multinational enterprises determining wages for, 105
in project-oriented companies, 176–177
rotator, 232–234
staffing and targeting, 11
stakeholders perspective of, 288–290
targeting local, 288
in workforce, 231–234
lagging indicators, in safety metrics, 153–154
Large Project Management in Oil and Gas, 169
laws, 283, 319, 326
benefits required by, 32–33
compensation, benefits and compliance with, 34
labor, 90
local content, 90, 330–332
multinational enterprises and compliance with, 112
staffing and complying with, 15–16
leaders, 76–77, 92–93
leadership, 92–93, 95
learning, 20–21
leases
auctions for, 320–321

for exploration and development, 319–320
land comparisons of United States, 320
in United States, 319–321
legislation, 280
liquefied natural gas (LNG), 56, 64, 167, 214, 253–265, 362
local content
benefits of, 330
costs of, 330–331
goal achievement by, 331–332
laws, 90, 330–332
locals. *See* host-country nationals
losses, 136
lost time incident rate (LTIR), 151
lump sum approach, 108

M

Macondo accident, 133–136, 319, 324
male-dominance, 222–227, 239, 248
males, 144, 222–227, 239, 240
management, 295, 365–370. *See also* human resource management; performance management
construction contracts based on, 215
contractors, 173
contracts and, 215
crew resource training for, 148
long-term revenue, 269–270
partnership, 173
of Petróleos Mexicanos, 318–319
project, 2, 169, 173, 195–196, 198
reservoir, 59–60
revenue plan for, 269–270
safety, 174
supply chain and procurement, 174
training for risk, 95, 173–174
management by objectives (MBO), 193
Marathon Petroleum, 135
margins, refining, 62
Maritime Union, 260
marketing, 61–64
McKinsey analysis, 260–261
measurements, 242
megaprojects, capital expense and, 168–169
membership, union, 281
mentoring, 241
Mercer, 228
mergers, 51, 68–69
merit pay, 32
methane, coal bed, 69
methodologies, 240
for change by unions, 280
for combining selection tools, 15
deciding on job screening, 14–15
employee selection, 87–90
global context recruiting, 84–87
government relations training, 335–338
of job screening, 14–15
poaching in recruiting, 13
project-oriented training, 189–191
for recruiting, 13
recruiting success and, 84–85
safety and health training, 148–149
selecting training and development, 20
staffing and recruiting, 11–13
stakeholder perspective training, 296–297
metrics, 153–154
project performance, 194–195
Mexican oil industry, renewal of, 317–319
Mexico, 317–319
Middle East, 76, 78
midstream activities, 60–61
Millennials, 235–236, 238, 240–241, 244–245, 297
mission, 47–48
MOL Group of Hungary, 366
monitoring
of compensation and benefits, 34–35, 112
of performance management, 34–35
performance management evaluation and, 26–27
project management evaluations and, 198
of training and development, 21–22, 97
monopoly, Petrobras, 317
multidimensional formats, 300
multinational corporation (MNC), 288, 302, 369–370
multinational enterprises (MNEs), 98–99, 101, 110, 284
compensation complications of, 103
labor market wage determinations by, 105
legal compliance in compensation and, 112
sources of labor for, 80–81, 84
staffing and, 90–91
stakeholder perspective and, 294
multinational partnerships, 78–79
multiple hurdles approach, 15, 144, 182–183, 237, 293, 335, 367
mutual assistance, 280

N

national oil companies (NOCs), 51, 63, 68–69, 95
global companies, 75–76
as global competitors, 363–364
government and, 311–319, 327–328

government and performance of, 313–315
government exploitation of its, 316–317
government intervention suffered by, 55, 71
governments and, 311–319
growing importance of, 53–56, 70
industry players and competitors', 53–56
oil and gas industry challenges and, 361–364
profitable and well-run, 56
as purely domestic, 76
revenue per employee for, 313–314
strategic interests of, 313, 315
National Research Council, 279
national security, 310–311
National Society of Black Engineers, 290
National Union of Petroleum and Natural Gas Workers, 283–284
nationals, permanent, 124
nationals, special-project, 124
nations
gas-producing, 66
leading oil-producing, 76–77
oil and gas industry's major producing, 44–45
natural gas, 65–67, 361
liquefied, 56, 64, 167, 214, 253–265, 362
natural gas to liquid (GTL), 65
natural resources, 1
net-to-net approach, 107
Nieto, Enrique Peña, 319
Nigeria, 283–284
Nigeria Labour Congress, 283
nongovernment organizations (NGOs), 268, 275, 278–279, 288–289, 294–295
North Dakota, 353–355
Norway, 324
Norwegian Continental Shell, 312
not in my backyard (NIMBY), 274

O

occupational disease, 139–140
occupational safety and health, 129–140
Occupational Safety and Health Administration (OSHA), 128, 150, 152, 159, 322–323
occupations, female, 223–225
oil. *See also* crude oil
benefits and costs of discovering, 270
government ownership of, 319–321
international companies for, 223
nations leading in production of, 76–77
price growth of crude, 42–43
refineries, 61–64

shale or tight, 352
oil and gas, 105–106, 169, 274
discovering reserves for, 42–43
federal agencies concerned with, 322–324
governments and importance of, 309–328, 346
growth of unconventional, 69–70, 351–360
Middle Eastern producers of, 76, 78
resources, 76–78
oil and gas companies, 268
code of ethics topics in, 285–286
recruiting in other, 235
by stock market capitalization, 51–52
world's largest publicly traded, 53–54
oil and gas industry. *See also* global oil and gas industry
Bakken Boom's share of employment in, 357–358
benefits prevalent in, 33
capital resources and, 1
challenges facing, 361–365
collective bargaining issues in, 282, 341
common jobs in, 11–12
compensation for imported labor in, 105–106
compensation in, 369–370
complexity of, 167–169
environmental expenditures of United States and, 271–272
fatality causes in, 141
females and, 222–227, 239
global industry of, 73–79
government regulations and, 322–328
government relations in, 332–333
governments as integral to, 311, 346
health and safety's importance in, 127–162
high-paying wages and, 28–29
hourly wage of sectors in, 28–29
importance of, 1, 70
increased complexity projects in, 167–169
industry differences from, 1–3, 370
influence of governments in, 309–328, 346
influence of stakeholders in, 267
job growth comparisons in, 9–10
largest contractor firms in, 212–213
males as dominant in, 222–227, 248
managing human resources in, 365–370
minority employment by sector, 289–290

national oil companies and challenges of, 361–364
nations of major production in, 44–45
natural resources and, 1
North Dakota's jobs in, 354–355
occupational disease and, 139–140
occupational safety and health in, 129–140
owners and investors in, 275–276
performance management in, 368–369
project development in, 170–174
project development phases in, 171–174
projected job openings in, 224–225
projects in, 167–170, 176–184, 214–215
rotators in, 217–221
safety and health in private sector *vs.*, 131
safety and health rates by value chain in, 131–132
safety hazards, 127–128
safety metrics in, 154
security and, 136–139
selection tools in, 367
skill shortages in, 253–265
skills required for challenges of, 364–365
sources for recruiting in, 13–14
staffing in, 365–367
stakeholders, 268–284, 305
technology challenges in, 362–363
training content in, 367–368
training requirements for, 18
unconventional sources for development of, 351–355
unions and, 283–284
value chain and, 58–64
wages in sectors of, 28–29
worker type and fatalities in, 142
workforce of, 211
world's energy and evolving, 361–370
world's most expensive projects in, 167–170
Oil and Gas Industry Guidance on Voluntary Sustainability Reporting, 275
oil and gas projects, 167–170, 176–184, 214–215
Oil and Gas Sector Guide on Implementing the UN Guiding Principles on Business and Human Rights, 274
oil field services, 57
oil majors. *See* integrated oil companies
oil sands, 43–44, 169
Oil Workers Union of the Mexican Republic, 318
O*net, 11
operations, global, 73–74
Organization of the Petroleum Exporting Countries (OPEC), 42, 47–49
other industries, 234–235
"Our Ethical, Environmental, and Social Responsibilities," 295
outcomes, safety, 148–149, 153, 158
outsourcing, 236
overruns, cost, 169, 215
owners, oil and gas industry investors and, 275–276

P

parent-country nationals (PCNs), 87, 91–93, 246
    advantages and disadvantages of staffing, 82–83, 217–218
    Chad-Cameroon Petroleum Development Project and, 121–122
    compensation and, 107, 110
    extended localization of, 108
    high costs and, 123–124
    terrorism and, 138–139
    training and, 146, 148
partnerships, 78–79, 173, 327–328
Patient Protection and Affordable Care Act, 160
pay level, 199. *See also* wages
    for better government relations, 343
    deciding on wages and, 30
    global context in, 103–108
    safety performance and, 156–157, 162
    from stakeholder perspective, 302
    in workforce, 243–244
pay-for-performance, 160
pay-mix, 200, 244, 303
    global context in, 108–109
    international assignees and, 109–110
    safety performance and, 157
    wages and deciding on, 30–31
peer-to-peer awards, 32
performance, 21. *See also* safety performance
    behavior and pay-for-, 160
    behavior evaluations for safety, 153
    benefits of safety, 161
    better government relations by measuring, 339–340
    contextual, 23
    evaluations and, 99–100, 193
    financial, 46–47
    involvement in project-oriented evaluations and, 197
    measurements for project managers', 194–197
    measuring global context, 99

measuring project-oriented company, 192–197
measuring workforce, 242
project managers' measures of, 195–196
projects and metrics for, 194–195
scores, 300
selecting for stakeholder relevant, 291–292
stakeholder perspective of measuring, 298–299
performance management
for better government relations, 338–342
career development and, 22–27
contextual performance in, 23
cultures differences and, 98–99
formats to use in, 24–25
in global environment, 98–102
government relations and involvement in, 340–341
key decisions in, 23–25
maximizing effectiveness of, 25–27, 101–102, 243
maximizing stakeholder perspective in, 301
measuring performance in, 23–24
monitoring and evaluating, 26–27
monitoring of, 34–35
in oil and gas industry, 368–369
in project-oriented companies, 192–198
providing appropriate feedback in, 25
safety through, 150–156
sources involved in, 24
from stakeholder perspective, 298–301
stakeholder perspective and involvement in, 299–300
ways training is related to, 23
of workforce, 241–243
performance-based pay
for better government relations, 343–344
bonuses as safety, 160
deciding on type of, 31–32
differences in plan types of, 31–32
global context in, 109–110
in project-oriented companies, 200–203
risks, 202–203
safety performance and types of, 157–161
safety recognition awards and, 160
from stakeholder perspective, 303
in workforce, 244–245
permanent nationals, 124

personality, tests, 144
personnel, 193
Petrobras, 79, 95, 315–317
petrochemicals, 67, 70
petrocracy, 318
Petróleos de Venezuela S.A. (PDVSA), 55, 315–317
Petróleos Mexicanos (Pemex), 315–319
Petroleum and Natural Gas Senior Staff Association of Nigeria, 283–284
Petronas, 119–126
PetroSkills, 366
Pew, Joseph N., 278
Pew, Mary, 278
Pew Charitable Trusts, 278
phases, 170–174
philosophy, 81–84
pipeline companies, 57
pipelines, 60–61, 119
piracy, 138–139
poaching, 13
policies, 216, 226–227, 239, 340
polycentric staffing philosophy, 81–82
predictions, for global oil and gas industry, 70
price, 42–43, 356–358
price volatility, 361–362
principles, 212–213, 280
private sector, 131, 327–328
privatization, 315–317
process safety, 132–136
procurements, 174
production
compensation costs for workers in, 103–104
firms for exploration and, 362
of global oil and gas industry, 47–48
nations in major gas-, 66
North Dakota crude oil, 353
Petróleos Mexicanos' declining, 318–319
upstream activities and, 58–60
production sharing agreement (PSA), 59, 78–79
productivity, 7, 261
profitability, 56, 60, 62, 352
Program Evaluation and Review Technique/Critical Path Method (PERT/CPM), 186
project development, 263–264
analysis phase in, 172
execution phase in, 173–174
feasibility phase in, 172
investment decision phase in, 173
in oil and gas industry, 170–174
Project Management Association of Japan, 182

Project Management Institute (PMI), 181
Project Management Professional (PMP), 181
project managers
    areas and bases of knowledge needed by, 187–189, 191
    certifications for, 181–182
    challenges of human resource management and, 170–205, 305, 351
    performance measurements for, 194–197
    performance measures of, 195–196
    recruiting for, 178–180
    skills related to, 179–180
    staffing projects and selecting, 181–182
    training, 185–191
project-oriented companies, 205
    combining tools in, 182–183
    compensation in, 199–204
    differences in human resources of, 175–176
    formats to use in, 197–198
    human resources in, 174–184
    human resources' related project aspects in, 175
    labor markets in, 176–177
    maximizing effectiveness of, 183–184, 191–192, 204
    maximizing project management in, 198
    measuring performance in, 192–197
    pay mix in, 200
    pay-level in, 199
    performance evaluation involvement in, 197
    performance management in, 192–198
    performance-based pay in, 200–203
    projects as training in, 187, 190
    project-specific training in, 184–187, 189
    recruiting for, 177–180
    selecting projects in, 180–181
    topics for training in, 186–189
    training candidates for, 184–186
    training in, 184–191
    training methods for, 189–191
    training to maximize, 191–192
    travel-related benefits in, 203–204
projects
    Gorgon energy, 253
    increased complexity of, 167–169, 195
    locations of, 254–256
    management of, 2, 169, 173, 195–196, 198
    megaprojects and capital expense, 168–169
    oil and gas, 167–170, 214–215
    oil and gas industry, 167–170, 176–184
    performance metrics for, 194–195
    recruiting for specific, 177–178
    sponsors for, 263
    staffing, 176–184
    as training in project-oriented companies, 187, 190
project-specific training, 184–187, 189
psychological assessments, 87

Q

quality, 194–195
quality-of-life, 277–278, 310
Queensland Curtis LNG project (QCLNG), 255

R

race, 292
railroads, 60 61
Raymond, Lee, 61–62
recessions, 222
recognition awards, 32, 203
recruiting
    applying testing tools in, 14, 87–88
    for better government relations, 333–334
    external, 12
    female associations for, 236
    global context method for, 84–87
    internal, 12, 236
    methods, 11–13, 84–87
    nature of assignment and, 84–85
    oil and gas industry's sources of, 13–14
    options for, 236
    other industry employee, 234–235
    other oil and gas company, 235
    poaching method of, 13
    for project managers, 178–180
    for project-oriented companies, 177–180
    safety and health valuing employees, 142–143
    selection tools used for, 14–15, 87–88
    for specific projects, 177–178
    staffing and, 9–17
    from stakeholder perspective, 290–291
    targeting females in, 236
    targeting Millennials in, 235–236, 238, 240–241, 244–245, 297
    in workforce, 234–236
refineries
    golden-age of, 62
    margins, 62
    oil refining, 61–64

profitability factors driven by, 62
questioning future of, 63
Texas City Refinery's explosion and, 133–135, 160
world's largest, 63
regimes, fiscal, 59
regional approach, 107–108
regulations
    drilling, 324–325
    foreign practice, 327
    labor, 326
    oil and gas industry and government, 322–328
    for pricing and taxes, 325–326
regulatory compliance, 340
reinvestments, 121–123
Reliance Industries, 63
renewal, of Mexican oil industry, 317–319
reports, 133–134, 275
reserves
    discovering oil and gas, 42–43
    global oil and gas industry, 47–48
    regions of selected company gas, 95–96
    regions of selected company oil, 95–96
    true, 42
reservoir management, 59–60
residential, communities, 260
resource curse, 49
Resource Efficiency, 196
resource-poor countries, 312–313, 315
resource-rich countries, 312, 315
resources, 1, 312–313, 315
    access, 362
    cultural, 283
    oil and gas, 76–78
    stakeholder allocations to, 286
retention, 220–221, 245
revenue, 269, 313–314
revenue management plan (RMP), 269–270
reverse discrimination, 292
rewards, 198
    awards as, 32, 160, 203, 366
    different cultures and, 101–102
    setting goals and giving rewards, 26
    stakeholder perspective and, 301
    total, 26
rights to explore, 59
risk, 124
    applicants not taking safety-related, 143–144
    assessing future, 143–144
    contracts and financial, 215
    females compared to males as taking less, 144
    management training, 95, 173–174

    multinational partnerships for avoidance of, 78–79
    occupational disease, 139–140
    performance-based pay, 202–203
    security, 136
    social media enhancing challenges and, 364
    training, 95
Rockefeller, John D., 41
rotations, 190, 265
rotators, 123, 244, 248
    causes of failed assignments of, 220–221
    current employment challenges of, 219
    defining, 217–218
    as employees, 217–219
    expected growth of global, 233–234
    firms providing benefits for, 245–246
    labor market and, 232–234
    number of worldwide, 233
    in oil and gas industry, 217–221
    as solution for staffing dilemma, 217–218
    special attention for retention of, 220–221
    top countries using, 232
Royal Dutch Shell, 75
royalties, 320–321
Russian territories, 279

S

safety, 124
    applicants not taking risks related to, 143–144
    categorizing metrics and, 153–154
    through compensation, 156–162
    contractors and, 216–217
    creating culture of, 140–141, 162
    different plan types of training, 146–147
    factors, 216–217
    findings on explosion and, 134–135
    hazards, 127–128
    in house or contracted staffing in, 142
    through human resource management, 140–141
    knowledge, 148–149
    lagging indicators in metrics of, 153–154
    management, 174
    maximizing effectiveness of, 144–145, 149, 156, 162
    measures of, 195–196
    metrics, 153–154
    nets, 304
    outcomes, 148–149, 153, 158
    performance, 148–149

through performance management, 150–156
performance of global oil and gas industry, 151–152
performance-based recognition awards for, 160
personal and process, 132–136
selection tools for multiple-hurdles approach to, 144
through staffing, 141–145
through training, 145–149
training for security and, 95
underreporting incidents in, 159
safety and health. *See also* occupational safety and health
content of training for, 146–148
maximizing effectiveness of, 149
oil and gas industry's occupational, 129–140
rates by value chain in oil and gas industry, 131–132
recruiting employees that value, 142–143
training methods and, 146–149
safety performance, 148–149
behavior evaluations for, 153
benefits and, 161
Chad-Cameroon Petroleum Development Project, 150–151
format for measuring, 155–156
of global oil and gas industry, 151–152
involvement decisions for, 155
maximizing effectiveness of, 156, 162
measures of, 152–155
measuring, 150–155
pay level and, 156–157, 162
pay-mix and, 157
performance-based pay types and, 157–161
safety training, 155
deciding who receives, 145–146
different plan types of, 146–147
improvement from, 148–149
level of engagement and, 148–149
maximizing effectiveness of health and, 149
salaries. *See* wages
sands, oil, 43–44
Saudi Basic Industries Corporation (SABIC), 76
Schlumberger, 331–332
science, technology, engineering, mathematics (STEM), 222–224, 228, 325, 327, 337
scorecard, 195, 300
scores, 300
screenings, 14–15
security, 95, 124, 128, 136–139, 277, 310–311
selection
for better government relations, 334
employee attributes in, 292
methodologies for employee, 87–90
project-oriented companies and project, 180–181
stakeholders' perspective of, 291–292
of workforce, 236–237
selection tools
better government relations by combining, 334–335
compensatory approach in, 15
methods for combining, 15
multiple hurdles *approach* in, 15
multiple-hurdles approach on safety-related, 144
oil and gas industry's, 367
perspective of stakeholders and combining, 293
recruitment and, 14–15, 87–88
seminars, 295
services, oil field, 57
Seven Sisters, as largest integrated oil companies, 51
sexual harassment, 226–227, 239
shale formation, 351–352
shale gas, 65–66, 69
shale oil, 352
Shell, Royal Dutch, 75
Shell Oil, 4, 52–53, 67, 83, 236, 279, 286
Gorgon's cost blowout and, 259
shift patterns, 2
short-term assignees, 92
skill-based pay, 27
skills, 19, 189, 239–240
Australian LNG industry's requirements for, 257–259
capital expense *vs.* sustainable, 262
company-sponsored "upskilling" for, 263
development of indigenous, 261–262
gaps in workforce, 227–230
global context training, 91–93
international assignment characteristics and, 89–90, 218
oil and gas industry required, 364–365
project management's essential, 169
project manager related, 179–180
proportion of categorized labor, 124
shortage of, 253–265
targeting those with stakeholder specific, 290
wages based on, 27
social media, 235, 275, 364

socialization, 94
societal impact, 358–360
society, 268–269, 273–275
Society for Human Resource Management, 32–33
Society of Hispanic Professional Engineers, 290
Society of Petroleum Engineers, 236, 239
Society of Women Engineers, 236
Soft Skills Council, 239
solar power, 70
Sonatrach, 138–139
South African Qualifications Authority, 182
special-project nationals, 124
Spindletop Charities, 248
Spindletop salt dome, 42
staffing, 3
    Australian LNG industry's, 257–259
    geocentric philosophy, 83–84
    in global environment, 80–91
    government relations and effectiveness of, 335
    host-country nationals and, 81–82
    improving government relations through, 329–335
    internal vs. external, 84–85
    key decisions for, 11–15
    labor market targeting in, 11
    legal compliance in, 15–16
    maximizing effectiveness of, 15–17, 90–91, 144–145, 183–184, 237–238, 293
    multinational enterprises and, 90–91
    in oil and gas industry, 365–367
    parent-country nationals and, 82–83, 217–218
    polycentric philosophy for, 81–82
    project managers and, 181–182
    projects, 176–184
    recruiting decisions for, 9–17
    recruiting methods for, 11–13
    rotators as solution for dilemma of, 217–218
    safety and in house or contracted, 142
    safety through, 141–145
    stakeholder perspective of, 288–293
    third-country nationals and, 83–84
    unconventional workforce, 230–238
stakeholder perspective
    benefits from, 303–304
    combining selection tools and, 293
    compensation from, 302–305
    corporate social responsibility and, 294
    evaluations and, 301
    formats to use from, 300–301
    goals and, 301
    of human resource management, 287, 295–296, 305
    involvement in performance management from, 299–300
    labor markets from, 288–290
    maximizing effectiveness from, 297, 301
    maximizing performance management from, 301
    measuring performance from, 298–299
    multinational enterprises and, 294
    pay levels from, 302
    pay mix from, 303
    performance management from, 298–301
    performance-based pay from, 303
    recruitment from, 290–291
    rewards and, 301
    selection from, 291–292
    staffing from, 288–293
    training courses from, 295
    training from, 293–297
    training methods from, 296–297
stakeholders, 172, 193–195, 197–202
    annual sustainability reports and, 275
    communities, society and, 268–271
    customers as, 277–278
    discrimination and, 280, 285, 293
    employees as group, 276–277
    HR practices for impacting, 7
    important, 286
    influence of oil and gas industry, 267
    maximizing effectiveness of, 293
    nongovernment organizations and, 278–279, 288–289, 294–295
    oil and gas industry, 268–284, 305
    proactiveness of, 2
    resource allocations to, 286
    satisfying multiple, 6–8
    selecting for performance relevant to, 291–292
    Shell and, 286
    surveys and, 277, 285, 297
    targeting of skills specific to, 290
    targeting underrepresented groups and, 288–290
    training courses related to, 295–296
    unions as, 279–284
Standard Oil Trust, 41–42
Statoil, 136–139, 312, 317
strategies. *See also* HR strategies
    Australian LNG project challenge and, 260–261
    business, 4–5
    compensation and benefits' implication, 28

employment structure for Australian LNG industry and, 261–265
functional, 5
human resource, 365–366
of human resource management, 3–6, 35
integrated oil companies interests and, 313, 315
labor shortages and retention, 220, 263–265
national oil companies interests and, 313, 315
training and development's implications of, 18
subdimensions, 300
success, 84–85, 89–90
supermajors, top 10 global, 49–50
supervision, 100
supply and demand, 228–230, 248
supply chain, 174
surveys, 89, 155, 159, 175, 179, 186, 259
human resource management, 24, 26, 34–35
stakeholders and, 277, 285, 297
United States Geological Survey, 42, 322, 324, 351
workforce, 219–221, 224, 227–228, 232–234, 237, 239–240, 243, 246

T

talent, 7–8
industrialized countries' shrinking, 364
retaining, 219–221, 245
shortage, 236, 248
supply and demand factors for shortage of, 228–230
taxes, 246–247, 318, 325–326
technology
combined cycle gas turbine, 66–67
for following up training, 97
global oil and gas industry's innovation and, 68
oil and gas industry challenges in, 362–363
territories, Russian, 279
terrorism, 136–139
testing tools, 14, 87–88
tests, 144
Texas City Refinery, 133–135, 160
third-country nationals (TCNs), 83–84, 87, 92–93, 107
3-D seismic data, 68, 79, 321
360 degree feedback, 24
tight gas, 69
tight oil, 352
time, 151, 194–195

tools, 14, 87–88, 182–183, 237. *See also* selection tools
top 10, global supermajors, 49–50
Total, 74–75, 295
total recordable incident rate (TRIR), 150–151
Trade Union Act, 283
Trade Union Congress, 284
Trade Union Ordinance, 283
trade unions, 283
trading, 53–54, 60–61
training, 23. *See also* safety training
for better government relations, 335–338
candidates for government relations, 336
candidates for workforce, 238–239
content of safety and health, 146–148
courses from stakeholder perspective for, 295
crew resource management, 148
cross-cultural, 94
effectiveness of, 20–22, 97, 240–241, 248, 297
global context in, 91–93
in global environment, 91–97
government relations and, 335–338
job specific, 147–148
methods for safety and health, 148–149
methods for workforce, 240
oil and gas industry content for, 18, 367–368
parent-country nationals and, 146, 148
for performance on jobs, 21
project managers', 185–191
project-oriented company, 184–192
project-oriented company candidates for, 184–186
project-oriented company topics for, 186–189
projects as, 187, 190
project-specific, 184–187, 189
risk management, 95, 173–174
safety and security, 95
safety through, 145–149
from stakeholder perspective, 293–297
stakeholder perspective methods for, 296–297
stakeholder related, 293–297
technology for follow-ups of, 97
of unconventional workforce, 238–241
ways of global context, 94–96
workforce, 238–241
training and development
effectiveness of, 20–22, 97, 240–241, 248, 297

employee satisfaction with, 18
focusing on employees by, 17–22
job performance and transfer of, 21
key decisions for, 19–20
learning in, 20–21
monitoring and evaluating, 21–22, 97
oil and gas industry requirements for, 18
selecting employees for, 19
selecting methods for, 20
selecting necessary and warranted areas for, 19–20
strategic implications of, 18
ways performance management related to, 23
transportation, 60–61, 63–64
travel, 203–204
travel expenses, 247
travelers, 92–93
Treasury Petroleum Fields (YPF), 312
turbines, gas, 66–67

U

unconventional oil, 69–70
Bakken Boom, employment and, 351–360
Bakken Boom's, 352–353
categories of, 352
employment and, 353–355
industry sources for development of, 351–355
shale oil or tight oil as, 352
underrepresentation, 288–290, 292
unions, 341
in global industry, 284
industrial, 283
Maritime, 260
methods for change by, 280
in Nigeria, 283–284
opposition to, 265
Petróleos Mexicanos', 318
prevalence in, 281
as stakeholders, 279–284
trade, 283
in United States, 283
United States (US), 11, 111, 135, 271–272, 278–279, 283, 352
Bureau of Labor Statistics, 152, 356
government influence and, 310–327
United States Geological Survey (USGS), 42, 322, 324, 351
universities, 290–291
"upskilling," company-sponsored, 263
upstream activities, 58–60, 123, 222–223
utilities, gas, 57

V

value chain
crude oil in, 60–61
government involvement in, 328, 346
oil and gas industry, 58–64
volatility, price, 361–362

W

wages, 358. *See also* compensation; compensation and benefits; pay-mix; performance-based pay; skill-based pay
Bakken Boom's average monthly, 356–357
danger pay in, 111
deciding on pay levels for, 30
merit pay in, 32
multinational enterprises and labor market, 105
oil and gas industry's, 28–29
pay-mix and, 30–31
skills and basis of, 27
well production, Bakken, 355–356
wellness programs, 160–161
wells, drilling, 59
Wi-Fi, 245
Williston Basin, 351–352
wind-powered energy, 70
work, principles and trends in, 212–213, 280
worker type, fatalities in oil and gas industry and, 142
workforce
aging, 221–221
benefits in, 245–247
candidates for training in, 238–239
characteristics of current, 221–230
combining tools in staffing, 237
compensation in unconventional, 243–248
discrimination and, 224, 226
drive-in and drive-out, 262
flown-in and flown-out, 262
involvement and formats in, 242
labor markets in, 231–234
male-dominated, 222–227, 239, 248
managing unconventional, 230
maximizing effectiveness of, 237–238, 240–241, 243, 247–248
measuring performance of, 242
methods of training for, 240
of oil and gas industry, 211
pay levels in, 243–244
pay mix in, 244
performance management of, 241–243
performance-based pay in, 244–245
recruiting in, 234–236

        selecting, 236–237
        skills gap in, 227–230
        skills to train in, 239–240
        staffing of unconventional, 230–238
        surveys, 219–221, 224, 227–228,
            232–234, 237, 239–240, 243, 246
        training methods for, 240
        training of unconventional, 238–241
        unique, 2
working conditions, 227
world
        most expensive oil and gas projects in,
            167–170
        number of rotators in, 233
        oil and gas industry and evolving,
            361–370
        oil and gas industry as most global in,
            73, 113
        refinery as largest in, 63
World Health Organization, 129

Y

young employees, 235–236, 238, 240–241,
    244–245, 297

International Journal of Human Resource Management, Journal of Global Mobility, Journal of Business Research, Worldatwork Journal, Workspan, and *Compensation and Benefits Review,* among others.

He is the coauthor of *Managing Human Resources,* 11th edition (Cengage Publishing) with Susan E. Jackson and Randall S. Schuler. He is the editor of the book *Managing Human Resources in North America* (Routledge) and the coeditor of the book *Global Compensation* (Routledge) with Luis R. Gomez-Mejia. He is on the editorial boards of six journals, including the *Journal of Management, Journal of Management Studies, Journal of Business Research, Human Resource Management, Human Resource Management Review,* and *International Journal of Human Resource Management.*

ANDREW INKPEN

Andrew Inkpen is the J. Kenneth and Jeanette Seward Chair in Global Strategy at Thunderbird School of Global Management at Arizona State University. Dr. Inkpen holds a B. Comm. degree from St. Mary's University, an MBA degree from the University of Western Ontario, and a PhD in Business Policy and International Business from the University of Western Ontario. Prior to entering academe, Dr. Inkpen worked in public accounting and qualified as a Chartered Accountant in Canada. He is codirector of the Thunderbird Center for Global Energy Studies.

Before joining the faculty at Thunderbird, Dr. Inkpen was on the faculty of Temple University. He has also taught at the University of Western Ontario, National University of Singapore, and Nanyang Technological University. Dr. Inkpen has taught a variety of courses, including Competitive Strategy, Global Strategy, and Corporate Strategy. Dr. Inkpen is actively involved in a variety of Executive Development Programs at Thunderbird. He is the Academic Director for the ExxonMobil General Leadership Program. This program is delivered 27 times a year in three global locations. He has also taught on many different oil and gas programs for clients such as Baker Hughes, Smith International, ExxonMobil Gas and Power Marketing, TNK-BP, and Integra. He has also worked other companies, including Ericsson, Pfizer, CEMEX, General Motors, LG Electronics, Teleflex, Integra, Goodyear, DENSO, Cisco

Systems, Solar Turbines, McDonald's, TRW, Alticor, Vitro, Warner-Lambert, Pharmacia & Upjohn, and Caremark.

Dr. Inkpen's research and teaching interests focus on the management of multinational firms, with a particular focus on strategic alliances, mergers and acquisitions, organizational learning, and global strategy. Articles by Dr. Inkpen have been published in various academic and practitioner publications including *Academy of Management Review*, *Academy of Management Executive*, *California Management Review*, *Strategic Management Journal*, *Journal of International Business Studies*, *Journal of Management Studies*, *Long Range Planning*, *Organizational Dynamics*, *Organization Science*, *Decision Sciences*, *Journal of Applied Behavioral Science*, and *European Management Journal*. He is the author of a book examining automotive supplier alliances titled *The Management of International Joint Ventures: An Organizational Learning Approach* (Routledge) and a book on global strategy called *Global Strategy: Value Creation and Advantage in the International Arena* (Oxford). He also coauthored *The Global Oil and Gas Industry: Management, Strategy, and Finance* (2011) with Michael H. Moffett. He is actively involved in teaching case development and has written more than 35 cases. He is a coauthor of the fifth edition of *International Management: Text and Cases* (Irwin/McGraw-Hill). He is on the editorial boards of seven journals, including the *Strategic Management Journal* (the leading academic journal in strategic management), *Journal of International Business Studies* (the leading journal in international business), *Organization Science*, *Journal of International Management*, *Management International Review*, and *Asia Pacific Journal of Management*.

MICHAEL H. MOFFETT

Michael H. Moffett is the Continental Grain Professor of Finance at the Thunderbird School of Global Management at Arizona State University. Formerly Associate Professor of Finance at Oregon State University, he has held teaching or research appointments at the University of Michigan, Ann Arbor, the Brookings Institution, Washington, D.C., the University of Hawaii at Manoa, the Aarhus School of Business (Denmark), the Helsinki School of Economics and Business Administration (Finland), the International Centre for Public Enterprises

(Yugoslavia), and the University of Colorado, Boulder. He is codirector of the Thunderbird Center for Global Energy Studies.

Professor Moffett's primary areas of teaching and research expertise are in multinational financial management, focusing on the financial demands of the global oil and gas industry and its continual development. Professor Moffett received a BA (Economics) from the University of Texas at Austin (1977), an MS (Resource Economics) from Colorado State University (1979), an MA (Economics) from the University of Colorado, Boulder (1983), and PhD (Economics) from the University of Colorado, Boulder (1985).

He has authored, coauthored, or contributed to a multitude of journal articles, books, and other publications. Professor Moffett is coauthor of several books in multinational business and finance, as well as the author of more than 50 case studies in international business, strategy, and financial management.

Professor Moffett is the coauthor of several books in multinational business and finance, *Multinational Business Finance*, 14th edition, with Arthur Stonehill and David Eiteman (Addison Wesley 2016), and *Fundamentals of Multinational Finance*, 5th edition, (Addison Wesley 2014). He also coauthor of *International Business*, 7th edition (Southwestern Publishing 2005), with Michael Czinkota and Ilkka Ronkainen, as well as *Global Business*, 4th edition (2002), and *Fundamentals of International Business* (2004). He is the author and coauthor of more than 50 case studies in international business, strategy, and financial management. He also coauthored *The Global Oil and Gas Industry: Management, Strategy, and Finance* (2011) with Andrew Inkpen.

He has acted as a consultant and educator with a multitude of global businesses including Adams, Allied Signal, American Express, AT&T, BP, Brasil Telecom of Brazil, Briggs & Stratton, Cemex de Mexico, Delphi, Delta Airlines, Discount Tire, Dow Chemical, East Asiatic Corporation of Denmark, EDS, Englehard, ExxonMobil, Fluor Corporation, Gate Gourmet, General Motors, Gudme Raaschou of Denmark, Honeywell, IBM, Kellogg's, Kimberley-Clarke, Legrand of France, Lincoln Electric, Mattel, Mobil Oil Corporation, ONGC of India, Parker Hannifin, Pfizer, Phelps Dodge Corporation, Solar Turbines, State Farm, Statoil of Norway, SK of Korea, Teleflex, Texaco, Vitro de Mexico, Warner Lambert, and Woodward Governor.